VARIATIONAL METHOD AND METHOD OF MONOTONE OPERATORS

IN THE THEORY OF NONLINEAR EQUATIONS

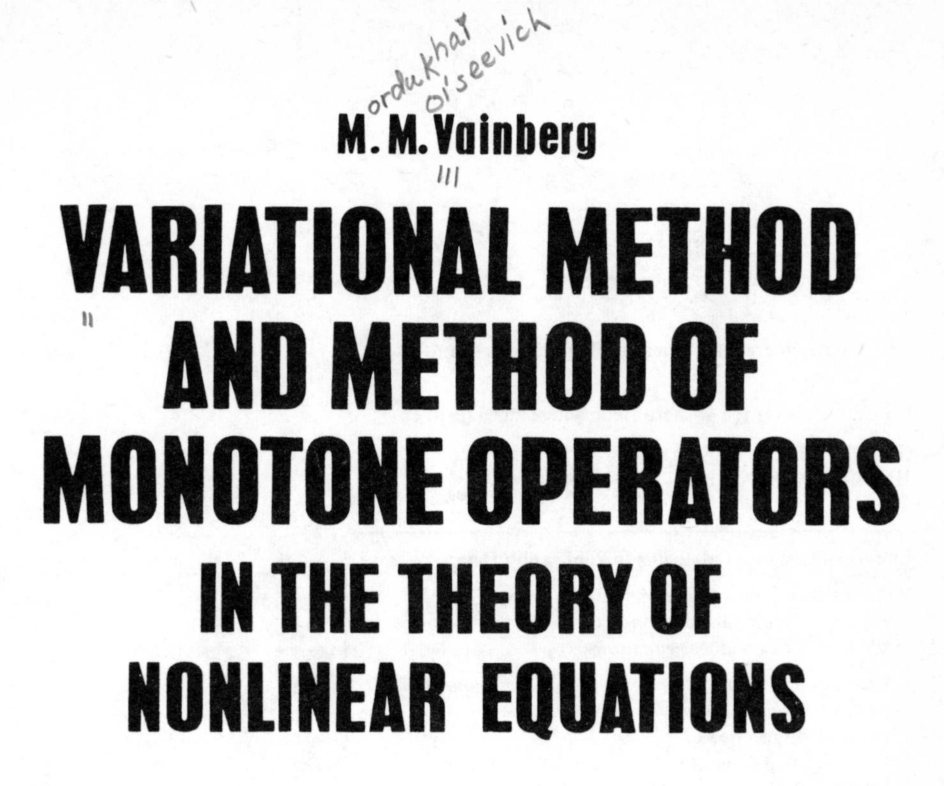

M. M. Vainberg

VARIATIONAL METHOD AND METHOD OF MONOTONE OPERATORS IN THE THEORY OF NONLINEAR EQUATIONS

Translated from Russian by A. Libin
Translation edited by D. Louvish

A HALSTED PRESS BOOK

JOHN WILEY & SONS
New York · Toronto

ISRAEL PROGRAM FOR SCIENTIFIC TRANSLATIONS
Jerusalem · London

Sole distributors for the Western Hemisphere and Japan

HALSTED PRESS, a division of
JOHN WILEY & SONS, INC., NEW YORK

Library of Congress Cataloging in Publication Data

Vaĭnberg, Mordukhaĭ Moiseevich.
Variational method and method of monotone operators in the theory of nonlinear equations.

Translation of *Variatsionnyĭ metod i metod monotonnykh operatorov v teorii nelineĭnykh uravneniĭ.*
"A Halsted Press book."
1. Differential equations, Nonlinear. 2. Calculus of variations. 3. Monotone operators. I. Title.
QA372.V3213 515'.35 73-16383
ISBN 0-470-89775-9

Distributors for the U.K., Europe, Africa and the Middle East

JOHN WILEY & SONS, LTD., CHICHESTER

Distributed in the rest of the world by

KETER PUBLISHING HOUSE JERUSALEM LTD.

ISBN 0 7065 1374 6
IPST cat. no. 22090

This book is a translation from Russian of
VARIATSIONNYI METOD I METOD MONOTONNYKH OPERATOROV V TEORII NELINEINYKH URAVNENII
Izdatel'stvo "Nauka"
Glavnaya Redaktsiya Fiziko-Matematicheskoi Literatury
Moscow, 1972

Printed in Israel

PREFACE

This book presents some recently developed methods for the study of nonlinear equations.

Fifteen years have elapsed since publication of the author's "Variational Methods for the Study of Nonlinear Operators." During this period variational methods for the study of nonlinear operators and equations have undergone intensive development, leading to the definition of new classes of nonlinear operators (monotone and accretive operators) and to a new method for the study of equations involving these operators, now known as the method of monotone operators.

Both the old and the new method have found wide application in various areas of pure and applied mathematics; we might mention, in particular, nonlinear integral equations, nonlinear equations of evolution and boundary-value problems for quasilinear partial differential equations.

Since the research relevant to the two methods is contained in short notes and periodical articles, both in the Soviet Union and in the West, the author has seen fit to present a systematic exposition of the new method and of further developments concerning the material of his previous book /9/. The present book is in this sense a sequel of the latter.

The manuscript was read by Professor M. I. Vishik, who also offered many invaluable comments.

M. Vainberg

CONTENTS

INTRODUCTION

This book presents two methods for the study of nonlinear equations. The first, the variational method, was discussed in a previous book of the author /9/, so we shall confine ourselves here to results achieved since the publication of /9/. The second, the method of monotone operators, is the fruit of the last fifteen years, so that our previous book included only a few preliminary results.

The reader will know that, when proving the solvability of an equation $\Phi(x)=0$ by the variational method, one uses the operator $\Phi(x)$ to construct a certain functional whose extremal or critical points serve as roots or pre-images of roots of $\Phi(x)$. There are several approaches. If the operator Φ has range in a normed space, the quest for a solution of the equation $\Phi(x)=0$ reduces to determining a point at which the functional $f(x)=\|\Phi(x)\|$ has an absolute minimum and, as we shall see, the solution may be obtained by determining a sequence minimizing the functional $f(x)$.

Another approach is to replace $\Phi(x)=0$ by an equivalent equation $F(x)=0$, where $F(x)$ is a potential operator from a vector space E to its dual. To be precise, let E be a real normed space and E^* its dual. An operator $F(x)$ from E to E^* is known as the gradient of a real functional $f(x)$ defined on E if

$$\lim_{t\to 0}\frac{f(x+th)-f(x)}{t}=\langle F(x),\ h\rangle,$$

where h is an arbitrary vector in E and $\langle y, h\rangle$ denotes the value of a linear functional $y\in E^*$ at h. An operator $F(x)$ which is the gradient of some functional $f(x)$ is called a potential operator and $f(x)$ is known as the potential of $F(x)$.

Since the roots of $F(x)$ are extremal points of its potential $f(x)$ and, if $f(x)$ is convex, the set of its extremal points is precisely the set of all roots of $F(x)$, one is interested in finding the extremal points of $f(x)$ or its minimum points, since its maxima are minima of the functional $(-1)f(x)$. Of great importance for studying the minima of functionals are analogs of the classical Weierstrass theorem.* For example, if a real-valued weakly lower semicontinuous

* [I.e., the theorem that a continuous function defined on a closed interval is bounded there and attains its g.l.b.]

functional is defined on a bounded convex closed subset of a reflexive Banach space E, it is bounded below there and attains its greatest lower bound. This proposition remains valid for $E = X^*$, where X is a separable normed space. For this reason, conditions for functionals to be weakly semicontinuous are of great interest for the variational method.

If $f(x)$ is a finite convex functional, it is weakly lower semicontinuous. If $f(x)$ is Gâteaux-differentiable, it is convex if and only if for any x and y

$$f(x) - f(y) - Df(y, x - y) \geqslant 0, \qquad (0.1)$$

where $Df(y, h)$ is its Gâteaux differential, which is continuous in h for fixed y. Another criterion for convexity of a differentiable functional $f(x)$ is

$$\langle F(x) - F(y), x - y \rangle \geqslant 0, \qquad (0.2)$$

where $F(x)$ is the gradient of $f(x)$. This is why inequality (0.2) or the equivalent inequality (0.1) are so widely applied in the variational treatment of nonlinear equations.

An operator $F(x)$ satisfying (0.2) is said to be monotone. It turns out that various theorems concerning the existence and uniqueness of solutions, established using the variational method for equations $F(x) = y$ or $x = BF(x)$, where $F(x)$ is a potential operator from E to E^* and B a linear operator from E^* to E, possess analogs when $F(x)$ is required to be monotone instead of potential. Existence-uniqueness proofs for solutions of equations with monotone operators employ a method distinct from that of the variational approach. This method, now known as the *method of monotone operators*, has found wide application in various problems of nonlinear analysis, such as nonlinear differential equations with unbounded operators in Hilbert and Banach spaces.

We now proceed to a brief chapter-by-chapter summary of the book. Before beginning, we note that much of the material is presented here for the first time in full detail, and moreover this is its first appearance in book form.

Chapter I (§§ 1 – 4) presents the parts of analysis in linear spaces needed later. In §1 we consider various types of continuity of mappings and introduce a few standard definitions. In particular, we define monotone, strictly monotone, strongly monotone and J-monotone operators, as well as the notions of hemicontinuity and demicontinuity. The fundamental proposition of this section is Theorem 1.2, which states that if a monotone operator is defined on an open set or a dense linear set in a Banach space it is hemicontinuous if and only if it is demicontinuous.

In §2 we consider the question of Gâteaux and Fréchet differentiability of mappings in normed spaces, define the derivative and gradient of a functional, introduce generalized Lagrange formulas for functionals and operators, establish the relationship between Gâteaux and Fréchet derivatives, examine differentiability properties of the norm and establish certain properties of the gradient of the norm. The fundamental propositions here concern the differentiability of the norm and the properties of its gradient.

In §3 we consider a few types of differentiability of mappings in linear topological spaces.

§4 contains a derivation of Taylor's formula with remainder both in the integral form and in Peano's version. Before doing this we introduce multilinear mappings and derivatives of higher orders, establish propositions pertaining to the continuity and symmetry of multilinear mappings, and examine the Riemann integral and some of its properties.

Chapter II (§§ 5—7) contains an exposition of certain problems from the theory of potential operators and differentiable mappings. §5 examines potential operators and monotone operators. The example of the Nemytskii operator shows that these two properties are independent. We cite several examples of Nemytskii operators: an operator which is both potential and monotone, an operator which is potential but not monotone, and an operator which is monotone but not potential. We then prove Theorem 5.1, stating that a necessary and sufficient condition for a potential operator to be monotone (strictly monotone) is that its potential be a convex (strictly convex) functional.

In §6 we first present conditions for operators to be potential operators, and then establish the equivalence of certain equations. Let F be a potential (or monotone) operator defined in a vector space E, and B a linear operator from the dual of E to E. Then the left-hand side of an equation of Hammerstein type $x - BF(x) = 0$ is not a potential (or monotone) operator. It turns out that this equation is equivalent to an equation $y - TF(Ay) = 0$ whose left-hand side is a potential (or monotone) operator in a real Hilbert space H, provided E and H satisfy certain conditions; the linear operators A and T depend in a certain way on B. This equivalence is proved in two cases (when E is a Hilbert space or a locally convex space) for potential operators.

In §7 we consider some properties of potential operators and differentiable mappings. It is shown, in particular (by a counterexample), that complete continuity of potential operators does not imply their strong continuity, and a criterion is established for strong continuity of differentiable mappings from one Banach space into another.

Chapter III (§§ 8—10) is of independent interest for the calculus of variations, as it studies the question of minima of nonlinear functionals and convergence of minimizing sequences. The material is later used to investigate nonlinear equations by the variational method. First, in §8, we consider the question of weak lower semicontinuity of nonlinear functionals. A criterion is established for a functional to be weakly lower semicontinuous, and quite natural sufficient conditions are derived for weak semicontinuity of differentiable functionals, essentially utilizing the convexity property of these functionals. The concept of a support functional (or subgradient) proves useful in this context. The main results of this section are Theorem 8.8, stating that any finite convex functional has a support functional at each point (this implies, in particular, a criterion for convexity of functionals), Theorem 8.10 (any finite convex functional is weakly lower semicontinuous) and Corollary 8.1, which furnishes a criterion for convexity of differentiable functionals. In the last part of the section we consider quasiconvex and semiconvex functionals.

In §9 we establish existence and uniqueness theorems for the minima of nonlinear functionals. One result (Theorem 9.1) is a necessary and sufficient condition for a minimum. Also proved are two generalized versions of the Weierstrass theorem (Theorems 9.2 and 9.3), for reflexive spaces and for duals of separable normed spaces, and theorems furnishing sufficient conditions for unconditional minima of functionals. In the last part of the section we consider the question of conditional extrema of functionals and prove the fundamental Theorem 9.11 on the conditional extremum of a functional defined in a linear topological space.

In § 10 we prove propositions on weak and strong convergence of minimizing sequences and ascertain when minimization problems for nonlinear functionals are well-posed. In this context a major role is played by increasing and strictly convex functionals, and also by certain properties of gradients of functionals.

Chapter IV (§§ 11 —13) is devoted to an examination of some methods for minimization of nonlinear functionals. § 11 considers the method of steepest descent. We first explain the idea of the method in Hilbert and Banach spaces and consider the relaxation properties of sequences obtained by steepest descent. Then, assuming the functional to be convex, we prove theorems concerning the convergence of a descent-type method which is also known as the method of gradient descent. These theorems are first established for differentiable functionals and then for twice differentiable functionals.

In § 12 we construct minimizing sequences by the Ritz method, given a real functional in a real separable normed space. Propositions are established on solvability of Ritz systems and it is determined when the Ritz approximations form a minimizing sequence. In the last part of the section we prove theorems on the (weak and strong) convergence of Ritz approximations to the minimum point of the functional under consideration.

The last section of Chapter IV invokes the Newton-Galerkin method for approximate solution of nonlinear equations in normed spaces in order to construct minimizing sequences. Although Newton approximations may be defined in the general case, useful propositions on convergence of Newton approximations to the minimum point of the functional are obtained only when the functional is convex.

In Chapter V we study the equations $F(x) = y$ and $x - BF(x) = 0$, where F is a potential operator and B a linear operator. The existence-uniqueness theorems of § 14 are proved for the first equation using mainly the results of Chapter III, while the proofs of analogous theorems for the second equation (§ 17) use both results concerning the first and theorems on the square root of a linear operator (§§ 15 and 16).

In § 15 we prove theorems on the square root of a bounded linear operator defined in a Banach space or locally convex space with range in another space of the same type. We first introduce a few notions and results from the theory of linear topological spaces, define the class of embeddable spaces E and give examples of such linear topological spaces. We then consider linear operators B from E to E' or from E' to E, where E is an embeddable space and E' the weak dual of E, and prove theorems pertaining to the properties of the square root of a restriction of the operator B and the representation of B as the product of two operators. Two cases are considered here, depending on whether B is a positive or indefinite operator.

In § 16 we study properties of the square root of a restriction of a linear operator B from a Banach space E to its dual E^* (or from E^* to E), without assuming the E is embeddable or that the operator is bounded. The operator B is assumed to be closed. The main results for a positive B are theorems on representation of B in the form $Bx = T^*Tx$ for every $x \in D(B)$, where T is a closed extension of the square root of the restriction of B to a linear subset of some Hilbert space, and T^* the adjoint of T. When B is indefinite, we establish other representations of the operator as the product of two closed operators. Theorems of this type are essential in proving the existence and uniqueness of solutions for equations of Hammerstein type, both with a potential operator F (§ 17) and with a monotone operator F (§§ 19, 20).

In § 17 we use the variational method to prove existence-uniqueness theorems for an equation of Hammerstein type $x - BF(x) = 0$, where F is a potential operator, in both Banach spaces and locally convex spaces. These theorems are proved for bounded, unbounded, positive and indefinite operators B. When B is unbounded we define generalized solutions, prove their existence and indicate sufficient conditions for a generalized solution to be an exact solution of the Hammerstein equation.

In Chapter VI (§§ 18—20) we study nonlinear equations with monotone operators. §18 begins with an exposition of the principal auxiliary propositions underlying the method of monotone operators and goes on to prove the fundamental existence-uniqueness theorems for equations of the type $F(x) = y$, where F is a monotone operator from a (real or complex) Banach space to its dual. We also define semimonotone and pseudomonotone operators and prove corresponding existence-uniqueness theorems.

In §19 the method of monotone operators is utilized to prove existence-uniqueness theorems for equations of type in Hammerstein Banach and locally convex spaces. We first establish Theorem 19.1, which states that the equation $x + BF(x) = 0$ is uniquely solvable in special Banach spaces, without assuming that the linear operator $B: E^* \to E$ is selfadjoint, and then prove various existence-uniqueness theorems for the equation $x - BF(x) = 0$, where F is a monotone operator from an embeddable vector space E to its dual E^* (or E'), and B a bounded selfadjoint operator from E^* (or E') to E. The basic idea of the proofs of the other theorems is to replace the equation $x - BF(x) = 0$, via the results of § 15, by an equivalent equation $y - TF(Ay) = 0$ whose left-hand side is a monotone operator in some Hilbert space, T and A certain linear operators. Also considered are certain properties of the Hammerstein operator. In particular, we prove that if the mapping $F: E \to E'$ is arbitrary and B is a positive bounded selfadjoint operator from E' to E, then F is monotone on the solutions of $x - BF(x) = 0$. Propositions of this kind are of interest for uniqueness theorems. We also prove a result (Theorem 19.9) concerning Hammerstein equations with a semimonotone operator F.

In § 20 we prove existence theorems for the equations $x + BF(x) = 0$ and $Bx + F(x) = 0$ in Banach and Hilbert spaces, on the assumption that F is monotone or semimonotone. B is not assumed to be bounded, the only requirement being that it be closed.

In Chapter VII (§§21 —23) we study equations with nonlinear accretive operators and prove theorems on the convergence of steepest-descent and Galerkin approximations to solutions of nonlinear equations. In §21 we define duality mappings from a normed space E to its dual E^* and study their properties. We prove, in particular, that a duality mapping is single-valued if E^* is strictly convex, that it is monotone, coercive and strongly monotone if E is strictly convex. With the help of the duality mapping we introduce a semi-inner product in normed spaces and define nonlinear accretive operators. We study certain properties of accretive operators and prove the fundamental Theorem 21.1 (nonlinear strongly accretive operators satisfying a Lipschitz condition are homeomorphisms), and existence theorems for nonlinear equations with such operators. In the second part of the section, adopting the hypotheses of Theorem 21.1, we show that the solution is the limit of its steepest-descent approximations in Hilbert space, L-spaces and sequence spaces.

In § 22 we prove that a duality mapping is uniformly continuous on any bounded set, consider the question of its weak continuity and prove a number of auxiliary propositions which are then used to prove fundamental existence theorems for equations with nonlinear accretive operators.

In § 23 we construct Galerkin approximations and derive sufficient conditions for their convergence to solutions of nonlinear equations with nonlinear accretive operators.

The eighth and last chapter applies the method of monotone operators to nonlinear integral equations, boundary-value problems for quasilinear partial differential equations and nonlinear differential equations in Banach spaces.

In all references to the author's book "Variational Methods for the Study of Nonlinear Operators" we shall employ the abbreviation VM.

Chapter I

SOME PROBLEMS OF ANALYSIS IN LINEAR SPACES

In the four sections of this chapter we present several problems of analysis to be used in the sequel. The most important of these are problems concerning the relations between the hemicontinuity and demicontinuity of mappings (§ 1), conditions for differentiability of the norm and properties of the gradient of the norm (§ 2), and Taylor's formula for operators. Some of the problems presented in this chapter have been discussed only in the periodical literature.

§ 1. PRELIMINARIES; VARIOUS TYPES OF CONTINUITY

Let E_x and E_y be linear spaces and G a mapping (not necessarily linear) from E_x to E_y, denoted $G: E_x \to E_y$ if G is defined on all of E_x. If E_x is a euclidean space of dimension one or higher and E_y is an abstract space (i. e., a general linear space), the mapping G is called an abstract function. If E_x is an abstract space and E_y the real line or the complex plane, G is called a functional. If both E_x and E_y are abstract spaces, G is called an operator. We shall assume that G is single-valued. The domain of definition of G is denoted by $D(G)$ and its range by $R(G)$.

1.1. Basic definitions

Definition 1.1. A mapping $G: X \to E_y$ where $X \subset E_x$, is said to be surjective if $R(G) = E_y$; it is injective if it is one-to-one, i. e., $G(x_1) = G(x_2)$ implies $x_1 = x_2$ for all $x_1, x_2 \in D(G) = X$; and it is bijective if it is both surjective and injective.

Note that if $G: X \to E_y$ is bijective, the inverse mapping of E_y onto X exists and is denoted by G^{-1}.

If E_x and E_y are linear topological spaces and $X \subset E_x$, we can speak of the continuity of a mapping $G: X \to E_y$ relative to the given topologies. In particular, if E_x and E_y are Banach spaces, continuity is defined in terms of the norm, i. e., G is continuous at a point x_0 if for positive ε there exists positive δ such that if $\|x - x_0\| < \delta$, $x \in X$, then

$$\| G(x) - G(x_0) \| < \varepsilon.$$

This definition is equivalent to the following: a mapping $G: X \to E_y$ is continuous at $x_0 \in X$ if for any sequence $\{x_n\} \subset X$ converging* to x_0 (i. e., $\|x_n - x_0\| \to 0$ as $n \to \infty$) the sequence $\{G(x_n)\} \subset E_y$ converges to $G(x_0)$. In the case of linear topological spaces, the continuity of G at a point x_0 may be defined with the help of transfinite sequences.

We shall write $x_n \to x_0$ $(x_\alpha \to x_0)$ if the sequence $\{x_n\}$ (transfinite sequence $\{x_\alpha\}$) converges to x_0 and $x_n \rightharpoonup x_0$ $(x_\alpha \rightharpoonup x_0)$ if this sequence converges weakly to x_0, i. e. if

$$l(x_n) \to l(x_0) \, (l(x_\alpha) \to l(x_0))$$

for any linear functional l.

Definition 1.2. A mapping $G: X \to E_y$ is a *homeomorphism* if it is bijective and both G and G^{-1} are continuous.

Naturally, if G is a homeomorphism of X onto E_y, then G^{-1} is a homeomorphism of E_y onto X.

Definition 1.3. A mapping $G: X \to E_y$, where $X \subset E_y$, is *bounded* if it maps every bounded set in X onto a bounded set in E_y; it is *compact* if it maps every bounded set in X onto a compact set in E_y; it is *completely continuous* if it is continuous and compact; it is *weakly continuous* if it maps every weakly convergent sequence onto a weakly convergent one and *strongly continuous* if it maps every weakly convergent sequence onto a convergent one.

Let E_x and E_y be linear spaces and $B(x, y)$ a bilinear form defined on E_x and E_y. We say that the bilinear form B is a pairing of the spaces E_x and E_y** or that E_x and E_y are paired (with respect to B) if the following conditions hold:

1. For every $x \neq 0$ in E_x there exists $y \in E_y$ such that $B(x, y) \neq 0$.
2. For every $y \neq 0$ in E_y there exists $x \in E_x$ such that $B(x, y) \neq 0$.

Henceforth we shall consider only pairings relative to a canonical bilinear form $\langle y, x \rangle$. We therefore assume that either E_x is a normed space and E_y is its dual E_x^* (in the usual sense) or

* This type of convergence is sometimes referred to as *strong*; we shall use the term *pointwise convergence*.

** See Bourbaki /1/.

that E_x is a separable locally convex space and E_y is its strong dual E'_x. In either case $\langle y, x\rangle$ is the value of the linear continuous functional y at the vector $x \in E_x$.

Definition 1.4. A linear operator A from a real space E_x to the real space E_x^* (or E'_x) is said to be positive if $\langle Ax, x\rangle \geqslant 0$, strictly positive if $\langle Ax, x\rangle \geqslant 0$ and $\langle Ax, x\rangle = 0$ only for $x=0$; strongly positive from E_x to E_x^* if $\langle Ax, x\rangle \geqslant c\|x\|^2$, where $c > 0$.

Definition 1.5. A mapping G from a real normed space E_x to E_x^* is said to be coercive if

$$\langle G(x), x\rangle \geqslant c(\|x\|)\|x\|, \qquad (1.1)$$

where $c(t)$ is a real-valued function of nonnegative t such that $c(t) \to \infty$ as $t \to \infty$.

This definition also applies to complex normed spaces if the left-hand side in (1.1) is replaced by $\operatorname{Re}\langle G(x), x\rangle$.

Definition 1.6. A mapping G from a normed space E_x to E_x^* is said to be monotone if

$$\operatorname{Re}\langle G(x) - G(y), x - y\rangle \geqslant 0 \qquad (\forall x, y \in D(G)); \qquad (1.2)$$

it is strictly monotone if equality can hold in (1.2) only when $x = y$; and strongly monotone if

$$\operatorname{Re}\langle G(x) - G(y), x - y\rangle \geqslant \|x - y\|\gamma(\|x - y\|) \qquad (1.3)$$

for any $x, y \in D(G)$, where $\gamma(t)$ is a real-valued nonnegative function defined for $t \geqslant 0$ such that $\gamma(t) \to +\infty$ as $t \to +\infty$ and $\gamma(t) = 0$ implies $t = 0$; G is uniformly monotone if (1.3) holds and $t\gamma(t)$ is an increasing real-valued function, vanishing at zero.

If $(-1)G(x)$ is a monotone operator, $G(x)$ is said to be dissipative or antimonotone.

Since we shall be concerned primarily with monotone operators in real spaces, the sign Re in (1.2) and (1.3) will henceforth be omitted.

Remark 1.1. Monotone operators may also be introduced via the concept of monotone sets. Let E_x be a real space. A set $E \subset E_x \times E_x^*$ is said to be monotone if, for any two points (x_1, y_1) and (x_2, y_2) in E, we have $\langle y_1 - y_2, x_1 - x_2\rangle \geqslant 0$. E is said to be maximally monotone if it is not a proper subset of another monotone set in $E_x \times E_x^*$. From this definition and from Definition 1.6 it follows that an operator G from E_x to E_x^* is monotone if its graph is a monotone set. The operator G is maximally monotone if its graph is a maximally monotone set.

One can consider monotone operators from E_x to E_y, if E_x and E_y are paired vector spaces over the same field.

We now introduce a more general notion of the monotonicity of an operator. Let E_x and E_y be normed spaces, J and G nonlinear operators with $D(J)=E_x$, $R(J)\subset E_y$, $D(G)\subset E_x$, $R(G)\subset E_y^*$.

Definition 1.7. The operator G is said to be J-monotone if

$$\operatorname{Re}\langle G(x)-G(y),\ J(x-y)\rangle\geqslant 0 \qquad (x,\ y\in D(G)).$$

Remark 1.2. If $E_y=E_x$ and J is the identity operator, Definitions 1.6 and 1.7 are identical, i. e., an operator is J-monotone if and only if it is monotone. If E_x is a reflexive Banach space, $E_y=E_x^*$, E_x^* is strictly convex and J is a duality mapping, then, as we shall show in §§ 21 and 22, the definition of a J-monotone operator is equivalent to the definition of a nonlinear accretive operator.

Definition 1.8. Let E_x and E_y be normed spaces. A mapping G from E_x to E_y is said to be demicontinuous at a point $x\in D(G)$ if for any sequence $x_n\to x\,(x_n\in D(G))$, we have $G(x_n)\rightharpoonup G(x)$; it is hemicontinuous at a point $x_0\in D(G)$ if for any vector x such that $x_0+tx\in D(G)$ for $0\leqslant t\leqslant\alpha$ $(\alpha=\alpha(x)>0)$ and for any sequence $t_n\to 0$ as $n\to\infty$ $(0<t_n\leqslant\alpha)$, we have $G(x_0+t_nx)\rightharpoonup G(x_0)$.

Demicontinuity of G clearly implies hemicontinuity, from which it follows, in turn, that $\langle h, G(x_0+tx)\rangle$ is continuous in t for any $h\in E_y^*$, if $x_0+tx\in D(G)$.

1.2. The relation between hemicontinuity and demicontinuity*

Definition 1.9. A mapping G from E_x to E_y, where E_x and E_y are normed spaces, is said to be locally hemibounded at a point x_0 if the conditions $x_0+tx\in D(G)$ for $0\leqslant t\leqslant\alpha$, $\alpha>0$, $t_n\to 0$ as $n\to\infty$, where $0\leqslant t_n\leqslant\alpha$, imply that the sequence $\{G(x_0)+t_nx)\}$ is bounded.

Lemma 1.1. Let G be a J-monotone operator satisfying the conditions:

1. J is a positive-homogeneous operator, i. e., $J(tx)=tJ(x)$ for $t>0$, and $R(J)=E_y$;

2. J is uniformly continuous in the unit ball $\|x\|\leqslant 1$;

3. $D(G)$ is an open set in E_x.

* These propositions are due to Kato /4/.

Then, if G is locally hemibounded at x_0, it is also locally bounded at x_0.

Proof. Let G be locally hemibounded at $x_0 \in D(G)$, and suppose that G is not locally bounded there. Then there exists a sequence $\{x_n\} \subset D(G)$ such that $x_n \to x_0$, but

$$0 < \|G(x_n)\| = r_n \to +\infty \text{ as } n \to \infty.$$

We shall show that this contradicts the assumption that G is locally hemibounded at x_0.

Let $\psi(s) = \sup\|Jx - Jy\|$, where $\|x\| \leqslant 1$, $\|y\| \leqslant 1$ and $\|x - y\| \leqslant s$. Since by assumption the mapping J is uniformly continuous, it follows that $\psi(s)$ is a nondecreasing function and that $\psi(s) \to 0$ as $s \to 0$. In view of the fact that J is positive-homogeneous, we have $\psi(s) < \infty$ for any $s > 0$. Thus

$$\|Jx - Jy\| \leqslant \psi(\|x - y\|), \quad \|x\| \leqslant 1, \quad \|y\| \leqslant 1. \tag{1.4}$$

Setting

$$t_n = \max\left[r_n^{-1}, \|x_n - x_0\|^{1/2}, (\psi(\|x_n - x_0\|))^{1/2}\right], \tag{1.5}$$

we obtain

$$t_n > 0, \; t_n r_n \geqslant 1, \quad \|x_n - x_0\| \leqslant t_n^2, \quad \psi(\|x_n - x_0\|) \leqslant t_n^2. \tag{1.6}$$

Now, by assumption, $r_n \to \infty$ and $\|u_n - u_0\| \to 0$, so that $\psi(\|x_n - x_0\|) \to 0$; thus, by (1.5),

$$t_n \to 0 \quad \text{as} \quad n \to \infty. \tag{1.7}$$

Let $u \in E_x$ and $u_n = x_0 + t_n u$. Since $x_0 \in D(G)$ and $D(G)$ is open, it follows that $u_n \in D(G)$ for sufficiently large n. By the J-monotonicity of G and the positive homogeneity of J, we may write

$$\begin{aligned} \operatorname{Re}\langle G(x_n), Ju\rangle \leqslant t_n^{-1} \operatorname{Re}\langle G(u_n), J(u_n - x_n)\rangle + \\ + t_n^{-1} \operatorname{Re}\langle G(x_n), J(t_n u) - J(u_n - x_n)\rangle. \end{aligned} \tag{1.8}$$

Let us estimate the right-hand side of (1.8). We first observe that

$$t_n^{-1}(u_n - x_n) = u - t_n^{-1}(x_n - x_0) \to u \quad \text{as} \quad n \to \infty,$$

because $t_n^{-1}\|x_n - x_0\| \leqslant t_n \to 0$. Thus, by the continuity and positive homogeneity of J, it follows that

$$t_n^{-1} J(u_n - x_n) = J(t_n^{-1}(u_n - x_n)) \to J(u)$$

and, since by assumption

$$\| G(u_n) \| = \| G(x_0 + t_n u) \|$$

is bounded as $n \to \infty$, the first term on the right of (1.8) is bounded as $n \to \infty$.

To estimate the second term on the right of (1.8), we note that $t_n u$ and $u_n - x_n$ converge to 0 as $n \to \infty$, and so, for sufficiently large n, they lie in the unit ball. In view of (1.4) and (1.6), we have

$$\| J(t_n u) - J(u_n - x_n) \| \leqslant \psi(\| t_n u - u_n + x_n \|) = \\ = \psi(\| x_n - x_0 \|) \leqslant t_n^2,$$

so that the second term on the right of (1.8) is majorized by $t_n \| G(x_n) \| = t_n r_n$. Hence,

$$\operatorname{Re} \langle G(x_n), Ju \rangle \leqslant C + t_n r_n, \tag{1.9}$$

where the constant C may depend on the vector u, but not on n. Since $t_n r_n \geqslant 1$ by (1.6), division of both sides of (1.9) by $t_n r_n$ gives

$$\varlimsup_{n \to \infty} \operatorname{Re} \langle (t_n r_n)^{-1} G(x_n), Ju \rangle < +\infty. \tag{1.10}$$

By assumption $R(J) = E_y$, so we may view $Ju = y$ as an arbitrary vector in E_y for which (1.10) holds. Replacing y by $(-y)$ (or by $\mp iy$ if E_y is a complex space) and using (1.10), we see that

$$\langle (t_n r_n)^{-1} G(x_n), y \rangle$$

is bounded when $n \to \infty$ for any fixed vector y in E_y. This contradicts the fact that

$$\| (t_n r_n)^{-1} G(x_n) \| = t_n^{-1} \to \infty,$$

completing the proof of the lemma.

1.3. Fundamental propositions

Theorem 1.1. Under the assumptions of Lemma 1.1, any hemicontinuous operator G is demicontinuous.

Proof. Let G be hemicontinuous at a point $x_0 \in D(G)$. Then it is locally hemibounded at this point, so we may use the same reasoning as in the proof of Lemma 1.1. In particular, if $x_n \to x_0$, then $r_n = \|G(x_n)\|$ is a bounded sequence. Set

$$t_n = \max\left[\|x_n - x_0\|^{1/2}, (\psi(\|x_n - x_0\|))^{1/2}\right].$$

Then all of (1.6) holds, except for the inequality $t_n r_n \geqslant 1$, and, moreover, (1.7) is also valid. The inequality (1.8) also holds for $u_n = x_n + t_n u$ and, as before, $t_n^{-1} J(u_n - x_n) \to J(u)$ as $n \to \infty$. Thus, the equality

$$G(u_n) = G(x_0 + t_n u) \to G(x_0),$$

which follows from the hemicontinuity of G, implies that the first term on the right of (1.8) converges to $\mathrm{Re}\langle G(x_0), Ju\rangle$. We have seen that the second term is majorized by $t_n r_n$. Since r_n is bounded and $t_n \to 0$, this term converges to zero.

Consequently,

$$\overline{\lim_{n\to\infty}} \mathrm{Re}\langle G(x_n) - G(x_0), Ju\rangle \leqslant 0.$$

Since $Ju = y$ may be an arbitrary vector of E_y, we see, as at the end of the proof of Lemma 1.1, that for any $y \in E_y$

$$\overline{\lim_{n\to\infty}} |\langle G(x_n) - G(x_0), y\rangle| = 0,$$

i. e., $G(x_n) \to G(x_0)$. Q. E. D.

Remark 1.3. When $E_y = E_x$ and J is the identity, i. e., G is a monotone operator, Lemma 1.1 and Theorem 1.1 also hold if $D(G)$ is a dense linear set in E_x.

From Theorem 1.1 and Remark 1.3 we have

Theorem 1.2. If $D(G)$ is a dense linear subset of E_x or an open subset of E_x and G is a monotone operator from E_x to E_x^*, then G is demicontinuous if and only if it is hemicontinuous.

As we have seen, if E_x is a reflexive Banach space, $E_y = E_x$, and J is a duality mapping from E_x to E_x^*, i. e.,

$$\langle Jx, x\rangle = \|Jx\|\|x\| = \|x\|^2,$$

then the notions of a J-monotone operator and a nonlinear accretive operator (§ 21) are equivalent. It turns out (see Lemma 22.1) that for a real space E_x having uniformly convex* dual E_x^*, a duality

* A Banach space E is uniformly convex if $x_n, y_n \in E$, $\|x_n\| \leqslant 1$, $\|y_n\| \leqslant 1$ and $\|x_n + y_n\| \to 2$ as $n \to \infty$ imply $\|x_n - y_n\| \to 0$. There are other definitions of uniform convexity (see Day /1/). Mil'man /1/ has shown that every uniformly convex Banach space is reflexive.

mapping is uniformly continuous on every bounded set. Hence the following corollary of Theorem 1.1:

Theorem 1.3. *Suppose E is a reflexive Banach space, E^* is uniformly convex, and J is a duality mapping from E to E^*. Then if a J-monotone (accretive) operator G from E to E is defined on an open set $D(G)$, hemicontinuity of G implies its demicontinuity.*

Remark 1.4. If the mapping $G: E \to E^*$, where E is a finite-dimensional Banach space, is hemicontinuous, then G is continuous on E. Indeed, by Theorem 1.2, G transorms every convergent sequence in E into a weakly convergent sequence in the finite-dimensional space E^*, but in finite-dimensional spaces strong and weak convergence are equivalent.

§ 2. GÂTEAUX AND FRÉCHET DIFFERENTIABILITY IN NORMED SPACES

2.1. Gâteaux differential and derivative

Let $F(x)$ be a nonlinear operator mapping a dense linear set $D(F) \subset E_x$ into E_y, where E_x and E_y are normed spaces.

If for a point $x \in D(F)$ the limit

$$\frac{d}{dt} F(x+th)\big|_{t=0} = \lim_{t\to 0} \frac{F(x+th)-F(x)}{t} = VF(x, h) \tag{2.1}$$

exists for any $h \in D(F)$, then $VF(x,h)$ is called the Gâteaux variation or Gâteaux differential at x of the operator $F(x)$. The operator $VF(x,h)$ is homogeneous in h, i. e., $VF(x, \alpha h) = \alpha VF(x,h)$, but it is not always additive. There are examples (see VM) for which

$$VF(x, h_1+h_2) \neq VF(x, h_1) + VF(x, h_2).$$

We shall assume that the Gâteaux differential, denoted by $DF(x,h)$, is a linear operator, that is, it is additive and homogeneous.* The differential $DF(x,h)$, being linear in h, can be written

$$DF(x, h) = P(x)h.$$

* Sometimes only $DF(x,h)$ is called the Gâteaux differential and $VF(x,h)$ the variation of F at x.

$P(x)$ is called the *derivative* of F at x, denoted by $F'(x)$, i. e.,

$$DF(x, h) = F'(x)h.$$

If for some fixed $x \in D(F)$ the derivative $F'(x)$ is a bounded operator, then it can be extended by continuity to an operator acting on all vectors $h \in E_x$. This extension, too, is denoted by $F'(x)$.

2.2. Fréchet differential and derivative

Let F be a nonlinear operator from $D(F) \subset E_x$ to E_y, where $D(F)$ is a dense linear subset of E_x. If at some point $x \in D(F)$

$$F(x+h) - F(x) = u(x, h) + \omega(x, h),$$

where $u(x, h)$ is a linear operator in $h \in D(F)$ and

$$\lim_{h \to 0} \frac{\omega(x, h)}{\| h \|} = 0,$$

then $u(x, h)$ is called the *Fréchet differential* of $F(x)$ at x and $\omega(x, h)$ the *remainder* of the differential. We assume that the operator $u(x, h)$ is bounded in h and denote it by $dF(x, h)$. Thus, we can write

$$dF(x, h) = F'(x)h.$$

Since $F'(x)$ is a bounded operator, it can be extended to an operator defined on all of E_x. This extension is also denoted by $F'(x)$. Our notations for the Gâteaux and Frêchet derivatives are the same, but one should remember that while the existence of the Fréchet derivative implies the existence of the Gâteaux derivative and the two are equal, the converse is not necessarily true.

Many properties of ordinary derivatives also hold for the derivatives of operators. For example, $(cF(x))' = cF'(x)$ for any constant c, and

$$(F_1(x) + F_2(x))' = F_1'(x) + F_2'(x).$$

If $F(x)$ is an operator from E_x to E_y and $\Phi(y)$ from E_y to E_z, and $F'(x)$ and $\Phi'(y)$ exist, then

$$[\Phi(F(x))]' = \Phi'(F(x))F'(x),$$

where the right-hand side is a product of linear operators, one from E_y to E_z and the other from E_x to E_y.

2.3. Derivative and gradient of a functional

Let $f(x)$ be a nonlinear functional defined on a dense linear subset $D(f)$ of a Banach space E, i. e., on a normed space dense in E. Suppose there exists a linear Gâteaux differential $Df(x, h)$ at a point $x \in D(f)$. Then, for fixed x and any $h \in D(f)$,

$$Df(x, h) = f'(x)h,$$

where $f'(x)$ is the Gâteaux derivative of the functional f at x. Since $Df(x, h)$ is a functional, $f'(x)$ is a linear functional for fixed x, whose domain of definition is all of $D(f)$. If for fixed x the derivative $f'(x)$ is a bounded linear functional, then it can be extended by continuity to a linear functional on all of E. This extension is called the *gradient* of the functional f at x, denoted by grad $f(x)$; $f(x)$ is called the *potential* of the operator $F(x) = \operatorname{grad} f(x)$. Clearly, grad $f(x) = F(x)$, as a continuous linear functional defined for all vectors $h \in E$, is an element of the dual E^* of E. If the gradient exists on some set $A \subset D(f)$, then grad $f(x)$ is an operator from A into E^*. It follows from the definition of grad $f(x)$ that

$$\langle \operatorname{grad} f(x), h \rangle = \frac{d}{dt} f(x + th)\Big|_{t=0} = \lim_{t \to 0} \frac{f(x + th) - f(x)}{t},$$

where $\langle y, x \rangle$ is the value of a linear functional $y \in E^*$ at $x \in E$. A few examples now follow.

Example 2.1. Let E be a real reflexive Banach space and B a positive bounded linear operator from E to E^*, i. e., $\langle Bx, x \rangle \geqslant 0$. Then the adjoint is also an operator from E to E^*. Consider the functional

$$f(x) = \langle Bx, x \rangle.$$

For this functional, we have

$$f(x + th) - f(x) = \langle Bx + tBh, x + th \rangle - \langle Bx, x \rangle = \\ = t\langle Bx, h \rangle + t\langle B^*x, h \rangle + t^2 \langle Bh, h \rangle.$$

Hence grad $f(x) = Bx + B^*x$.

Example 2.2. Let H be a real Hilbert space. Then

$$\operatorname{grad}(x, x) = 2x; \quad \operatorname{grad}\|x\| = \frac{x}{\|x\|} \qquad (x \neq 0).$$

Example 2.3.* Let $\{x(u)\}$ denote the set of real-valued, absolutely continuous functions on the interval [0, 1] such that $x(0) = x(1) = 0$ and $x'(u)$ is square-integrable on [0,1]. Consider the functional

$$f(x) = \frac{1}{2}\int_0^1 (x'(u))^2\, du.$$

We have

$$\frac{d}{dt} f(x + th)\,\Big|_{t=0} = \int_0^1 x'(u)\, h'(u)\, du.$$

We make the family $\{x(u)\}$ a Hilbert space H by defining an inner product via the formula

$$[x, y] = \int_0^1 x'(u)\, y'(u)\, du \qquad (\forall x, y \in H).$$

Then

$$\frac{d}{dt} f(x + th)\,\Big|_{t=0} = [x, h],$$

so that

$$\operatorname{grad} f(x) = x.$$

If we view $\{x(u)\}$ as a linear set in $L^2(0, 1)$, then

$$Df(x, h) = \int_0^1 x'(u)\, h'(u)\, du,$$

and the Gâteaux derivative is not a bounded functional for arbitrary fixed $x \in D(f)$. However, if $x'(u)$ is absolutely continuous and $x''(u)$ is square-integrable on [0,1], then

* See Mikhlin /3/.

$$f'(x)h = Df(x, h) = \int_0^1 x'(u)h'(u)\,du =$$
$$= -\int_0^1 x''(u)h(u)\,du = (-x'', h),$$

and for these fixed points $f'(x)$ is a bounded functional, so that

$$F \equiv \operatorname{grad} f = -\frac{d^2}{du^2}.$$

Thus, the domain of the operator F consists solely of those functions on $D(f)$ having abolutely continuous first derivatives and square-integrable second derivatives on $[0,1]$.

This example shows that the form of the gradient and its domain depend on the space in which the functional is considered.

2.4. Generalized Lagrange formula and the Lipschitz inequality

Lemma 2.1. Let $f(x)$ be a functional having a linear Gâteaux differential at every point in a convex subset ω of a linear space E. Then the generalized Lagrange formula holds at any points $x,\ x+h \in \omega$, i.e.,

$$f(x+h) - f(x) = Df(x+\tau h, h), \qquad 0 < \tau < 1. \tag{2.2}$$

Proof. Set $\varphi(t) = f(x+th)$; then

$$\varphi'(t) = \frac{d}{dt} f(x+th) = Df(x+th, h),$$

so that

$$f(x+h) - f(x) = \varphi(1) - \varphi(0) = \varphi'(\tau) = Df(x+\tau h, h),$$

where $0 < \tau < 1$. Q. E. D.

Remark 2.1. Let E be a normed space and let $f(x)$ be defined on a set $X \supset \omega$ which is dense in E. Then, if the Gâteaux derivative of $f(x)$ is a bounded linear functional at each $x \in \omega$, we can write (2.2) as

$$f(x+h) - f(x) = \langle \operatorname{grad} f(x+\tau h), h \rangle, \qquad 0 < \tau < 1. \tag{2.3}$$

We now show that in some sense the generalized Lagrange formula remains valid for operators.

Lemma 2.2. Let $F(x)$ be an operator from a dense linear set X in a normed space E_x to a Banach space E_y, and suppose $F(x)$ has a linear Gâteaux differential at each point of a convex set $\omega \subset X$. Then the generalized Lagrange formula holds at all $x, x+h \in \omega$, i.e.,

$$\langle F(x+h) - F(x), e\rangle = \langle DF(x+\tau h, h), e\rangle, \qquad (2.4)$$
$$0 < \tau < 1, \quad \tau = \tau(e),$$

where e is an arbitrary unit vector in the space E_y^*.

Proof. Consider the functional $\varphi(x, z) = \langle F(x), z\rangle$, $z \in E_y^*$. It has a linear Gâteaux differential at $x \in \omega$, since

$$\frac{1}{t}[\varphi(x+th, z) - \varphi(x, z)] = \left\langle \frac{1}{t}[F(x+th) - F(x)], z\right\rangle,$$

and therefore

$$D\varphi((x, z), h) = \langle DF(x, h), z\rangle.$$

The functional $\varphi(x, z)$ satisfies the generalized Lagrange formula

$$\varphi(x+h, z) - \varphi(x, z) = D\varphi((x+\tau h, z), h), \qquad 0 < \tau < 1,$$

so that

$$\langle F(x+h) - F(x), z\rangle = \langle DF(x+\tau h, h), z\rangle.$$

Q. E. D.

Note that this proof uses the existence of $DF(x, h)$ on ω only in the weak sense.

Lemma 2.2. leads to the Lipschitz inequality for operators.

Lemma 2.3. Let $F(x)$ be a nonlinear operator from X to E_y. where X is a dense linear set in a normed space E_x, and suppose $F(x)$ has a bounded Gâteaux differential $DF(x, h)$ on a convex set $\omega \in X$. Then the Lipschitz inequality holds for all x $x, x+h \in \omega$, i.e.,

$$\| F(x+h) - F(x) \| \leqslant \| F'(x+\tau h) \| \| h \|, \quad 0 < \tau < 1. \qquad (2.5)$$

Proof. By a corollary of the Hahn-Banach Theorem,* for fixed x and h there exists a unit vector $e \in E_y^*$ such that

* See, for example, Hille and Phillips /1/, Theorem 2.7.4, or Dunford and Schwartz /1/, p. 65, Corollary 14.

$$\langle F(x+h)-F(x),\ e\rangle=\| F(x+h)-F(x)\| .$$

This fact and equality (2.4) imply (2.5).

2.5. Relation between Gâteaux and Fréchet derivatives

As noted above, the existence of the Fréchet derivative implies that of the Gâteaux derivative and their equality. The converse is generally false. For example, we proved in Vainberg /3/, pp. 91—92, that the Nemytskii operator in L^2 may have a linear Gâteaux differential everywhere but its Fréchet differential exists nowhere. However, we have the following assertion, which is analogous to the elementary theorem on differentiability of a function of several variables when all its partial derivatives are continuous.

Theorem 2.1. *Let $F(x)$ be a nonlinear operator mapping a dense linear set X in a normed space E_x into a Banach space E_y. Suppose $F(x)$ has a bounded linear Gâteaux differential $DF(x,h)=$ $=F'(x)h$ at every point of $S=U\cap X$, where U is a neighborhood of $x\in X$ in E_x. Then if $F'(x)$ is continuous at a point x_0 (in the topology of $L(E_x,E_y)$) with respect to the set S, the Fréchet differential exists and*

$$dF(x_0,\ h)=DF(x_0,\ h).$$

In other words, a continuous Gâteaux derivative is a Fréchet derivative.

Proof. Let $x_0+h\in S$ and

$$\omega(x_0,\ h)=F(x_0+h)-F(x_0)-F'(x_0)h,$$

where, by assumption, $F'(x_0)h=DF(x_0,\ h)$. Then for any unit vector $e\in E_y^*$ we have

$$\langle\omega(x_0,\ h),\ e\rangle=\langle F(x_0+h)-F(x_0),\ e\rangle-\langle F'(x_0)h,\ e\rangle;$$

using formula (2.4), we obtain

$$\langle\omega(x_0,\ h),\ e\rangle=\langle[F'(x_0+\tau h)-F'(x_0)]h,\ e\rangle,\ 0<\tau<1.$$

By a corollary of the Hahn-Banach Theorem, the vector e can be chosen so that for fixed h

$$\langle\omega(x_0,\ h),\ e\rangle=\|\omega(x_0,\ h)\| .$$

We now have the inequality

$$\| \omega(x_0, h) \| \leqslant \| F'(x_0 + \tau h) - F'(x_0) \| \| h \|$$

or, by our assumptions,

$$\lim_{h \to 0} \frac{\| \omega(x_0, h) \|}{\| h \|} = 0,$$

i. e., $DF(x_0, h) = dF(x_0, h)$. Q. E. D.

2.6. Differentiability and gradient of the norm

In subsection 2.3 (Example 2) we saw that the norm in a Hilbert space is differentiable and found its gradient. In general, the norm in a Banach space is differentiable if and only if the unit sphere $S = \{x : \|x\| = 1\}$ is smooth, i. e., S has a support plane at each of its points.*

As an example, let us evaluate the differential of the norm in $L^p(G)$, where $p > 1$ and G is a measurable set of finite or infinite measure in a finite-dimensional euclidean space.** For $p = 2$ this is a Hilbert space, and so we distinguish the following two cases:

1. $1 < p < 2$. Consider the function

$$\varphi(\xi) = (|1 + \xi|^p - 1 - p\xi) |\xi|^{-p}$$

defined on the real line. Since

$$\lim_{\xi \to \pm\infty} \varphi(\xi) = 1, \quad \lim_{\xi \to 0} \varphi(\xi) = 0,$$

it follows that φ is bounded and

$$C_1 |\xi|^p \leqslant |1 + \xi|^p - 1 - p\xi \leqslant C_2 |\xi|^p \qquad (C_1, C_2 = \text{const}).$$

Set $\xi = th(u)/x(u)$, where $h(u), x(u) \in L^p(G)$ and $x(u) \neq 0$. Upon integration we obtain

$$C_1 \| th \|^p \leqslant \| x + th \|^p - \| x \|^p - \\ - pt \int_G h(u) |x(u)|^{p-1} \operatorname{sign} x(u)\, du \leqslant C_2 \| th \|^p.$$

* See subsection 2.7, Corollary 2.1.

** See Mazur /1/.

Dividing these inequalities by t and letting $t \to 0$, we have

$$Dg(x, h) = p \int_G |x(u)|^{p-1} \operatorname{sign} x(u) h(u) \, du,$$

where $g(x) = \|x\|^p$.

2. Let $p > 2$. We consider the function

$$\psi(\xi) = (|1 + \xi|^p - 1 - p\xi)(|\xi|^p + \xi^2)^{-1}$$

and, by repeating the previous argument, obtain the same result. Let

$$f(x) = \|x\|, \quad g(x) = \|x\|^p = [f(x)]^p.$$

We have the equality

$$Dg(x, h) = p[f(x)]^{p-1} Df(x, h),$$

which we combine with previous results to yield

$$Df(x, h) = \|x\|^{1-p} \int_G |x(u)|^{p-2} x(u) h(u) \, du.$$

This linear functional is bounded, because $(p^{-1} + q^{-1} = 1)$ $|(x(u)|^{p-1} \in L^q(G)$, whence

$$\operatorname{grad} \|x\| = \|x\|^{1-p} |x(u)|^{p-2} x(u), \qquad x \neq 0. \tag{2.6}$$

In view of the fact that grad $\|x\|$ is an operator from L^p to L^q and is continuous at every nonzero $x \in L^p$, it follows from Theorem 2.1 that the norm in L^p is Fréchet-differentiable.

In the same fashion one can prove that the norm in the sequence space $l^p (p > 1)$ is Fréchet-differentiable and

$$\operatorname{grad} \|x\| = \|x\|^{1-p} z;$$
$$z = \left(|x_1|^{p-2} x_1, |x_2|^{p-2} x_2, \ldots\right) \in l^q,$$

where $p^{-1} + q^{-1} = 1$.

We now consider the general case. Let E be a normed space, $x \in E$ and $f(x) = \|x\|$. Let us first note that the norm is not differentiable at zero, since $\|th\|/t$ has no limit as $t \to 0$. Let the norm be differentiable. Then we have the inequality $t > 0$

$$(\|x+th_1+th_2\|-\|x\|)\,t^{-1}=$$
$$=(\|(x+2th_1)+(x+2th_2)\|-2\|x\|)(2t)^{-1}\leqslant$$
$$\leqslant(\|x+2th_1\|-\|x\|)(2t)^{-1}+(\|x+2th_2\|-\|x\|)(2t)^{-1}$$

and, letting $t\to 0$, we obtain

$$Vf(x,\ h_1+h_2)\leqslant Vf(x,\ h_1)+Vf(x,\ h_2).$$

Since this inequality holds for all vectors h_1 and h_2, we may replace h_1 and h_2 by $-h_1$ and $-h_2$, respectively, and, by the homogeneity of $V(x,\ h)$ with respect to h,

$$Vf(x,\ h_1+h_2)\geqslant Vf(x,\ h_1)+Vf(x,\ h_2),$$

i. e.,

$$Vf(x,\ h_1+h_2)=Vf(x,\ h_1)+Vf(x,\ h_2),$$

or

$$Vf(x,\ h)=Df(x,\ h).$$

Moreover, since

$$(\|x+th\|-\|x\|)\,t^{-1}\leqslant\|h\|\operatorname{sign}t,$$

it follows that

$$|Df(x,\ h)|\leqslant\|h\|.$$

Thus, we have proved

Lemma 2.4. If the norm of a normed space is Gâteaux-differentiable then the derivative exists at each nonzero point of the space and it is a bounded linear functional.

By this lemma, Gâteaux-differentiability of the norm implies the existence of its gradient

$$\operatorname{grad}\|x\|,\qquad x\neq 0.$$

This gradient maps the Banach space E into its dual E^*, since, for every fixed $x\in E\ (x\neq 0)$, grad $\|x\|$ is a bounded linear functional on E.

We cite some of the properties of the gradient of the norm.*

* See Vainberg /12/, /13/.

Lemma 2.5. If the norm in a normed space E *is Gâteaux-differentiable, then* grad $\|x\|$ *is an odd operator, defined for all* $x \neq 0$, *and*

$$\|\operatorname{grad}\|x\|\,\| = 1, \ \langle \operatorname{grad}\|x\|, x\rangle = \|x\|$$
$$u \ \operatorname{grad}\|\alpha x\| = \operatorname{sign}\alpha \operatorname{grad}\|x\|$$

for all real $\alpha \neq 0$.

Proof. By definition,

$$\langle \operatorname{grad}\|x\|, h\rangle = \lim_{t\to 0}(\|x+th\| - |x|)\,t^{-1}.$$

Letting $h = x$, we have for $|t| < 1$

$$\langle \operatorname{grad}\|x\|, x\rangle = \|x\|.$$

Hence

$$\|x\| = \langle \operatorname{grad}\|x\|, x\rangle \leqslant \|x\|\|\operatorname{grad}\|x\|\,\|$$

and

$$\|\operatorname{grad}\|x\|\,\| \geqslant 1. \tag{2.7}$$

If $t > 0$ and $\|h\| = 1$, we write

$$\langle \operatorname{grad}\|x\|, h\rangle = \lim_{t\to 0}(\|x+th\| - \|x\|)\,t^{-1} \leqslant 1,$$

and since

$$\|\operatorname{grad}\|x\|\,\| = \sup_{\|h\|=1}\langle \operatorname{grad}\|x\|, h\rangle,$$

it follows that

$$\|\operatorname{grad}\|x\|\,\| \leqslant 1.$$

By this inequality and (2.7),

$$\|\operatorname{grad}\|x\|\,\| = 1.$$

Finally, for $\alpha \neq 0$,

$$\langle \operatorname{grad}\|\alpha x\|, h\rangle = \lim_{t\to 0}(\|\alpha x + th\| - \|\alpha x\|)\,t^{-1} =$$
$$= \lim_{t\to 0}\left(\left\|x + \frac{t}{\alpha}h\right\| - \|x\|\right)\left(\frac{t}{\alpha}\operatorname{sign}\alpha\right)^{-1} = (\operatorname{sign}\alpha)\langle \operatorname{grad}\|x\|, h\rangle,$$

i. e.,

$$\operatorname{grad}\|\alpha x\| = (\operatorname{sign}\alpha)\operatorname{grad}\|x\|.$$

Q. E. D.

As we have seen (Theorem 2.1), a continuous Gâteaux derivative is a Fréchet derivative. It turns out that for norms the converse is also valid.

Lemma 2.6. The norm is Fréchet-differentiable if and only if its gradient is continuous.

Proof. Sufficiency follows from Theorem 2.1. We prove necessity.* By the homogeneity of the norm and the properties of the gradient, it suffices to prove continuity at any point on the unit sphere. Let the sequence $\{x_k\} \subset E$ converge to x ($\|x_k\| = \|x\| = 1$). Then

$$\lim_{k\to\infty} \langle \operatorname{grad}\|x_k\|, \quad x_k - x\rangle = 0,$$

since $\|\operatorname{grad}\|x_k\|\,\| = 1$. Taking into account that $\langle \operatorname{grad}\|x_k\|, x_k\rangle = \|x_k\|$, we see that for any $\delta > 0$ there exists k_0 such that for $k \geqslant k_0$

$$1 - \langle \operatorname{grad}\|x_k\|, x\rangle < \delta,$$

i. e., all vectors grad $\|x_k\|$ lie in the set

$$K(x, \delta) = \{f \in E^*: \|f\| = 1, \langle f, x\rangle \geqslant (1-\delta)\|x\|\}.$$

But, as shown by Shmul'yan /1/, the diameter of $K(x, \delta)$ converges to zero as $\delta \to 0$, so that

$$\lim_{k\to\infty} (\operatorname{grad}\|x_k\| - \operatorname{grad}\|x\|) = 0.$$

Q. E. D.

2.7. Directional derivative

As in subsection 2.1, let $F(x)$ be an operator mapping $D(F) \subset E_x$ into E_y. Then if the limit

$$V_+F(x, h) = \lim_{t\to+0} \frac{F(x+th) - F(x)}{t} \tag{2.8}$$

* See Kadets /1/.

exists for every unit vector $h \in D(F)$, it is called the variation of F at x in the direction h. Clearly, the directional variation coincides with the variation if

$$V_+F(x, h) = -V_+(x, -h).$$

As an example, consider a sublinear functional $p(x)$ on a real normed space E_x, that is, a real-valued functional which is subadditive and positive-homogeneous:

$$p(x+y) \leqslant p(x) + p(y) \quad (\forall x, y \in E_x);$$
$$p(\alpha x) = \alpha p(x) \quad \text{for} \quad \alpha > 0.$$

Lemma 2.7. A sublinear functional has a directional variation at every point.

Proof. Let $b \geqslant a > 0$. Then

$$p(bx + aby) \leqslant p(ax + aby) + p((b-a)x),$$

or

$$b[p(x+ay) - p(x)] \leqslant a[p(x+by) - p(x)],$$

i. e.,

$$\frac{1}{a}[p(x+ay) - p(x)] \leqslant \frac{1}{b}[p(x+by) - p(x)].$$

Thus, the function

$$g(t) = \frac{1}{t}[p(x+th) - p(x)]; \quad t > 0; \quad x, h \in E_x$$

decreases with decreasing t, and since

$$p(x) \leqslant p(x+th) + p(-th),$$

it follows that

$$g(t) \geqslant -p(-h).$$

But $g(t)$ is bounded below, and so the limit

* See Dunford and Schwartz /1/, p. 445. Since a sublinear functional $p(x)$ is the Minkowski functional of a convex set $K \subset E_x$ (see Naimark /1/, p. 30), $V_+p(x, h)$ is known as the tangent function of the set K.

$$\lim_{t \to +0} g(t) = V_+(p)(x, h)$$

exists. Q. E. D.

Set $V_-p(x, h) = -V_+p(x, -h)$. Lemma 2.7 can be applied to the functional $p(x) = \|x\|$, because the norm in the space E_x is a sublinear functional. If for any h and fixed $x \neq 0$

$$V_-p(x, h) = V_+p(x, h), \tag{2.9}$$

it follows by Lemma 2.4 that grad $\|x\|$ exists for $x \neq 0$.

Let us proceed to verify the validity of (2.9). To this end, consider the real-valued function

$$\varphi(t) = \|x + th\| \qquad (-\infty < t < +\infty),$$

which takes finite values. It is convex because, for $\alpha \geqslant 0$, $\beta \geqslant 0$ and $\alpha + \beta = 1$, we have

$$\varphi(\alpha t_1 + \beta t_2) = \|\alpha(x + t_1 h) + \beta(x + t_2 h)\| \leqslant \alpha\varphi(t_1) + \beta\varphi(t_2).$$

But any convex function $\varphi(t)$ satisfies

$$\frac{\varphi(t_1) - \varphi(t)}{t_1 - t} \leqslant \frac{\varphi(t_2) - \varphi(t)}{t_2 - t} \qquad (t_1 < t < t_2),$$

implying the existence of $\varphi'_-(t)$, $\varphi'_+(t)$ and the inequality $\varphi'_-(t) \leqslant \varphi'_+(t)$.

Consider a straight line through the point $M = (t, \varphi(t))$, with slope α such that

$$\varphi'_-(t) \leqslant \alpha \leqslant \varphi'_+(t).$$

This straight line is a support line of the curve $\varphi(t)$ at M. The support line at M is unique if and only if $\varphi'_-(t) = \varphi'_+(t)$, i. e., the derivative $\varphi'(t)$ exists. Obviously, (2.9) means that $\varphi'_-(0) = \varphi'_+(0)$, so that (2.9) holds if and only if the curve $\varphi(t)$ has a unique support at the point $M = (0, \varphi(0))$. We now state an additional test for the validity of (2.9).

Let x_0 be a boundary point of the ball $D = \{x \in E_x : \|x\| \leqslant r,\ r > 0\}$. Then there is a support hyperplane to D through this point,* i. e., there exists a linear functional $f \in E_x^*$ such that $f(x_0) = \|x_0\| = r$ and $f(x) \leqslant \|x\|$ for $x \in D$.

* See, for example, Naimark /1/, p. 37.

*Lemma 2.8.** *The equality (2.9), where* $x = x_0$ $(\|x_0\| = r > 0)$ *and* h *is an arbitrary unit vector in* E_x, *is valid if and only if the support hyperplane to the ball* D *at* x_0 *is unique.*

Lemmas 2.4, 2.5 and 2.7 imply

Corollary 2.1. The gradient of the norm exists at every nonzero point of the space E_x *if and only if the unit sphere* $S \subset E_x$ *is smooth, i.e., there is a unique support hyperplane at every point of the sphere.***

We note furthermore that, for a finite real-valued convex functional $p(x)$ defined on E_x, $V_- p(x,h)$ and $V_+ p(x, h)$ also exist. To see this, it suffices to consider (for fixed $x \in E_x$) the convex function

$$f(t) = p(x + th), \qquad t \in (-\infty, +\infty),$$

for which $f'_-(t)$ and $f'_+(t)$ exist, and to use the fact that $f'_-(0) = V_- p(x, h)$ and $f_+(0) = V_+ p(x, h)$. We shall return to this question in subsection 8.4.

2.8. Schwartz differentiability

Let $F(x)$ be an operator mapping a dense linear set $D(F) \subset E_x$ into E_y. Then if the limit

$$V_s F(x, h) = \lim_{t \to 0} \frac{F(x + th) - F(x - th)}{2t}, \qquad x \in D(F)$$

exists for any vector $h \in D(F)$, $V_s F(x, h)$ is called the Schwartz variation of $F(x)$.

If $F(x)$ has a directional variation at a fixed point x, then the Schwartz variation also exists at this point. Indeed, for $t > 0$

$$\lim_{t \to 0} \frac{E(x + th) - F(x - th)}{2t} =$$

$$= \frac{1}{2}\left[\lim_{t \to 0} \frac{F(x + th) - F(x)}{t} - \lim_{t \to 0} \frac{F(x + t(-h)) - F(x)}{t}\right] =$$

$$= \frac{1}{2}[V_+ F(x, h) + V_- F(x, h)]$$

(see subsection 2.7), and for $t < 0$, $\tau = -t$,

$$\lim_{t \to 0} \frac{F(x + th) - F(x - th)}{2t} =$$

$$= \lim_{\tau \to 0} \frac{F(x + \tau h) - F(x - \tau h)}{2\tau} = \frac{1}{2}[V_+ F(x, h) + V_- F(x, h)].$$

* See Dunford and Schwartz /1/, p. 449.

** See Banach /1/, p. 144, James /1/, Cudia /1/, Sunderesan /1/.

It thus follows from the results of the previcus subsection that every sublinear or convex functional defined in E_x has a Schwartz variation at each point.

The Schwartz variation is a homogeneous operator, i. e.,

$$V_sF(x, \alpha h) = \alpha V_sF(x, h).$$

It need not, however, be additive in h.

§ 3. DIFFERENTIABILITY IN LINEAR TOPOLOGICAL SPACES

3.1. Variation and generalized Lagrange formula

Let E_x and E_y be real linear topological spaces and $F(x)$ a mapping from E_x to E_y. The directional variation $V_+F(x, h)$, Schwartz variation $V_sF(x, h)$, and Gâteaux variation $VF(x, h)$ are defined as in normed spaces. V_sF and VF are homogeneous functions of h; for example $VF(x, \alpha h) = \alpha VF(x, h)$ for any real α. As in normed spaces, $VF(x, h)$ need not be additive in h. To obtain sufficient conditions for $VF(x,h)$ to be additive in h, we first prove a generalized Lagrange formula for functionals.

Lemma 3.1. If a real functional $f(x)$ has variation $Vf(x,h)$ on a convex set $K \subset E_x$, then for all x, $x+h \in K$, there exists $\tau \in (0, 1)$ such that

$$f(x+h) - f(x) = Vf(x+\tau h, h). \tag{3.1}$$

Proof. Let $\varphi(t) = f(x+th)$, $t \in (0, 1)$. By hypothesis, the real function $\varphi(t)$ is continuous on $[0,1]$. We have

$$\varphi'(t) = \lim_{\Delta t \to 0} \frac{f(x+th+h\,\Delta t) - f(x+th)}{\Delta t} = Vf(x+th, h),$$

whence, by the classical Lagrange theorem,

$$f(x+h) - f(x) = \varphi(1) - \varphi(0) = \varphi'(\tau) = Vf(x+\tau h, h),$$

where $0 < \tau < 1$.

3.2. Conditions for linearity of the Gâteaux variation

*Lemma 3.2.** *Let $f(x)$ be a functional defined in some neighborhood U of a point $x_0 \in E_x$ and suppose $f(x)$ satisfies the following conditions:*

1. $Vf(x, h)$ exists for all $x \in U$ and is continuous at x_0 for every fixed $h \in E$.

2. $Vf(x_0, h)$ is continuous at $h=0$.

Then $Vf(x_0, h)$ is a continuous linear functional with respect to h.

Proof. It suffices to prove additivity. Fix h_1 and h_2. Select $\delta(\varepsilon) > 0$ such that, for $|t| < \delta$, we have $x_0 + th_1 \in U$, $x + th_1 + th_2 \in U$ and

$$\left.\begin{aligned} &\left|Vf(x_0, h_1) - \frac{1}{t}(f(x_0 + th_1) - f(x_0))\right| < \frac{\varepsilon}{4}, \\ &\left|Vf(x_0, h_2) - \frac{1}{t}(f(x_0 + th_2) - f(x_0))\right| < \frac{\varepsilon}{4}, \\ &\left|Vf(x_0, h_1 + h_2) - \frac{1}{t}(f(x_0 + th_1 + th_2) - f(x_0))\right| < \frac{\varepsilon}{4}; \end{aligned}\right\} \tag{3.2}$$

$$\left.\begin{aligned} &\left|Vf(x_0 + th_2 + \tau_1 th_1, h_1) - Vf(x_0, h_1)\right| < \frac{\varepsilon}{8}, 0 < \tau_1 < 1, \\ &\left|Vf(x_0 + \tau_2 th_1, h_1) - Vf(x_0, h_1)\right| < \frac{\varepsilon}{8}, \quad 0 < \tau_2 < 1. \end{aligned}\right\} \tag{3.3}$$

It follows from (3.3) and (3.1) that

$$\left|\frac{1}{t}\left[f(x_0 + th_1 + th_2) + f(x_0) - f(x_0 + th_2) - f(x_0 + th_1)\right]\right| =$$
$$= |Vf(x_0 + th_1 + \tau_1 th_1, h_1) - Vf(x_0 + \tau_2 th_1, h_1)| < \frac{\varepsilon}{4}.$$

Hence, by (3.2),

$$|Vf(x_0, h_1 + h_2) - Vf(x_0, h_1) - Vf(x_0, h_2)| < \varepsilon$$

and, finally, since the choice of $\varepsilon > 0$ was arbitrary, , we have

$$Vf(x_0, h_1 + h_2) = Vf(x_0, h_1) + Vf(x_0, h_2).$$

Q. E. D.

*Lemma 3.3.*** *Let $F(x)$ be an operator from E_x to E_y, defined on some neighborhood U of a point $x \in E_x$ and let the following conditions be satisfied:*

* See Vainberg /2/, /3/.

** See Marinescu /1/, Vainberg /9/.

1. $VF(x, h)$ exists at every point $x \in U$ and is continuous at x_0 for every fixed $h \in E$.

2. $VF(x_0, h)$ is continuous at $h=0$.

Then $VF(x_0, h)$ is a linear operator from E_x to E_y, continuous with respect to h.

Proof. Let E_y' be the strong dual of E_y and let $z \in E_y'$. Then the functional

$$f(x) = z(F(x)) = \langle F(x), z\rangle$$

has Gâteaux variation

$$Vf(x, h) = \langle VF(x, h), z\rangle$$

satisfying the hypotheses of Lemma 3.2, so that

$$\langle VF(x_0, h_1 + h_2), z\rangle = \langle VF(x_0, h_1), z\rangle + \langle VF(x_0, h_2), z\rangle.$$

Hence, since $Z \in E_y'$ is an arbitrary element, it follows that*

$$VF(x_0, h_1 + h_2) = VF(x_0, h_1) + VF(x_0, h_2).$$

Thus, by condition 2 of the lemma, $VF(x_0, h)$ is a continuous linear operator.

3.3. Gâteaux differential and derivative; gradient of a functional

Let F be a mapping from E_x to E_y having variation $VF(x_0, h)$ at a point x_0. If the variation is linear in h, it is denoted by $DF(x_0, h)$ and called the Gâteaux differential of the mapping F at x_0. As a linear operator, $DF(x_0, h)$ can be expressed as

$$DF(x_0, h) = F'(x_0)h,$$

where $F'(x_0)$ is an element in the space of all linear mappings from E_x into E_y, known as the Gâteaux derivative of the operator F at x_0. Thus the correspondence $x \to F'(x)$ defines a mapping of E_x into the space of linear mappings from E_x to E_y.

If $f(x)$ is a given functional on E_x, we can speak of its Gâteaux derivative $f'(x)$. If $Vf(x, h)$ is a continuous linear functional in h, the derivative $f'(x)$ is called the gradient of the functional.

* We consider linear topological spaces having "enough" continuous linear functionals.

Thus

$$\operatorname{grad} f(x) = f'(x),$$

if $f'(x)h$ is a continuous linear functional for every fixed x.

Let E'_x be the dual of E_x. If an operator $\Phi(x)$ from E_x to E'_x is the gradient of some functional $f(x)$, we call $\Phi(x)$ a potential operator and $f(x)$ is called its potential.

3.4. On the various notions of differentiability

In normed spaces, Gâteaux and Fréchet differentiability are the main types of differentiability; moreover, any continuous Gâteaux derivative is also a Fréchet derivative.

More than twenty different notions of differentiability are known in linear topological spaces.* We examine some of them here. The different notions of differentiability depend on the topology τ of the space $E_{xy} = \mathscr{L}(E_x, E_y)$ of continuous linear mappings from E_x to E_y. We consider the most important topologies.** Let γ be a system of bounded subsets of E_x such that any finite union of sets in γ belongs to γ and the union of all sets in γ is dense in E_x. A neighborhood of zero $W(M, V) \subset L(E_x, E_y)$ is defined as the set of linear operators $A \in L(E_x, E_y)$ such that $A(M) \subset V$, where $M \in \gamma$ and V is a neighborhood of zero in E_y. If the space E_y is locally convex and $\{V\}$ is a fundamental system of convex neighborhoods of zero in E_y, then the set $\{W(M, V)\}$ forms a fundamental system of convex neighborhoods in the space $L(E_x, E_y)$ where M ranges over γ. This topology is called the *γ-topology* on $L(E_x, E_y)$, it is separable and locally convex. The most important cases are the following:

1. If γ is the set of all finite subsets of E_x, the γ-topology in $L(E_x, E_y)$ is the topology of pointwise convergence (the weakest topology).

2. If γ is the set of compact subsets of E_x, the γ-topology is the topology of compact convergence.

3. If γ is the set all bounded subsets of E_x, the γ-topology in $L(E_x, E_y)$ is known as the topology of bounded convergence (this is the strongest γ-topology in $L(E_x, E_y)$]).

Definition 3. We say that a mapping $F(x)$ from E_x into E_y is *τ-differentiable at a point $x_0 \in E_x$ in the sense*

* See Averbukh and Smolyanov /1/, /2/, for a discussion and classification of all these notions, including the equivalence of some of them to differentiability of mappings.

** See Bourbaki /1/.

of VE* if, in some neighborhood U of x_0, F is Gâteaux-differentiable and its derivative $F'(x)$ is continuous at x_0 as a mapping from E_x into the space $L(E_x, E_y)$ of continuous linear mappings with respect to the topology τ.

Definition 3.2. A mapping $F(x)$ from E_x into E_y is differentiable at a point $x \in E_x$ in the sense of Sebastião e Silva** if there exists a continuous linear operator $A \in L(E_x, E_y)$ such that

$$F(x+h) - F(x) = Ah + \omega(x, h),$$

where $\omega(x, th)/t \to 0$ as $t \to 0$, $t \neq 0$, uniformly, in $h \in B$ for any $B \in \gamma$.

By selecting different systems γ of subsets of E_x, we obtain various types of differentiability in the sense of Sebastião e Silva.

§4. MULTILINEAR MAPPINGS AND TAYLOR'S FORMULA

4.1. Multilinear mappings

Let $E_1, E_2, \ldots, E_k$ be k linear spaces and let $E = E_1 \times E_2 \ldots \times E_k$. A mapping $u_k(x_1, x_2, \ldots, x_k)$ from E to a linear space E_y is said to be k-linear if it is linear in each $x_i \in E_i$. If the spaces are normed, then a k-linear operator u_k is said to be bounded if

$$\|u_k(x_1, x_2, \ldots, x_k)\| \leqslant M \|x_1\| \ldots \|x_k\|. \tag{4.1}$$

The smallest possible constant M in (4.1) is called the norm of u_k, denoted by $\|u_k\|$.

Theorem 4.1. A k-linear operator u_k defined in normed spaces is continuous if and only if it is bounded.

Proof. Necessity. Since u_k is continuous at the point $(0, 0, \ldots, 0)$, it follows that if

$$(x_1, x_2, \ldots, x_k) \in D = \{x \in E: \|x_i\| \leqslant r;\ i = 1, 2, \ldots, k\},$$

then

$$\|u_k(x_1, x_2, \ldots, x_k)\| \leqslant 1.$$

* See Averbukh and Smolyanov /2/, Vainberg and Engel'son /2/.

** See Sebastião e Silva /1/, Averbukh and Smolyanov /1/, /2/.

Consider an arbitrary point $(x_1, x_2, \ldots, x_k)$. If $x_i = 0$ for some i, then $u_k(x_1, x_2, \ldots, x_k) = 0$, and thus (4.1) holds for any $M \geqslant 0$. Let $x_i \neq 0$ for every $i = 1, 2, \ldots, k$; then

$$\left\| u_k \left(\frac{rx_1}{\|x_1\|}, \frac{rx_2}{\|x_2\|}, \ldots, \frac{rx_k}{\|x_k\|} \right) \right\| \leqslant 1,$$

hence

$$\|u_k(x_1, x_2, \ldots, x_k)\| \leqslant r^{-k} \|x_1\| \|x_2\| \ldots \|x_k\|.$$

Sufficiency. Let (4.1) hold. We fix a point $(a_1, a_2, \ldots, a_k)$ and prove continuity of u_k at this point. For arbitrary $\varepsilon > 0$, set

$$\|a_i\| + 1 \leqslant a, \quad 0 < \delta < 1, \quad \delta \leqslant (kM)^{-1} a^{1-k} \varepsilon.$$

Then, if $\|x_i - a_i\| < \delta$ $(i = 1, 2, \ldots, k)$,

$$\begin{aligned} &\|u_k(x_1, x_2, \ldots, x_k) - u_k(a_1, a_2, \ldots, a_k)\| \leqslant \\ &\leqslant \|u_k(x_1 - a_1, x_2, \ldots, x_k) + u_k(a_1, x_2 - a_2, x_3, \ldots, x_k) + \ldots \\ &\qquad \ldots + u_k(a_1, a_2, \ldots, x_k - a_k)\| \leqslant kM\delta a^{k-1} \leqslant \varepsilon, \end{aligned}$$

i. e., u_k is continuous at $(a_1, a_2, \ldots, a_k)$. In considering k-linear operators, we shall assume that $E_1, E_2, \ldots, E_k$ are all copies of the same space, i. e., $x_i \in E_x$ $(i = 1, 2, \ldots, k)$.

Examples of k-linear operators are the following operators in the L-spaces:

$$\begin{aligned} &I_k(x_1, x_2, \ldots, x_k) = \\ &\quad = \int_a^b \ldots \int_a^b K_k(s, t_1, \ldots, t_k) x_1(t_1) \ldots x_k(t_k) \, dt_1 \ldots dt_k. \end{aligned}$$

If we set $x_1 = x_2 = \ldots = x_k = x$ in a k-linear operator from E_x to E_y, we get a homogeneous operator $P(x) = u_k(x, \ldots, x)$ of degree k, since $P(\alpha x) = \alpha^k P(x)$.

The sum of a finite number of homogeneous operators all of degree k is called a homogeneous polynomial of degree This k-linear operator is said to be symmetric if its values do not change under any permutation of the arguments. A symmetric k-linear operator is called a k-linear form. If K_k is a symmetric function of $t_1, t_2, \ldots, t_k$, then $I_k(x, x, \ldots, x)$ is an example of a k-linear form. If u_k is a k-linear form, then $u_k(x, x, \ldots, x)$ is called a power and denoted by $u_k x^k$.

4.2. Higher-order derivatives

Let E_x and E_y be linear topological spaces, $F(x)$ a mapping from E_x to E_y, and $F'(x)$ its Gâteaux derivative, defined in a neighborhood U of a point x_0. If $F'(x)$, viewed as a mapping of U into the space of linear operators from E_x to E_y, has a derivative $(F'(x))'$, the latter is denoted by $F''(x)$ and called the second Gâteaux derivative.

Higher order derivatives are defined in analogous-fashion. Suppose that the Gâteaux derivatives $F'(x)$, $F''(x)$, ..., $F^{(k)}(x)$ exist in some neighborhood U of a point x_0. Let us denote by $D = D(F'(x_0))$ the domain of the linear operator $F'(x_0)$ and assume that D is constant when x_0 ranges over U. Then, for all fixed $h \in D$, the mapping $F'(x)h$ from E_x to E_y has a derivative $(F'(x)h)' = F''(x)h$, which is a linear operator for each fixed $x \in U$.

Suppose the domain of this operator is D, so that for all fixed h, $g \in D$, $[F''h]g = F''(x)hg$ is a mapping of U into E_y, and for fixed $x \in U$ the mapping $F''(x)hg$ is bilinear in h, $g \in D$. Of course, if the spaces are normed and $F'(x)$ and $F''(x)h$ are bounded linear operators for every $x \in U$, then D is all of E_x and the bilinear operator $F^n(x)hg$ is defined for all h, $g \in E_x$. Assuming $F^{(i)}(x)h_1h_2 \ldots$ $\ldots h_{i-1}$ to be a linear operator with domain D for fixed $h_1, h_2, \ldots, h_{i-1}$ and any fixed $x \in U$, we see that $F^{(i)}(x)h_1h_2 \ldots h_i$ is an i-linear operator $(1 < i \leqslant k)$ with domain D.

Theorem 4.2. *Suppose the mapping $F(x)$ from a linear topological space E_x to a linear topological space E_y is k times Gâteaux-differentiable in some neighborhood U of a point x_0. Let $F^{(k)}(x)h_1h_2 \ldots h_k$ be defined for all $h_i \in D$, where D is a linear subset of E_x. If for any fixed $h_1, h_2, \ldots, h_k \in D$ the mapping $F^{(k)}(x)h_1h_2 \ldots h_k$ of U into E_y is hemicontinuous, then the k-linear operator $F^{(k)}(x_0)$ $h_1h_2 \ldots h_k$ is symmetric.*

Proof. $(k = 2)$. We consider the expression

$$\Delta = \langle F(x_0 + ah_1 + bh_2) - F(x_0 + ah_1) - F(x_0 + bh_2) + F(x_0), e\rangle,$$

where a, b are arbitrary positive numbers such that $x_0 + ah_1 + bh_2 \in U$ and e is an arbitrary element of the strong dual E'_y of E_y. As in the proof of Theorem 5.1 in VM, we obtain

$$\langle F''(x_0 + \tau_1 bh_2 + \tau_2 ah_1) h_1h_2, e\rangle = \langle F''(x_0 + \tau_3 ah_1 + \tau_4 bh_2) h_2h_1, e\rangle,$$

where $0 < \tau_i < 1$, $i = 1, 2, 3, 4$.

Using the hemicontinuity and letting $a \to 0$ and then $b \to 0$ we obtain

$$\langle F''(x_0) h_1 h_2, e\rangle = \langle F''(x_0) h_2 h_1, e\rangle,$$

whence since $e \in E'_y$ is an arbitrary element, it follows that

$$F''(x_0) h_1 h_2 = F''(x_0) h_2 h_1.$$

Q. E. D.

If the mapping is Fréchet differentiable, the hypotheses of the theorem may be weakened.

Suppose E_x and E_y are normed spaces, $F(x)$ is a mapping of some neighborhood U of $x_0 \in E_x$ into E_y, and $F'(x)$ is its Fréchet derivative, defined for all $x \in U$. Moreover, suppose that for any fixed $x \in U$ the derivative is a continuous linear operator, i. e., $F'(x)$ is a mapping of U into $E_{xy} = \mathscr{L}(E_x, E_y)$. If $F'(x)$ has a Fréchet derivative, it is denoted by $F''(x)$. Suppose that for any fixed $x \in U$ and $h \in E_x$ the linear operator $F''(x)h$ is continuous. Higher-order derivatives are defined in similar fashion, and again $F^{(k)}(x)$ is a k-linear operator, i. e., $F^{(k)}(x) h_1, h_2 \ldots h_k$ is a bounded k-linear operator for any fixed $x \in U$.

Theorem 4.3. If a mapping $F(x)$ from a normed space E_x to a normed space E_y is k times Fréchet-differentiable at a point x_0, then the k-linear operator $F^{(k)}(x_0) h_1 h_2 \ldots h_k$ is symmetric.

Proof. ($k = 2$). Since $x_0 \in U$, we can find $r > 0$ such that, whenever $\|h\| = r$ and $\|g\| = r$, we have $x_0 + th + tg \in U$ for $0 < t \leqslant 1$. Let us consider two abstract functions from $[0, 1]$ to E_y:

$$\varphi(t) = F(x_0 + th + g) - F(x_0 + th),$$
$$\psi(t) = F(x_0 + h + tg) - F(x_0 + tg).$$

We have

$$\varphi(1) - \varphi(0) = \psi(1) - \psi(0) =$$
$$= F(x_0 + h + g) - F(x_0 + h) - F(x_0 + g) + F(x_0).$$

Hence

$$\varphi'(t) = [F'(x_0 + th + g) - F'(x_0 + th)]\, h =$$
$$= [F'(x_0 + th + g) - F'(x_0)]\, h - [F'(x_0 + th) - F'(x_0)]\, h.$$

By Fréchet differentiability

$$F'(x_0 + th + g) - F'(x_0) = F''(x_0)(th + g) + o(th + g),$$
$$F'(x_0 + th) - F'(x_0) = F''(x_0)\, th + o(th),$$

whence

$$\varphi'(t) = F''(x_0)\, gh + [o(th+g) + o(th)]\, h.$$

Similarly,

$$\psi'(t) = F''(x_0)\, hg + [o(h+tg) + o(tg)]\, g.$$

Using the generalized Lagrange formula, we have

$$\begin{aligned} J_e &= \langle \varphi(1) - \varphi(0), e\rangle = \langle \varphi'(t_1), e\rangle = \\ &= \langle F''(x_0)\, gh, e\rangle + \langle [o(t_1h+g) + o(t_1h)]\, h, e\rangle, \\ J_e &= \langle \psi(1) - \psi(0), e\rangle = \langle F''(x_0)\, hg, e\rangle + \\ &\qquad + \langle o(h+t_2g) + o(t_2g)]\, h, e\rangle, \end{aligned}$$

where $t_1,\ t_2 \in (0, 1)$ and e is an arbitrary vector in E_y^*. Replacing h and g by τh and τg, we have

$$\begin{aligned} \langle F''(x_0)\, gh - F''(x_0)\, hg, e\rangle &= \frac{1}{\tau} \langle o(\tau h + t_2\tau g)\, g + \\ &+ o(t_1\tau h + \tau g)\, h + o(t_1\tau h)\, h + o(t_2\tau g)\, g, e\rangle. \end{aligned}$$

Letting $\tau \to 0$, we obtain

$$\langle F''(x_0)\, gh - F''(x_0)\, hg, e\rangle = 0$$

or, since $e \in E_y^*$ is arbitrary,

$$F''(x_0)\, gh = F''(x_0)\, hg.$$

Q. E. D.

Note that the assertion of this theorem holds for mappings of one linear topological space into another if they are k times differentiable in the sense of VE or of Sebastião e Silva.*

4.3. On the Riemann integral

Let $f(t)$ be a function of a scalar variable $t \in [a, b]$, taking values in a vector space E.

We consider a partition of $[a, b]$ by points

$$t_0 = a < t_1 < t_2 < \ldots < t_n = b,$$

* Averbukh and Smolyanov /1/.

and select one point τ_k in each interval $\Delta t_k = t_{k+1} - t_k$. Consider the Riemann sum

$$\sigma = \sum_{k=0}^{n-1} f(\tau_k)\,\Delta t_k.$$

If σ has a limit as $\max \Delta t_k \to 0$, it is known as the Riemann integral of the abstract function $f(t)$ and denoted by

$$\int_a^b f(t)\,dt. \tag{4.2}$$

*Theorem 4.4.** *If a function $f(t)$ taking values in a Banach space is bounded and almost everywhere continuous on $[a, b]$, it is integrable on $[a, b]$.*

Let $f(t)$ be differentiable on $[a, b]$. We are interested in the validity of the equality

$$\int_a^b f'(t)\,dt = f(b) - f(a). \tag{4.3}$$

As we know from any standard course of analysis, (4.3) holds if the real-valued function $f(t)$ has a Riemann-integrable derivative $f'(t)$ at every point of $[a, b]$, i. e., the derivative is bounded and almost everywhere continuous.

Let $f(t)$ be a function, with values in a Banach space E; suppose that $f'(t)$ exists at every point of $[a, b]$ and is bounded and a. e. continuous.

Consider the function

$$\varphi(t) = \langle f(t), e\rangle,\ e \in E^*.$$

Then $\varphi'(t) = \langle f'(t), e\rangle$ and $\varphi'(t)$ is bounded and almost everywhere continuous. But $\varphi'(t)$ satisfies (4.3), whence

$$\int_a^b \langle f'(t), e\rangle\,dt = \langle f(b) - f(a), e\rangle.$$

Since for any continuous functional $e \in E^*$

$$\int_a^b \langle f'(t), e\rangle\,dt = \left\langle \int_a^b f'(t)\,dt, e\right\rangle,$$

it follows that (4.3) is valid. We have thus proved

* Graves /1/.

Theorem 4.5. If $f(t)$ is differentiable at each point of $[a, b]$ and the derivative is bounded and almost everywhere continuous, then (4.3) is valid.

The simplest properties of the Riemann integral remain valid for the integral (4.2):

$$1) \quad \int_a^b cf(t)\,dt = c\int_a^b f(t)\,dt, \quad c = \text{const};$$

$$2) \quad \int_a^b f(t)\,dt = \int_a^{t_1} f(t)\,dt + \int_{t_1}^b f(t)\,dt, \quad a < t_1 < b;$$

$$3) \quad \int_a^b [f_1(t) + f_2(t)]\,dt = \int_a^b f_1(t)\,dt + \int_a^b f_2(t)\,dt;$$

$$4) \quad \left\| \int_a^b f(t)\,dt \right\| \leqslant \int_a^b \|f(t)\|\,dt \quad (a < b),$$

for a normed vector space E.

We shall also need the following analog of integration by parts.

Theorem 4.6. Let $\varphi(t)$ be an abstract function from $[a, b]$ to E and $g(t)$ a real-valued function of a real variable $t \in [a, b]$. If $\varphi'(t)$ and $g'(t)$ exist at every point $t \in [a, b]$ and they are bounded and almost everywhere continuous on $[a, b]$, then

$$\int_a^b \varphi'(t)\,g(t)\,dt + \int_a^b \varphi(t)\,g'(t)\,dt = \varphi(b)\,g(b) - \varphi(a)\,g(a).$$

Indeed, setting $\varphi(t)g(t) = f(t)$, we see that the assertion follows from Theorem 4.5.

4.4. Integral over a segment

Let $\Phi(x)$ be a mapping of a convex set $\omega \subset E_x$ into $E_{xy} = L(E_x, E_y)$, so that, for every $h \in E_x$, $\Phi(x)h$ is an operator from E_x to E_y. Consider a segment $[x_1, x_2] \subset \omega$, i. e. the set $x = x_1 + t(x_2 - x_1)$, where $0 \leqslant t \leqslant 1$.

By definition,

$$\int_{x_1}^{x_2} \Phi(x)\,dx = \int_0^1 \Phi[x_1 + t(x_2 - x_1)](x_2 - x_1)\,dt.$$

This integral is called the integral over the segment $[x_1, x_2]$. By Theorem 4.4, it exists if E_y is a Banach space and the abstract function $f(x)=\Phi[x_1+t(x_2-x_1)](x_2-x_1)$ is bounded and almost everywhere continuous in $t\in[0,1]$.

If a mapping $F(x)$ from E_x to E_y is k times differentiable and the mapping

$$\Phi(x)=F^{(k)}(x)(x_2-x)^{k-1},\quad x=x_1+t(x_2-x_1),$$

takes E_x into $\mathscr{L}(E_x, E_y)$, then by definition

$$\int_{x_1}^{x_2} F'(x)\,dx=\int_0^1 F'[x_1+t(x_2-x_1)](x_2-x_1)\,dt,$$

$$\int_{x_1}^{x_2} F^{(i)}(x)(x_2-x)^{i-1}\,dx=\int_0^1 F^{(i)}[x_1+t(x_2-x_1)](x_2-x_1)^i(1-t)^{i-1}\,dt$$
$$(i=1,2,\ldots,k).$$

4.5. Taylor's formula with remainder in integral form

Theorem 4.7. Let $F(x)$ be a mapping of a convex set in a linear space E_x into a Banach space E_y, which satisfies the following condition: for all $x_1, x_2\in\omega$ the abstract function

$$\varphi(t)=F[x_1+t(x_2-x_1)]$$

is n times differentiable at every point $t\in[0,1]$, $\varphi^{(n)}(t)$ is bounded and almost everywhere continuous on $[0,1]$. Then for any $x_1, x_2\in\omega$, we have the following analog of Taylor's formula:

$$F(x_2)=F(x_1)+\sum_{k=1}^{n-1}\frac{1}{k!}F^{(k)}(x_1)(x_2-x_1)^k+r_n(x_1,x_2), \tag{4.4}$$

where

$$r_n(x_1,x_2)=\frac{1}{(n-1)!}\int_{x_1}^{x_2}F^{(n)}(x)(x_2-x)^{n-1}\,dx.$$

Proof. The boundedness of $\varphi^{(n)}(t)$ implies continuity of $\varphi^{(k)}(t)$ for $k\leqslant n-1$ and hence its boundedness on $[0,1]$. For $n=1$, Taylor's formula takes the form

$$F(x_2)=F(x_1)+\int_{x_1}^{x_2}F'(x)\,dx=$$
$$=F(x_1)+\int_0^1 F'[x_1+t(x_2-x_1)](x_2-x_1)\,dt. \tag{4.5}$$

Since

$$F'[x_1 + t(x_2 - x_1)](x_2 - x_1) = \frac{d}{dt} F[x_1 + t(x_2 - x_1)],$$

it follows by Theorem 4.5 that

$$\int_0^1 \frac{d}{dt} F[x_1 + t(x_2 - x_1)]\,dt = F(x_2) - F(x_1),$$

so that (4.5) holds. Thus, the theorem holds for $n = 1$. We proceed by induction on n. Suppose (4.5) is valid for $i = m - 1$ and let us prove it holds for $i = m$. Let

$$F(x_2) = F(x_1) + \sum_{k=1}^{m-2} \frac{1}{k!} F^{(k)}(x)(x_2 - x_1)^k + r_{m-1}(x_1, x_2).$$

Consider the expression

$$r_m(x_1, x_2) = \frac{1}{(m-1)!} \int_{x_1}^{x_2} F^{(m)}(x)(x_2 - x)^{m-1}\,dx =$$

$$= \frac{1}{(m-1)!} \int_0^1 F^{(m)}[x_1 + t(x_2 - x_1)](x_2 - x_1)^m (1 - t)^{m-1}\,dt. \qquad (4.6)$$

But

$$F^{(m)}[x_1 + t(x_2 - x_1)](x_2 - x_1)^m =$$
$$= \frac{d}{dt} F^{(m-1)}[x_1 + t(x_2 - x_1)](x_2 - x_1)^{m-1} = \frac{d}{dt}\psi(t),$$

and $(1 - t)^{m-1}$ is continuously differentiable. Hence, we may apply Theorem 4.6 to the last integral in (4.6) and write

$$r_m(x_1, x_2) =$$
$$= \left[\frac{1}{(m-1)!} F^{(m-1)}[x_1 + t(x_2 - x_1)](x_2 - x_1)^{m-1}(1 - t)^{m-1}\right]_0^1 -$$
$$- \frac{1}{(m-1)!} \int_0^1 F^{(m-1)}[x_1 + t(x_2 - x_1)] \times$$
$$\times (x_2 - x_1)^{m-1}[-(m-1)(1 - t)^{m-2}]\,dt =$$
$$= -\frac{1}{(m-1)!} F^{(m-1)}(x_1)(x_2 - x_1)^{m-1} +$$
$$+ \frac{1}{(m-2)!} \int_0^1 F^{(m-1)}[x_1 + t(x_2 - x_1)](x_2 - x_1)^{m-1}(1 - t)^{m-2}\,dt,$$

i. e.,

$$r_m(x_1, x_2) = -\frac{1}{(m-1)!} F^{(m-1)}(x_1)(x_2 - x_1)^{m-1} + r_{m-1}(x_1, x_2).$$

Combining this with the preceding results, we have

$$F(x_2) = F(x_1) + \sum_{k=1}^{m-1} \frac{1}{k!} F^{(k)}(x)(x_2 - x_1)^k + r_m(x_1, x_2).$$

Q. E. D.

Note that if $\varphi^{(n)}(t)$ is continuous on [0,1], Theorem 4.7 also holds if E_y is a linear topological space.*

4.6. Taylor's formula with remainder in Peano's form

As is well known from elementary analysis, if a real-valued function $f(t)$ of a real variable is n times differentiable at a point a, then

$$f(a+h) = f(a) + hf'(a) + \frac{h^2}{2} f''(a) + \ldots$$
$$\ldots + \frac{h^n}{n!} f^{(n)}(a) + o(h^n).$$

It turns out that this formula remains valid for abstract functions, i. e., when $f(t)$ is a function of a real variable with values in a normed space E. We shall prove this formula under the additional assumption that $f'(t)$ is continuous in some neighborhood of a. Of course, for $n > 2$ the existence of $f^{(n)}(a)$ implies the continuity of $f'(a)$ in some neighborhood of a.

Lemma 4.4. If an abstract function $\varphi(t)$ is n times differentiable at $t = 0$ and $\varphi'(t)$ is continuous in some interval $[0, \tau]$, then for $0 \leqslant t < \alpha \leqslant \tau$ we have Taylor's formula

$$\varphi(t) = \varphi(0) + \frac{t}{1}\varphi'(0) + \ldots + \frac{t^n}{n!}\varphi^{(n)}(0) + o(t^n). \qquad (4.7)$$

Proof (by induction). For $n = 1$, we have by the definition of the derivative $\varphi'(0)$

$$\varphi(t) = \varphi(0) + t\varphi'(0) + o(t).$$

* Taylor's formula for normed spaces was proved by Graves /1/. For linear topological spaces it was proved later; see, e.g., Sebastião e Silva /1/, Averbukh and Smolyanov /1/.

Suppose that for $1 \leqslant k \leqslant n-1$

$$\varphi(t) = \varphi(0) + t\varphi'(0) + \ldots + \frac{t^k}{k!}\varphi^{(k)}(0) + o(t^k).$$

Then for the abstract function $\varphi'(t)$ we also have

$$\varphi'(t) = \varphi'(0) + t\varphi''(0) + \ldots + \frac{t^k}{k!}\varphi^{(k+1)}(0) + o(t^k). \tag{4.8}$$

We now write

$$\varphi(t) = \varphi(0) + t\varphi'(0) + \ldots + \frac{t^{k+1}}{(k+1)!}\varphi^{(k+1)}(0) + r(t).$$

Since $\varphi(t)$ is differentiable on $(0, \tau)$, it follows that $r(t)$ is also differentiable there, so that

$$\varphi'(t) = \varphi'(0) + t\varphi''(0) + \ldots + \frac{t^k}{k!}\varphi^{(k+1)}(0) + r'(t).$$

Hence, by (4.8),

$$r'(t) = o(t^k) = t^k o(1). \tag{4.9}$$

Using Lemma 2.7 of VM, which states that if $x(t)$ is an abstract function which is continuously differentiable on (a, b), then

$$\int_\alpha^\beta x'(t)\,dt = x(\beta) - x(\alpha)$$

for $\alpha, \beta \in (a, b)$, we obtain from (4.9)

$$r(t) - r(0) = \int_0^t z^k o(1)\,dz = \frac{t^{k+1}}{k+1} o(1) = o(t^{k+1}).$$

Since $r(0) = 0$, it follows that $r(t) = o(t^{k+1})$. Thus

$$\varphi(t) = \varphi(0) + t\varphi'(0) + \ldots + \frac{t^{k+1}}{(k+1)!}\varphi^{(k+1)}(0) + o(t^{k+1}).$$

Q. E. D.

We have proved (4.7) for $t \geqslant 0$. The proof for $t \leqslant 0$ is similar. With the help of Lemma 4.4, we shall prove

Theorem 4.8. Suppose a mapping $F(x)$ of a convex set ω in a normed space E_x into a normed space E_y satisfies the following condition: the abstract function

$$\varphi(t) = F[x_1 + t(x_2 - x_1)],$$

where $x_1, x_2 \in \omega$, is n times differentiable from the right at $t = 0$ and $\varphi'(t)$ is continuous on $[0, \alpha)$, where α is a positive number. Then we have Taylor's formula

$$F(x_1 + h) = F(x_1) + \sum_{k=1}^{n} \frac{1}{k!} F^{(k)}(x_1) h^k + r(x_1, h), \qquad (4.10)$$

where $h = t(x_2 - x_1)$ and $\lim\limits_{h \to 0} \frac{r(x_1, h)}{\| h \|^n} = 0.$

Proof. Since the abstract function $\varphi(t)$ satisfies the hypotheses of Lemma 4.4, we can write

$$\varphi(t) = \varphi(0) + t\varphi'(0) + \ldots + \frac{t^n}{n!}\varphi^{(n)}(0) + \rho(x_1, x_2, t), \qquad (4.11)$$

where $\rho(x_1, x_2, t) = o(t^n)$ as $t \to 0$. Setting

$$\rho(x_1, x_2, t) = r(x_1, h),$$

we have

$$0 = \lim_{t \to 0} \frac{r(x_1, h)}{t^n} = \| x_2 - x_1 \|^n \lim_{h \to 0} \frac{r(x_1, h)}{\| h \|^n},$$

i. e.,

$$\lim_{h \to 0} \frac{r(x_1, h)}{\| h \|^n} = 0.$$

Furthermore, since

$$\varphi^{(k)}(0) = F^{(k)}(x_1)(x_2 - x_1)^k \qquad (k = 1, 2, \ldots, n),$$

formula (4.11) takes the form (4.10). This completes the proof.*

* Taylor's formula with remainder in Peano form has been considered by Averbukh and Smolyanov /1/ for linear topological spaces.

Chapter II

POTENTIAL OPERATORS AND DIFFERENTIABLE MAPPINGS

The general theory of potential operators which we presented for the case of Banach spaces in VM has been extended by Engel'son /2, 3/ to the case of locally convex linear topological spaces. Here we present some new material on monotone potential operators and differentiable mappings, established after the publication of VM.

§ 5. POTENTIAL AND MONOTONE POTENTIAL OPERATORS

5.1. Examples of potential operators

Let E be a vector space and E^* its dual; let A be a subset of E. Recall that an operator $F(x)$ from A to E^* is said to be potential if there exists a functional $f(x)$, defined on E, such that for all $x \in A$

$$F(x) = \operatorname{grad} f(x).$$

The functional $f(x)$ is called the potential of the operator $F(x)$.

We present a few examples of potential operators defined in L-spaces.*

Example 5.1. Let B be a measurable set of finite or infinite measure in euclidean s-space, $g(u,x)$ a real function continuous in $u \in (-\infty, +\infty)$ for almost all fixed $x \in B$ and measurable in $x \in B$ for any u. Then if

$$|g(u, x)| \leqslant a(x) + b|u|^{p-1}, \tag{5.1}$$

where $a(x) \in L^q(B)$, $b > 0$, $p > 1$, $p^{-1} + q^{-1} = 1$, the Nemytskii

* See Vainberg /3, 9/.

operator* h, defined by

$$hu = g(u(x), x),$$

is a continuous potential operator from $L^p(B)$ to $L^q(B)$ and its potential is

$$f(u) = f_0 + \int_B dx \int_0^{u(x)} g(v, x)\,dv,$$

where $f_0 = \text{const}$.

Example 5.2. Consider the Nemytskii operator in a space of vector-functions. Let B be a measurable set of positive (finite or infinite) Lebesgue measure in euclidean s-space and let $g_i(u_1, u_1, \ldots, u_n, x)$ be real functions $(i = 1, 2, \ldots, n)$, jointly continuous in $u_i \in (-\infty, +\infty)$ for almost every fixed $x \in B$ and measurable in $x \in B$ for any fixed $u_i (i = 1, 2, \ldots, n)$. Then, if

$$|g_i(u_1, u_2, \ldots, u_n)| \leqslant a_i(x) + b \sum_{k=1}^{n} |u_k|^{p-1} \quad (i = 1, 2, \ldots, n),$$

where $a_i(x) \in L^q(B)$, $b > 0$, $p > 1$, $p^{-1} + q^{-1} = 1$, then the Nemytskii operator $h = (h_1, h_2, \ldots, h_n)$,

$$h_i u = g_i(u_1(x), u_2(x), \ldots, u_n(x), x)$$

is defined and continuous from the direct sum of n copies of $L^p(B)$, denoted by $L_{p,n}(B)$, to the dual space $L_{q,n}(B)$.

This operator is potential if and only if** there exists a function $G(u_1, u_2, \ldots, u_n, x)$ such that

$$g_i(u_1, u_2, \ldots, u_n, x) = \frac{\partial}{\partial u_i} G(u_1, u_2, \ldots, u_n, x). \quad (5.2)$$

Example 5.3. Let E be a real reflexive Banach space with dual E^*, and A a linear operator from E to E^* with dense domain $D(A)$. Then the adjoint A^* is also an operator from E to E^*. Assume that the intersection of $D(A)$ and $D(A^*)$ is dense in E.

* Inequality (5.1) is a necessary and sufficient condition for h to be defined and continuous from L^p to L^q (see VM, §19, where the history of this problem is also discussed).

** VM, § 21. The author's results on the Nemytskii operator in the L-spaces were summarized in /9/. For other spaces, see Shimogaki /1, 2/, Shragin /1–15/, Vainberg and Shragin /1–5/ and others. See also Krasnosel'skii, Zabreiko et al. /1/, where additional bibliography is given.

Consider the quadratic functional

$$f(x) = \langle Ax, x \rangle,$$

for which (see Example 2.1)

$$Df(x, h) = \langle Ax + A^*x, h \rangle,$$

where x and h are arbitrary vectors in $C = D(A) \cap D(A^*)$. Since C is dense in E, we can extend the linear functional $Df(x, h)$ by continuity, giving

$$\operatorname{grad} f(x) = Ax + A^*x \qquad (\forall x \in C).$$

Consequently, if $A = A^*$ on C (such an operator is naturally called symmetric), then $Ax = 2^{-1} \operatorname{grad} f(x)$, i. e., A is a potential operator from C to E^*.

5.2. Examples of monotone potential operators

Let E be a real vector space, with strong dual E', and D a subset of E. Recall that an operator $F(x)$ from D to E' is monotone if for all $x, y \in D$

$$\langle F(x) - F(y), x - y \rangle \geqslant 0 \tag{5.3}$$

and it is strictly monotone if (5.3) is a strict inequality for $x \neq y$.

We present some examples of monotone potential operators.

Example 5.4. Consider the Nemytskii operator h from L^p to L^q (see Example 5.1). Inequality (5.3) becomes here

$$\int_B [g(u(x), x) - g(v(x), x)](u(x) - v(x))\, dx \geqslant 0, \tag{5.4}$$

where $u(x)$, $v(x)$ are arbitrary functions in L^p. This inequality holds if and only if $g(u, x)$ is nondecreasing in u for almost all fixed $x \in B$. For if $g(u, x)$ is not nondecreasing on some set σ of positive measure, we can find two functions $u(x)$, $v(x)$ with support in σ, such that for $x \in \sigma$

$$u(x) - v(x) > 0; \quad g(u(x), x) - g(v(x), x) < 0;$$

$u(x),\ v(x) \in L^p$. For these functions we have

$$\int_B [g(u(x), x) - g(v(x), x)](u(x) - v(x))\,dx =$$
$$= \int_\sigma [g(u(x), x) - g[v(x), x](u(x) - v(x)]\,dx < 0,$$

so that h is not a monotone operator. If the function $g(u,x)$ is nondecreasing in u for almost every $x \in B$, then, for any two functions $u(x)$, $v(x) \in L^p$, inequality (5.4) is valid. Thus, if $g(u,x)$ generates a Nemytskii operator from L^p to L^q $(p > 1, p^{-1} + q^{-1} = 1)$, this operator is potential, but it is monotone only if $g(u,x)$ is nondecreasing in u for almost all fixed $x \in B$.

Example 5.5. Consider the Nemytskii operator in a space of vector-functions (see Example 5.2). In this case the monotonicity condition becomes

$$\langle hu - hv, u - v\rangle = \sum_{i=1}^{n} \langle h_i u - h_i v, u_i - v_i\rangle =$$
$$= \sum_{i=1}^{n} \int_B [g_i(u_i(x), u_2(x), \ldots, u_n(x), x) -$$
$$- g_i(v_1(x), v_2(x), \ldots, v_n(x), x)](u_i(x) - v_i(x))\,dx. \tag{5.5}$$

This inequality is valid for any $u(x)$, $v(x) \in L_{p,n}$ if for almost every $x \in B$ and arbitrary numbers $u_1, u_2, \ldots, u_n; v_1, \ldots, v_n$

$$\sum_{i=1}^{n} [g_i(u_1, u_2, \ldots, u_n, x) - g_i(v_1, \ldots, v_n)](u_i - v_i) \geqslant 0. \tag{5.6}$$

On the other hand, if (5.5) holds, one can prove, reasoning as in the previous example, that (5.6) is a necessary condition for monotonicity of the operator h. Thus, if the Nemytskii operator is defined from $L_{p,n}$ to $L_{q,n}$ $(p > 1, p^{-1} + q^{-1} = 1)$, then it is potential and monotone if and only if inequalities (5.2) and (5.6) hold. Conditions (5.2) and (5.6) are obviously independent, so there exist Nemytskii operators which are potential but not monotone from $L_{p,n}$ to $L_{q,n}$, and others which are monotone but not potential.

Example 5.6. Let the operators A and A^* satisfy the conditions of Example 5.3 and let $C = D(A) \cap D(A^*)$. If A is positive on $D(A)$, i. e., if

$$\langle Ax, x\rangle \geqslant 0 \qquad (\forall x \in D(A)), \tag{5.7}$$

then it is positive and hence also monotone on $C \subset E$. As we have seen (Example 5.3), this operator is potential on C only if $A^* = A$.

Thus, a linear operator A from E to E^* is potential and monotone on C if condition (5.7) holds and $A^* = A$. These conditions are clearly independent, so that linear operators from E to E^* may be monotone but not potential, and vice versa.

Example 5.7. Let E be a real normed space with Gâteaux-differentiable norm (see Corollary 2.1). Consider the functional

$$\varphi(x) = \| x \|^{\alpha+1} \quad (\alpha > 0).$$

Since

$$\frac{d}{dt}\varphi(x + th) = (\alpha + 1)\| x + th \|^{\alpha} \frac{d}{dt}\| x + th \|,$$

it follows that

$$U_\alpha(x) = \operatorname{grad} \| x \|^{\alpha+1} = (\alpha + 1)\| x \|^{\alpha} \operatorname{grad} \| x \|.$$

We claim that $U_\alpha(x)$ is a monotone operator.* In fact, using the equality (see Lemma 2.5)

$$\langle \operatorname{grad} \| x \|, x \rangle = \| x \|,$$

we have

$$\begin{aligned} &\langle U_\alpha(x) - U_\alpha(y), x - y \rangle = \\ &= (\alpha + 1)[\| x \|^{\alpha+1} + \| y \|^{\alpha+1}] - \langle U_\alpha(x), y \rangle - \langle U_\alpha(y), x \rangle \geqslant \\ &\geqslant (\alpha + 1)[\| x \|^{\alpha+1} + \| y \|^{\alpha+1} - \| x \|^{\alpha}\| y \| - \| y \|^{\alpha}\| x \|] = \\ &= (\alpha + 1)(\| x \| - \| y \|)(\| x \|^{\alpha} - \| y \|^{\alpha}) \geqslant 0. \end{aligned}$$

Thus, the potential operator

$$U_\alpha(x) = (\alpha + 1)\| x \|^{\alpha} \operatorname{grad} \| x \|,$$

from E to E^* is monotone for any $\alpha \geqslant 0$. In view of the condition

$$\langle U_\alpha(x), x \rangle = (\alpha + 1)\| x \|^{\alpha+1},$$

$U_\alpha(x)$ is a coercive operator for $\alpha > 0$. Note also that $U_\alpha(x)$ is a positive-homogeneous operator of degree α.

* See Vainberg /1/.

5.3. Conditions for monotonicity of potential operators

Let ω be an open convex set in a real vector space E, and $F(x)$ a potential operator defined on ω with potential $f(x)$.

*Theorem 5.1.** *A potential operator* $F(x)$ *defined on* ω *is monotone (strictly monotone) if and only if its potential* $f(x)$ *is a convex (strictly convex) functional on* ω.

Proof. Necessity. We first recall that a functional $f(x)$ is convex on ω if, for any $x, y \in \omega$ and any $\lambda \in (0, 1)$,

$$f(\lambda x + (1-\lambda) y) \leqslant \lambda f(x) + (1-\lambda) f(y).$$

Let $F(x)$ be a monotone (strictly monotone) operator on ω and let x, y be arbitrary vectors in ω. We write

$$\begin{aligned}\Lambda = \lambda f(x) + (1-\lambda) f(y) - f(\lambda x + (1-\lambda) y) = \\ = \lambda [f(x) - f(\lambda x + (1-\lambda) y)] + \\ + (1-\lambda)[f(y) - f(\lambda x + (1-\lambda) y)].\end{aligned}$$

In view of the fact that $\operatorname{grad} f(x) = F(x)$, we obtain from the Lagrange formula that

$$\begin{aligned}\Delta = \lambda \langle F(\lambda x + (1-\lambda) y + t_1 (1-\lambda)(y-x)), (1-\lambda)(y-x)\rangle + \\ + (1-\lambda)\langle F(\lambda x + (1-\lambda) y + t_2 \lambda (x-y)), \lambda (x-y)\rangle,\end{aligned}$$

where $0 < t_1, t_2 < 1$. Setting $u = \lambda x + (1-\lambda) y + t_2\lambda(x-y)$, $v = \lambda x + (1-\lambda) y + t_1 (1-\lambda)(y-x)$, $\gamma = t_1(1-\lambda) + t_2\lambda > 0$, and taking into account that $u, v \in \omega$, we have

$$\Delta = \frac{\lambda(1-\lambda)}{\gamma} \langle F(u) - F(v), u - v\rangle \geqslant 0,$$

i. e.,

$$\lambda f(x) + (1-\lambda) f(y) - f(\lambda x + (1-\lambda) y) \geqslant 0.$$

Consequently the monotonicity (strict monotonicity) of $F(x)$ implies the convexity (strict convexity) of $f(x)$.

Sufficiency. Consider the real-valued function of a real argument

$$\varphi(t) = f(y + th), \quad 0 \leqslant t \leqslant 1,$$

* Vainberg /18/; compare Kachurovskii /9/, Levitin and Polyak /1/.

where $h = x - y$, with x, y distinct (arbitrary but fixed) vectors in ω. By the Gâteaux differentiability and convexity (strict convexity) of $f(x)$, the function $\varphi(t)$ is convex (strictly convex) and differentiable on $[0,1]$, so that $\varphi'(t)$ is increasing* (strictly increasing). But

$$\varphi'(t) = \frac{d}{dt} f(y + th) = \langle F(y + th), h \rangle,$$

whence

$$\langle F(x) - F(y), x - y \rangle = \varphi'(1) - \varphi'(0) \geqslant 0.$$

Q. E. D.

We add a direct proof of the sufficiency part of Theorem 5.1 in the case of monotonicity, without using the properties of real convex functions of a real variable. By the definition of a potential operator,

$$\langle F(x), h \rangle = \lim_{t \to 0} \frac{f(x + th) - f(x)}{t};$$

$$\langle F(y), h \rangle = \lim_{t \to 0} \frac{f(y) - f(y - th)}{t},$$

from which we have, for $h = x - y \neq 0$,

$$\langle F(x) - F(y), x - y \rangle = \lim_{t \to 0} \frac{f(x + th) + f(y - th) - f(x) - f(y)}{t}.$$

But for $t < 0$, $1 + t > 0$ and $h = x - y$, we have, by the convexity of $f(x)$, that

$$\begin{aligned} f(x + th) + f(y - th) - f(x) - f(y) &= \\ = f((1 + t)x - ty) + f((1 + t)y - tx) - f(x) - f(y) &\leqslant \\ \leqslant (1 + t) f(x) - tf(y) + (1 + t) f(y) &- \\ - tf(x) - f(x) - f(y) &\equiv 0. \end{aligned}$$

Since $t < 0$,

$$\frac{1}{t}[f(x + th) + f(y - th) - f(x) - f(y)] \geqslant 0,$$

and, letting $t \to 0$, we find that

$$\langle F(x) - F(y), x - y \rangle \geqslant 0.$$

* See, for example, Bourbaki /2/, p. 49, Proposition 8.

§ 6. CONDITIONS FOR OPERATORS TO BE POTENTIAL

Conditions for operators to be potential were presented in VM. Here we shall quote the basic facts and confine ourselves to making a few remarks.

6.1. Integral along a polygonal line and the line integral

Let E_x be a normed space, E_y a Banach space and E_{xy} the space of all continuous linear mappings from E_x to E_y. Let $\Phi(x)$ be a mapping from an open convex set* $\omega \subset E_x$ into E_{xy}. We assume that for all fixed $x, h \in E_x$, the function

$$\varphi(t) = \Phi(x + th)\,h, \quad 0 \leqslant t \leqslant 1,$$

with values in E_y, is continuous in t. We shall call such a mapping radially continuous. In this case (see subsections 4.3, 4.4) the integral along a segment $[x_1, x_2] \subset \omega$

$$\int_{x_1}^{x_2} \Phi(x)\,dx = \int_0^1 \Phi[x_1 + t(x_2 - x_1)](x_2 - x_1)\,dt$$

exists. Let $0 = t_0 < t_1 < t_2 < \ldots < t_k < \ldots < t_n = 1$ and let x_k $(k = 0, 1, 2, \ldots, n)$ be points in ω. Set

$$x(t) = x_{k-1} + (t - t_{k-1})(t_k - t_{k-1})^{-1}(x_k - x_{k-1})$$
$$(k = 1, 2, \ldots, n;\ t \in [t_{k-1}, t_k]).$$

The abstract function $x(t)$ determines a segment $[x_{k-1}, x_k] = l_k \subset \omega$. Set $L = \bigcup_{k=1}^{n} l_k$. Then by definition

$$\int_L \Phi(x)\,dx = \int_{x_0}^{x_n} \Phi(x)\,dx =$$
$$= \sum_{k=1}^{n} \int_{t_{k-1}}^{t_k} \Phi\left[x_{k-1} + \frac{t - t_{k-1}}{t_k - t_{k-1}}(x_k - x_{k-1})\right] \frac{x_k - x_{k-1}}{t_k - t_{k-1}}\,dt.$$

* We may require instead that ω be simply-connected instead of convex (see VM, p. 49).

If $y(t)$ is a continuous abstract function with values in ω such that $x_0 = y(0)$, $x_n = y(1)$ and $\{L^{(i)}\}$ is a sequence of polygonal lines converging uniformly to the curve $y(t)$, then we define

$$\int_0^1 \Phi[y(t)]\, dy(t) = \lim_{i\to\infty} \int_{L^{(i)}} \Phi(x)\, dx.$$

6.2. On the line integral's independence of the path of integration

Theorem 6.1. Let $\Phi(x)$ be a radially continuous mapping from an open convex set $\omega \subset E_x$ to E_y. The integral*

$$\int_{x_0}^{x} \Phi(x)\, dx$$

is independent of the path of integration in ω if and only if $\Phi(x)$ is the Gâteaux derivative of an operator $\Psi(x)$ mapping ω into E_y, i.e., $\Phi(x) = \Psi'(x)$.

Proof. Necessity. Set

$$\Psi(x) = \int_{x_0}^{x} \Phi(x)\, dx, \qquad x,\ x_0 \in \omega. \tag{6.1}$$

Since by assumption the inegral does not depend on the path of integration, it follows that the operator $\Psi(x)$ is well defined on ω. Consider a point $(x+\Delta x) \in \omega$, where $\Delta x = h\,\Delta t$, $x = x(t) \subset \omega$ and h is a fixed vector. Set $x(\tau) = x(t) + (\tau - t)h$. Then

$$I = \frac{1}{\Delta t}[\Psi(x + h\,\Delta t) - \Psi(x)] = \frac{1}{\Delta t}\int_t^{t+\Delta t} \Phi(x(\tau))\, h\, d\tau,$$

$$\Phi(x(t))\, h = \frac{1}{\Delta t}\int_t^{t+\Delta t} \Phi(x(t))\, h\, d\tau,$$

whence

$$I - \Phi(x(t))\, h = \frac{1}{\Delta t}\int_t^{t+\Delta t} [\Phi(x(\tau)) - \Phi(x(t))]\, h\, d\tau.$$

* Compare Gavurin /1/ and Vainberg /9/, §6.

Using the usual estimates, we get

$$\|I-\Phi(x(t))h\| \leqslant \\ \leqslant \sup_{0 \leqslant \tau - t \leqslant \Delta t} \|[\Phi(x(t)+(\tau-t)h)-\Phi(x(t))]h\|.$$

Hence, since $\Phi(x)$ is radially continuous, we have

$$\lim_{\Delta t \to 0} I = \Phi(x(t))h,$$

i. e., $\Phi(x)=\Psi'(x)$.

Sufficiency. Let $\Phi(x)=\Psi'(x)$. Then, for any points x_0, $x \in \omega$ and any polygonal line L defined by

$$x(t)=x_{k-1}+(t-t_{k-1})(t_k-t_{k-1})^{-1}(x_k-x_{k-1}), \\ t \in [t_{k-1}, t_k] \; (k=1, 2, \ldots, n; \; t_0=0, \; t_n=1; \; x_n=x),$$

we obtain, using the previous results,

$$\int_{x_0}^{x} \Phi(x)\,dx = \\ = \sum_{k=1}^{n} \int_{t_{k-1}}^{t_k} \Phi\left[x_{k-1}+\frac{t-t_{k-1}}{t_k-t_{k-1}}(x_k-x_{k-1})\right]\frac{x_k-x_{k-1}}{t_k-t_{k-1}}\,dt = \\ = \sum_{k=1}^{n} \int_{t_{k-1}}^{t_k} \Psi'\left[x_{k-1}+\frac{t-t_{k-1}}{t_k-t_{k-1}}(x_k-x_{k-1})\right]\frac{x_k-x_{k-1}}{t_k-t_{k-1}}\,dt = \\ = \sum_{k=1}^{n} \int_{t_{k-1}}^{t_k} \frac{d}{dt}\Psi\left[x_{k-1}+\frac{t-t_{k-1}}{t_k-t_{k-1}}(x_k-x_{k-1})\right]dt = \\ = \sum_{k=1}^{n} (\Psi(x_k)-\Psi(x_{k-1})) = \Psi(x)-\Psi(x_0).$$

Since this is valid for any polygonal line with endpoints x_0, x, it follows that

$$\int_L \Phi(x)\,dx = \Psi(x)-\Psi(x_0),$$

i. e., the line integral depends only on the endpoints of L. This completes the proof.

This theorem implies that the primitive of the mapping $\Phi(x)$ is determined by (6.1) up to an additive constant, so that the general form of the primitive is

$$\Psi(x) = \Psi_0 + \int_{x_0}^{x} \Phi(x)\, dx \tag{6.2}$$

or, alternatively,

$$\Psi(x) = \Psi_0 + \int_0^1 \Phi(x_0 + t(x - x_0))(x - x_0)\, dt, \tag{6.3}$$

where Ψ_0 is an arbitrary element in E_y.

6.3. Integral form of condition for potential operators

Let E_x be a normed space and E_y the real line. Then E_{xy} is the space of continuous linear functionals defined on E_x, i. e., E_x^*. Consider a mapping $F(x)$ from an open convex set $\omega \subset E_x$ to E_x^*. We shall assume that $F(x)$ is a radially continuous operator, i. e., the function

$$\varphi(t) = \langle F(x + th), h\rangle, \qquad 0 \leqslant t \leqslant 1$$

is continuous in t for any $x, h \in \omega$. For operators from E_x to E_x^*, radial continuity is not the same, of course, as hemicontinuity (see §1).

In order to determine whether an operator $F(x)$ from E_x to E_x^*, is potential, i. e., to test for the existence of a Gâteaux-differentiable functional $f(x)$ such that $\operatorname{grad} f(x) = F(x)$, we shall use Theorem 6.1. The following proposition is a corollary of that theorem.

Theorem 6.2. *Let $F(x)$ be a radially continuous operator from an open convex set $\omega \subset E_x$ to E_x^*. Then $F(x)$ is potential if and only if, for any polygonal line $l \subset \omega$, the line integral*

$$\int_l \langle F(x), dx\rangle$$

is independent of the path of integration.

Note that under the hypotheses of this theorem, it follows from (6.3) that the potential $f(x)$ of the operator $F(x)$ has the form

$$f(x) = f_0 + \int_0^1 \langle F(x_0 + t(x - x_0)), x - x_0\rangle\, dt, \tag{6.4}$$

where $\langle y, x\rangle$ denotes the value of a linear functional $y \in E_x^*$ at the vector $x \in E_x$.

Note also that if $F(x)$ is a potential operator defined on E_x, (6.4) is valid for points x_0, $x \in E_x$ such that the function $\langle F(x_0 + t(x - x_0)),\ x - x_0\rangle$ is bounded on $t \in [0,1]$ or integrable with respect to t on $[0,1]$. In fact, setting

$$\varphi(t) = f(x_0 + t(x - x_0)),$$

we get

$$\varphi'(t) = \langle F(x_0 + t(x - x_0)),\ x - x_0\rangle$$

and, with the help of the formula for the primitive (see, for example, Natanson /1/), we obtain

$$\varphi(1) - \varphi(0) = \varphi(0) + \int_0^1 \varphi'(t)\,dt,$$

or

$$f(x) = f(x_0) + \int_0^1 \langle F(x_0 + t(x - x_0)),\ x - x_0\rangle\,dt.$$

Example 6.1. Let E be a real reflexive Banach space and B a bounded linear operator from E to E^*. We seek conditions for B to be potential. Toward this end we consider, by Theorem 6.2, the line integral

$$\int_{x_1}^{x_2} \langle Bx,\ dx\rangle = \int_0^1 \langle B(x_1 + t(x_2 - x_1)),\ x_2 - x_1\rangle\,dt =$$
$$= \frac{1}{2}\langle Bx_2,\ x_2\rangle - \frac{1}{2}\langle Bx_1,\ x_1\rangle + \frac{1}{2}\langle (B - B^*)\,x_1,\ x_2\rangle$$

for arbitrary x_1, $x_2 \in E$. This integral can be represented as a difference of the values taken by some functional at points x_1 and x_2, i. e., it is independent of the path of integration, if and only if $\langle (B - B^*)x_1, x_2\rangle = 0$ for any x_1, $x_2 \in E$, i. e., $B^* = B$. In this case (compare Example 2.1), the potential of B has the form

$$f(x) = f_0 + \int_0^1 \langle Btx,\ x\rangle\,dt = f_0 + \frac{1}{2}\langle Bx,\ x\rangle.$$

Example 6.2. Let D be a measurable set of positive (finite or infinite) measure in euclidean s-space and $G(u_1, u_2, \ldots, u_n, x)$ a real function defined on $u_i \in (-\infty, +\infty)$ $(i = 1, 2, \ldots, n)$, $x \in D$. Suppose that for almost every $x \in D$ the function G has partial derivatives

$$g_i(u_1, u_2, \ldots, u_n, x) = \frac{\partial}{\partial u_i} G(u_1, u_2, \ldots, u_n, x),$$

which are jointly continuous in $(u_1, u_2, \ldots, u_n)$ for almost every $x \in D$, measurable in x for fixed $u = (u_1, u_2, \ldots, u_n)$, and satisfy the inequalities

$$|g_i(u_1, u_2, \ldots, u_n, x)| \leqslant a_i(x) + b \sum_{k=1}^{n} |u_k|^{p-1} \quad (i = 1, 2, \ldots, n), \tag{6.5}$$

where $p > 1$, $b > 0$, $a_i(x) \in L^q(D)$, $p^{-1} + q^{-1} = 1$. Under these conditions (see VM, § 19) the Nemytskii operator

$$h = (h_1, h_2, \ldots, h_n),$$

$$h_i u = g_i(u_1(x), u_2(x), \ldots, u_n(x), x)$$

is a continuous operator from $L_{p,n}(D)$ to $L_{q,n}(D)$. To determine whether it is a potential operator, we take two arbitrary vectors $u(x) = (u_1(x), u_2(x), \ldots, u_n(x))$, $u(x) + v(x) = (u_1(x) + v_1(x), \ldots, u_n(x) + v_n(x))$ and consider the line integral

$$I = \int_u^{u+v} \langle hw, w \rangle = \int_0^1 \langle h(u + tv), d(u + tv) \rangle =$$
$$= \int_0^1 dt \sum_{i=1}^{n} \int_D g_i(u_1(x) + tv_1(x), \ldots$$
$$\ldots, u_n(x) + tv_n(x), x)\, v_i(x)\, dx =$$
$$= \int_0^1 dt \int_D \sum_{i=1}^{n} g_i(u_1(x) + tv_1(x), \ldots$$
$$\ldots, u_n(x) + tv_n(x), x)\, v_i(x)\, dx =$$
$$= \int_0^1 dt \int_D \frac{d}{dt} G(u_1(x) + tv_1(x), \ldots, u_n(x) + tv_n(x), x)\, dx.$$

Using Fubini's theorem and setting

$$\Phi(u)=\int_D G(u_1(x),\ u_2(x),\ \ldots,\ u_n(x),\ x)\,dx,$$

we obtain

$$I=\int_D G(u_1(x)+v_1(x),\ \ldots,\ u_n(x)+v_n(x),\ x)\,dx-$$
$$-\int_D G(u_1(x),\ \ldots,\ u_n(x),\ x)\,dx=\Phi(u+v)-\Phi(u).$$

This equality shows that the line integral does not depend on the path, and thus, by Theorem 6.2, h is a potential operator. Using formula (6.4), we find the potential of the operator h:

$$f(u)=f_0+\int_0^1\langle h(0+th),\ u\rangle\,dt=$$
$$=f_0+\int_D G(u_1(x),\ \ldots,\ u_n(x),\ x)\,dx-\int_D G(0,\ \ldots,0,\ x)\,dx=$$
$$=\int_D G(u_1(x),\ u_2(x),\ \ldots,\ u_n(x),\ x)\,dx+C\qquad(C=\text{const}).$$

6.4. Conditions for potential operators (differential form)

Let E be a vector space, E' its strong dual, and ω an open convex set in E.

Theorem 6.3. *Let $F(x)$ be an operator from E to E' which is Gâteaux-differentiable at every point of ω, the Gâteaux differential $DF(x,h)$ being hemicontinuous in x. Then $F(x)$ is potential in ω if and only if the bilinear functional $\langle DF(x,h_1),h_2\rangle$ is symmetric, i.e.,*

$$\langle DF(x,\ h_1),\ h_2\rangle=\langle DF(x,\ h_2),\ h_1\rangle.$$

We omit the proof, which is similar to that of Theorem 5.1. in VM.

Example 6.3. Let $F(x)$ be a positive-homogeneous potential operator of degree $k>0$ defined in a real space E. Then, by formula (6.4), its potential is of the form

$$f(x)=f_0+\frac{1}{k+1}\langle F(x),\ x\rangle.$$

The converse also holds:* If this last formula is valid for a Gâteaux-differentiable potential operator $F(x)$ with gradient $f(x)$, then $F(x)$ is positive-homogeneous of degree k. If we set $f(0) = f_0 = 0$, the fact that $F(x)$ is homogeneous of degree k implies that the functional $f(x)$ is homogeneous of degree $1+k$. Moreover, if $F(x)$ is an odd (even) operator, $f(x)$ is an even (odd) functional.

Remark 6.1. Suppose $F(x)$ is a potential operator defined in a real linear space E, positive-homogeneous of degree $k > 0$; then the functional $\varphi(x) = \langle F(x), x\rangle$ has the following properties.

1. If $\varphi(x_0) > 0$, $\varphi(x)$ increases along the ray tx_0, $t \geqslant 0$.
2. If $\varphi(x_0) < 0$, $\varphi(x)$ decreases along the ray tx_0, $t \geqslant 0$.
3. If $F(x)$ is an odd (even) functional and $\varphi(x_0) \geqslant 0$, $\varphi(x)$ increases along tx_0 and $-tx_0$ $(t \geqslant 0)$($\varphi(x)$ increases along tx_0, decreases along $-tx_0$).
4. If $\varphi(x)$ assumes both positive and negative values on a ball $\|x\| \leqslant r$ in a normed space E and assumes its supremum and infimum on this ball, then $\varphi(x)$ attains its supremum and infimum on the sphere $\|x\| = r$.

6.5. Reduction to equations with potential operators

Let $\Phi(x)$ be a nonlinear operator from a normed space E into itself. Then the problem of the existence of solutions for the equation

$$\Phi(x) = 0 \tag{6.6}$$

reduces to finding the minimum of the functional $f(x) = \|\Phi(x)\|$ and the existence of solutions can be proved by constructing a minimizing sequence. As we shall see (§ 21), various propositions can be obtained in this manner. An alternative approach is to construct a functional whose critical points are the solutions of (6.6) or their pre-images. Potential operators are very important in the application of this method. In fact, if $\Phi(x)$ is a potential operator defined in a linear space and $\varphi(x)$ is its potential, then (6.6) is simply $\operatorname{grad}\varphi(x) = 0$, implying that the solutions of equation (6.6) are precisely the critical points of the functional $\varphi(x)$.

If $\Phi(x)$ is not potential, one must replace (6.6) by an equivalent equation $\Psi = 0$, where Ψ is a potential operator. This problem has been solved for equations of Hammerstein type

$$x - AF(x) = 0, \tag{6.7}$$

* See Vainberg /14/.

where A is a linear operator and F a potential operator in a real Hilbert space or a real reflexive space (Banach or locally convex). *

The left-hand side of equation (6.7) (see VM, Example 5.2) is not a potential operator. In this subsection we examine equation (6.7) in a Hilbert space H and prove that it can be reduced to an equivalent equation with potential operators.

Let A be a bounded positive selfadjoint operator in H and F a potential operator in H with potential ; let $A^{1/2}$ be the positive square root of A. Consider the equation

$$y - A^{1/2}F(A^{1/2}y) = 0, \qquad y \in H, \tag{6.8}$$

whose left-hand side can be written

$$\operatorname{grad}\left(\frac{1}{2}(y, y) - f(A^{1/2}y)\right),$$

since $\operatorname{grad}(y, y) = 2y$ and

$$\begin{aligned}(\operatorname{grad} f(A^{1/2}y), h) = \lim_{t \to 0} \frac{f(A^{1/2}y + tA^{1/2}h) - f(A^{1/2}y)}{t} = \\ = (F(A^{1/2}y), A^{1/2}h) = (A^{1/2}F(A^{1/2}y), h),\end{aligned}$$

i. e.,

$$\operatorname{grad} f(A^{1/2}y) = A^{1/2}F(A^{1/2}y).$$

The left-hand side of equation (6.8) is thus a potential operator. $A^{1/2}F(A^{1/2}y)$ is called the s u p e r p o s i t i o n o p e r a t o r.

We claim that equations (6.7) and (6.8) are equivalent, i. e., there exists a one-to-one correspondence between their solution sets. Indeed, let y_i be a solution of (6.8). Then $x_i = A^{1/2}y_i$ is a solution of (6.7). If y_1 and y_2 are two distinct solutions of (6.8), then

$$\begin{aligned}0 < (y_1 - y_2, y_1 - y_2) = (A^{1/2}F(A^{1/2}y_1) - A^{1/2}F(A^{1/2}y_2), y_1 - y_2) = \\ = (F(A^{1/2}y_1) - F(A^{1/2}y_2), A^{1/2}y_1 - A^{1/2}y_2) = \\ = (F(x_1) - F(x_2), x_1 - x_2),\end{aligned}$$

so that $x_1 = A^{1/2}y_1$ and $x_2 = A^{1/2}y_2$ are two distinct solutions of (6.7). Note that F is strictly monotone on the solution set of equation (6.7).

* See Vainberg /9,16/ and Engel'son /2/.

Conversely, let x_i be a solution of equation (6.7). Set $y_i = A^{1/2}F(x_i)$. Then, by (6.7), $x_i = A^{1/2}y_i$ and, using both equalities, we get

$$y_i - A^{1/2}F(A^{1/2}y_i) = 0,$$

i. e., $y_i = A^{1/2}F(x_i)$ is a solution of equation (6.8). If x_1 and x_2 are two distinct solutions of equation (6.7), then, since $y_i = A^{1/2}F(x_i)$, the solutions y_1 and y_2 of (6.6) are also distinct; otherwise the solutions $x_i = A^{1/2}y_i$ $(i = 1, 2)$ would coincide.

Consider the general case. Let A be a bounded selfadjoint operator in H, and suppose that

$$\alpha = \inf(Ax, x) < 0, \quad \beta = \sup(Ax, x) > 0, \quad \|x\| = 1.$$

Let E_t be a resolution of the identity* for the operator A. Then

$$E(\Delta) = E_\beta - E_0 = P_1$$

is a projection onto an invariant subspace $H_1 \subset H$ which reduces A. Let P_2 denote the projection onto the subspace

$$H_2 = H \ominus H_1$$

and, for every $y \in H$, set

$$((y))^2 = \|P_1 y\|^2 - \|P_2 y\|^2, \quad \varphi(y) = 2f(By) - ((y))^2,$$

where f is the potential of the operator F and B the principal square root of A, i. e.,

$$B = A_+^{1/2} - A_-^{1/2}; \quad A_+ = 2^{-1}(|A| + A), \quad A_- = 2^{-1}(|A| - A).$$

Setting $T = A_+^{1/2} + A_-^{1/2}$ and noting that $(P_1 - P_2)B = T$, we see that multiplication of the equality $\operatorname{grad}\varphi(y) = 0$, i. e.,

$$BF(By) - (P_1 - P_2)y = 0 \tag{6.9}$$

by $(P_1 - P_2)$ gives

$$TF(By) - y = 0. \tag{6.10}$$

* See, for example, Akhiezer and Glazman /1/.

We claim that equations (6.10) and (6.7) are equivalent. Indeed, let y_i be a solution of (6.10). Multiplying (6.10) by B and using the equality $BT = A$, we see that $x_i = By_i$ is a solution of (6.7). If y_1 and y_2 are two distinct solutions of (6.10), then

$$0 < (y_1 - y_2,\ y_1 - y_2) = (TF(By_1) - TF(By_2),\ y_1 - y_2),$$

so that $x_1 = By_1 \neq By_2 = x_2$.

Conversely, let x_1 and x_2 be distinct solutions of (6.7), i. e., $x_i = AF(x_i) = BTF(x_i)$, $i = 1, 2$. If $y_i = TF(x_i)$, then

$$x_i = By_i \qquad (i = 1,\ 2), \tag{6.11}$$

and thus $y_i = TF(By_i)$, i. e., y_i are solutions of (6.10). These solutions are distinct, since otherwise we would have by (6.11) that $x_1 = x_2$, a contradiction.

6.6. Reduction of equations of Hammerstein type to equivalent equations in Hilbert space

Let E be a real reflexive locally convex subspace of a real Hilbert space H. We assume* that $E \subset H$, H is a dense subspace of the strong dual E' of E, the topologies of E and H and of H and E' are respectively compatible, and that the bilinear functional $\langle y, x\rangle$, $x \in E$, $y \in E'$, is equal to (y, x) — the inner product in H — for $y \in H$. Examples: the spaces $E = L^p(D)$ and $H = L^2(D)$, where D is a finite measurable subset of euclidean s-space and $p > 2$, or rigged Hilbert spaces.** Consider the equation

$$x - BF(x) = 0, \quad x \in E, \tag{6.12}$$

where F is an operator from E to E' and B a bounded linear operator from E' to E. Let B_H be the restriction of B to H. Suppose B_H is a positive selfadjoint operator in H. Then (Theorem 15.1) the positive square root $A = B_H^{1/2}$ has a bounded extension T from E' to H whose adjoint T' is a bounded operator from H to E, with $B = T'T$ and $T' = A$.

Consider the equation

$$y - TF(Ay) = 0 \quad (y \in H). \tag{6.13}$$

As before, equations (6.12) and (6.13) are considered equivalent if their solution sets coincide.

* See Vainberg and Engel'son /1/, and also §15.

** See Gel'fand and Vilenkin /1/; also Gel'fand and Kostyuchenko /1/.

*Lemma 6.1.** *Equations (6.12) and (6.13) are equivalent; each solution y of (6.13) determines a solution $x = Ay$ of (6.12) and each solution x of (6.12) a solution $y = TF(y)$ of (6.13).*

The proof is the same as in the previous subsection. Let y_i be a solution of equation (6.13), i. e., let

$$y_i - TF(Ay_i) = 0.$$

Applying A to both sides and using the equality $AT = B$, we see that $x_i = Ay_i$ is a solution of equation (6.12) in E. Let y_1 and y_2 be distinct solutions of equation (6.13). Then

$$0 < (y_1 - y_2,\ y_1 - y_2) = (TF(Ay_1) - TF(Ay_2),\ y_1 - y_2) =$$
$$= \langle F(Ay_1) - F(Ay_2),\ Ay_1 - Ay_2 \rangle = \langle F(x_1) - F(x_2),\ x_1 - x_2 \rangle.$$

Thus $x_1 = Ay_1$ and $x_2 = Ay_2$ are distinct solutions of (6.12). Conversely, let x_i be a solution of (6.12). We have

$$x_i - ATF(x_i) = 0.$$

If we set $TF(x_i) = y_i$, then $x_i = Ay_i$. These two equalities imply

$$y_i = TF(Ay_i),$$

i. e., y_i is a solution of (6.13). Now, to every two distinct solutions x_1 and x_2 of (6.12) there correspond two distinct solutions of (6.13), since $y_1 = y_2$ implies $x_1 = Ay_1 = Ay_2 = x_2$. Q. E. D.

Note that the proof of Lemma 6.1 makes no use of the assumption that $F(x)$ is a potential operator.

Let $F(x)$ be a potential operator and $f(x)$ its potential. Define a functional

$$\varphi(y) = \frac{1}{2}(y,\ y) - f(Ay)$$

on H. We have

$$(\operatorname{grad} \varphi(y),\ h) = (y,\ h) - \lim_{t \to 0} t^{-1}[f(Ay + tAh) - f(Ay)] =$$
$$= (y,\ h) - \langle F(Ay),\ Ah \rangle = (y - A'F(Ay),\ h),$$

where h is an arbitrary vector in H. Hence, since $A' = T$, it follows that

$$\operatorname{grad} \varphi(y) = y - TF(Ay),$$

* See Vainberg /16, 18/.

i. e., the left-hand side of (6.13) is a potential operator defined in H. Thus, if F is a potential operator from E to E', equation (6.12) is reducible to an equivalent equation with a potential operator in the Hilbert space H.

We now consider a more general case. Let B_H be an indefinite operator in . Set

$$B_H^+ = 2^{-1}(|B_H| + B_H), \quad B_H^- = 2^{-1}(|B_H| - B_H),$$
$$A = (B_H^+)^{1/2} + (B_H^-)^{1/2}, \quad C = (B_H^+)^{1/2} - (B_H^-)^{1/2},$$

and assume that B_H^+ and B_H^- have extensions B^+ and B^-, respectively, which are bounded operators from E' to E. Then (Theorem 15.2) the operators A and C have extensions V and W, respectively, which are bounded operators from E' to H, where $V' = A$, $W' = C$, and $B = AW$. Consider the equation

$$y - WF(Ay) = 0 \quad (y \in H). \tag{6.14}$$

Lemma 6.2.* *Equations (6.12) and (6.14) are equivalent; each solution y of (6.14) determines a solution $x = Ay$ of (6.12) and each solution x of (6.12) a solution $y = WF(x)$ of (6.14).*

Proof. Let y_i be a solution of equation (6.14), i. e.,

$$y_i - WF(Ay_i) = 0.$$

Set $x_i = Ay_i$; since $AW = B$, we may apply A to the previous equality, to get

$$x_i - BF(x_i) = 0,$$

i. e., x_i is a solution of equation (6.12). If y_1 and y_2 are distinct solutions of (6.14), then

$$0 < (y_1 - y_2, y_1 - y_2) = (WF(Ay_1) - WF(Ay_2), y_1 - y_2),$$

so that $x_1 = Ay_1 \neq Ay_2 = x_2$. Conversely, let x_1, x_2 be distinct solutions of (6.12) i. e.,

$$x_i = BF(x_i) = AWF(x_i), \quad i = 1, 2.$$

Setting $y_i = WF(x_i)$, we get $x_i = Ay_i$, so that

$$y_i = WF(Ay_i), \quad i = 1, 2,$$

* See Vainberg /16, 18/.

i. e., y_i are solutions of (6.14). These solutions are distinct, else $x_1 = Ay_1 = Ay_2 = x_2$, contradicting the assumption $x_1 \neq x_2$. Q. E. D.

The proof of Lemma 6.2 does not assume that $F(x)$ is a potential operator. Suppose now that F is a potential operator with potential f. Let E_t be a resolution of the identity for the operator B_H. We have by assumption that

$$\alpha = \inf (B_H y, y) < 0, \quad \beta = \sup (B_H y, y) > 0, \quad \| y \| = 1,$$

so that (in the notation of the previous subsection) the space H is the direct sum of two orthogonal subspaces H_1 and H_2.

Set $((y))^2 = \|P_1 y\|^2 - \|P_2 y\|^2$, $y \in H$, where P_1 and P_2 are the projections of H onto H_1 and H_2, respectively. Then $\operatorname{grad} ((y))^2 = = 2(P_1 y - P_2 y)$.

Now consider the functional

$$\varphi (y) = \frac{1}{2} ((y))^2 - f(Ay), \quad y \in H,$$

for which

$$\operatorname{grad} \varphi (y) = (P_1 - P_2) y - A' F(Ay),$$

where A' is the adjoint of $A = V'$. Consider the equation

$$(P_1 - P_2) y = VF(Ay) = 0, \qquad y \in H, \tag{6.15}$$

whose left-hand side is a potential operator by construction. This equation is equivalent to (6.14). Indeed, applying the operator $(P_1 - P_2)$ and noting that $(P_1 - P_2)^2 = I$, $(P_1 - P_2) V = W$, we obtain

$$y - WF(Ay) = 0. \tag{6.16}$$

Since $P_1 - P_2$ vanishes only at the zero of the space H, it follows that it is invertible, so that equations (6.14) and (6.15) are equivalent. But we have seen that equations (6.14) and (6.12) are also equivalent, so we reach the following conclusion:

If F is a potential operator in H, then equation (6.12), when considered in the space E, is equivalent to equation (6.15), whose left-hand side is a potential operator in the Hilbert space.

The above method for reducing equation (6.12) to an equation with potential operators in Hilbert space was discussed in VM for L-spaces and in Engel'son /2, 3/ for the case of locally convex spaces.

§ 7. SOME GENERAL PROPERTIES OF POTENTIAL OPERATORS AND DIFFERENTIABLE MAPPINGS

General properties of potential operators and differentiable mappings were set forth in Vainberg /3, 9/. Here we present the basic facts with the addition of some propositions to be proved later.

7.1. Biorthogonal basis

Let E be a Banach space with dual E^*. Two sequences $\{x_n\} \subset E$ and $\{y_n\} \subset E^*$ form a countable biorthogonal basis if

$$y_i(x_k) = \langle y_i, x_k \rangle = \begin{cases} 0, & i \neq k \\ 1, & i = k \end{cases} \quad (i, k = 1, 2, 3, \ldots),$$

and any vectors $x \in E$ and $y \in E^*$ may be expressed in the form

$$x = \sum_{k=1}^{\infty} a_k x_k; \quad y = \sum_{k=1}^{\infty} b_k y_k,$$

where

$$\left\| x - \sum_{k=1}^{n} a_k x_k \right\| \to 0, \quad \left\| y - \sum_{k=1}^{n} b_k y_k \right\| \to 0$$

as $n \to \infty$.

It follows from the definition that

$$a_k = \langle y_k, x \rangle = y_k(x), \quad b_k = \langle y, x_k \rangle = y(x_k),$$

where, as usual, $y_k(x)$ and $y(x_k)$ are the values of the linear functional $y(x)$ for given x, y, x_k, y_k.

Let $E^{(n)}$ be the finite-dimensional subspace of E spanned by the vectors $x_1, x_2, \ldots, x_n$ and P_n the projection onto $E^{(n)}$. Then, for any $x \in E$,

$$P_n x = \sum_{k=1}^{n} \langle y_k, x \rangle x_k.$$

The subspace E_n^* and projections Q_n of E^* onto E_n^* may be similarly introduced.

7.2. On the strong continuity of potential operators

Let E_x and E_y be normed spaces. Recall that an operator $\Phi(x)$ from E_x to E_y is *strongly continuous* if it transforms every sequence $\{x_n\}$ converging weakly to $x_0 \in E_x$ into a sequence $\{\Phi(x_n)\}$ converging strongly to $\Phi(x_0)$. The following proposition was proved in VM.*

Theorem 7.1. Let E be a reflexive Banach space with a countable biorthogonal basis, ω an open convex set in E. A continuous operator Φ from ω to E^* is strongly continuous if and only if, for any $\varepsilon > 0$, there exists a positive integer $n_0(\varepsilon)$ such that for any $n \geqslant n_0$ and any $x \in \omega$

$$\| \Phi(P_n x) - \Phi(x) \| < \varepsilon,$$

where P_n is the projection of E onto $E^{(n)} \subset E$.

This theorem is clearly valid for potential operators. We shall cite, however, a proposition on the strong continuity of potential operators which does not assume the existence of a biorthogonal basis. We first recall the definition of a completely compact operator. An operator $\Phi(x)$ from E_x to E_y is said to be completely compact on $A \subset E_x$ if it is compact on A and uniformly continuous on every bounded subset $\sigma \subset A$.

*Theorem 7.2.** Let ω be an open convex set in a reflexive Banach space and $F(x)$ a potential operator defined on ω. Then $F(x)$ is strongly continuous on ω if and only if it is completely compact.*

The assertion of Theorem 7.2 is well known for linear operators. In fact, if A is a completely continuous linear operator from a Banach space E_x to a Banach space E_y, it transforms every weakly convergent sequence into a strongly convergent one, so that it is strongly continuous. Moreover, a continuous linear operator is uniformly continuous, so that complete continuity coincides with complete compactness.

Theorem 7.3.† Let E be a reflexive Banach space with countable biorthogonal basis, ω an open convex set in E and $f(x)$ a weakly continuous functional on ω. Then the operator $F(x) =$ $=\operatorname{grad} f(x)$ is strongly continuous in ω if and only if it is uniformly continuous in ω.

We present an additional proposition (see VM, Theorem 7.3) concerning the compactness of potential operators in reflexive spaces.

* See VM, Theorem 7.1 and Tsitlanadze /1/.

** See VM, Theorem 7.4.

† See VM, Theorem 7.7.

Theorem 7.4. Let E be a reflexive Banach space, ω an open convex set in E, and $f(x)$ a continuous functional in E which is weakly continuous in ω. Then the operator $F(x) = \operatorname{grad} f(x)$ is compact in ω if $F(x)$ is uniformly continuous in ω.

7.3. Relation between complete and strong continuity of potential operators

It is known (see VM, Theorems 1.2 and 1.4) that a strongly continuous operator defined in a reflexive space is not only completely continuous but even completely compact. In addition, as we have seen (Theorem 7.2), for potential operators in reflexive spaces complete compactness is equivalent to strong continuity. The question naturally arises of whether complete continuity of potential operators in reflexive spaces is equivalent to their strong continuity. This turns out not to be the case, as shown by the following example, due to Aizengendler. *

Example 7.1. For any vector $x = \{\xi_i\}$ in the real Hilbert space l^2, consider the vector $F(x) = y = \{\eta_1, \eta_2, \ldots\}$, where

$$\eta_i = \begin{cases} \frac{1}{i^2} \sin \frac{1}{\xi_i}, & \xi_i \neq 0 \\ 0, & \xi_i = 0 \end{cases} \qquad (i = 1, 2, \ldots).$$

The operator $F(x)$ maps l^2 into l^2. Consider the set $\omega \subset l^2$ of all vectors $x = \{\xi_i\}$ with positive coordinates $\xi_i > 0$. This set is open and convex. We claim that $F(x)$ is completely continuous on ω. Indeed, let $x^0 \in \omega$ be a fixed point, $x \in \omega$ an arbitrary one. Then

$$\| F(x) - F(x^0) \|^2 = \sum_{k=1}^{\infty} \frac{1}{k^4} \left(\sin \frac{1}{\xi_k} - \sin \frac{1}{\xi_k^0} \right)^2.$$

Let $\varepsilon > 0$ be arbitrary. Since the above series converges, there exists a positive integer $m = m(\varepsilon)$ such that

$$\sum_{k=m+1}^{\infty} \frac{1}{k^4} \left(\sin \frac{1}{\xi_k} - \sin \frac{1}{\xi_k^0} \right)^2 < \frac{\varepsilon^2}{4}.$$

Since $\sin \frac{1}{t}$ is continuous for $t > 0$, there exists $\delta_k = \delta_k(\varepsilon) > 0$ ($k = 1, 2, \ldots, m$) such that

$$\left| \sin \frac{1}{\xi_k} - \sin \frac{1}{\xi_k^0} \right| < \frac{\varepsilon^2}{4mM}, \quad M = \sum_{k=1}^{m} \frac{1}{k^4},$$

* Aizengendler /1/.

provided $|\xi_k-\xi_k^0|\leqslant\delta_k$. Set $\delta=\min(\delta_1, \ldots, \delta_m; \xi_1^0, \ldots, \xi_m^0)$; then $\| F(x)-F(x^0)\| < \varepsilon$ for $\| x-x_0 \| < \delta$, i. e., $F(x)$ is continuous at x^0. The operator $F(x)$ is compact, since it maps l^2 into the Hilbert cube $0\leqslant\eta_i\leqslant\frac{1}{i^2}$. Thus $F(x)$ is completely continuous in ω.

We now prove that $F(x)$ is a potential operator in ω. Let x_1, x_2 be arbitrary points of ω. Then (subsection 4.4) the integral along the segment $[x_1, x_2]\subset\omega$ is

$$\int_{x_1}^{x_2} (F(x), dx) = \int_0^1 (F(x_1+t(x_2-x_1)), x_2-x_1)\, dt =$$
$$= \int_0^1 \left(\sum_{k=1}^{\infty} \left[\sin \frac{1}{\xi_k^{(1)}+t(\xi_k^{(2)}-\xi_k^{(1)})} \right] \frac{\xi_k^2-\xi_k^{(1)}}{k^2} \right) dt =$$
$$= \sum_{k=1}^{\infty} \frac{1}{k^2} \int_{\xi_k^{(1)}}^{\xi_k^{(2)}} \sin\frac{1}{z}\, dz.$$

If we set

$$\int \sin\frac{1}{z}\, dz = \varphi(z)+C, \quad \Phi(x)=\sum_{k=1}^{\infty} \frac{1}{k^2}\varphi(\xi_k),$$

then

$$\int_{x_1}^{x_2} (F(x), dx) = \sum_{k=1}^{\infty} \frac{1}{k^2}\varphi(\xi_k^{(2)}) - \sum_{k=1}^{\infty} \frac{1}{k^2}\varphi(\xi_k^{(1)}) =$$
$$= \Phi(x_2)-\Phi(x_1).$$

Hence, by Theorem 6.2, $F(x)$ is a potential operator in ω.

Thus $F(x)$ is a completely continuous potential operator. We now show that it is not strongly continuous. By Theorem 7.2, it will suffice to show that $F(x)$ is not uniformly continuous in ω. Indeed, consider the sequence of pairs $x_n^{(1)}$, $x_n^{(2)}\in\omega$, where

$$x_n^{(1)} = \left\{ \frac{1}{n\pi}, \xi_2, \xi_3, \ldots \right\}; \quad x_n^{(2)} = \left\{ \frac{2}{\pi(2n+1)}, \xi_2, \xi_3, \ldots \right\}$$

and ξ_i are fixed. For any n,

$$\| F(x_n^2)-F(x_n^{(1)})\| = \left| \sin n\pi - \sin\frac{2n+1}{2}\pi \right| = 1,$$

but on the other hand $\lim_{n\to\infty}\|x_n^{(2)}-x_n^{(1)}\|=0$. It follows that $F(x)$ is not strongly continuous.

7.4. On the strong and complete continuity of differentiable mappings

Let E_x and E_y be normed spaces, D_r the open ball of radius r with center at zero, and $F(x)$ a mapping from E_x to E_y. The operator $F(x)$ is said to be Fréchet-differentiable at the point $x\in E_x$ if (see subsection 2.2)

$$F(x+h)-F(x)=F'(x)h+\omega(x, h), \tag{7.1}$$

where $F'(x)$ is a bounded linear operator from E_x to E_y and

$$\lim_{h\to 0}\frac{\omega(x, h)}{\|h\|}=0.$$

The operator $F(x)$ is said to be uniformly Fréchet-differentiable on D_r if, for any $\varepsilon>0$, there exists $\delta(\varepsilon)>0$ such that for $\|h\|<\delta(\varepsilon)$

$$\|\omega(x, h)\|<\varepsilon\|h\| \qquad (\forall x,\ x+h\in D_r).$$

Theorem 7.5 (see VM, Theorem 4.4). *Let $F(x)$ be a completely continuous Fréchet-differentiable mapping from E_x to E_y. If the derivative $F'(x)$, as a mapping from E_x to E_{xy}, is compact on a convex subset $\sigma\subset E_x$, then the operator $F(x)$ is strongly continuous on σ.*

Theorem 7.6 (see VM, Theorem 4.8). *Suppose that:*

1. For every fixed $x\in E_x$ the derivative $F'(x)$ is a completely continuous linear operator.

2. The derivative is compact as a mapping from E_x to E_{xy}.

Then $F(x)$ is a completely continuous operator from E_x to E_y.

Remark 7.1. As noted by Palmer,* under the hypotheses of Theorem 7.6, $F(x)$ is a strongly continuous operator. In fact, by condition 2 of Theorem 7.5 and the complete continuity of $F(x)$, the strong continuity of $F(x)$ follows from Theorem 7.5.

Theorem 7.7.** *The Fréchet derivative of a completely continuous operator from E_x to E_y is a completely continuous linear operator.*

* Palmer /1/.

** See, for example, VM, Theorem 4.7.

As shown by Bonic,* the converse is not valid, as can be seen from the following example.

Let $(e_1, e_2, \ldots)$ be the standard basis in the space c_0 of real null sequences, so that for any $x \in c_0$ we have

$$x = S_n x + R_n x,$$

where

$$S_n x = \sum_{k=1}^{n} x_k e_k, \quad R_n x = \sum_{k=n+1}^{\infty} x_k e_k.$$

In the space c_0, we have $\|x\| = \sup |x_k|$;** a set $E \subset c_0$ is compact if and only if it is bounded and, for any $\varepsilon > 0$, there exists a positive integer n_0 such that $\|R_n x\| < \varepsilon$ for $n \geqslant n_0$ and any $x \in E$. Consider the mapping

$$F(x) = \sum_{k=1}^{\infty} x_k^2 e_k$$

from c_0 into c_0. The operator $F(x)$ is Gâteaux-differentiable and

$$DF(x, h) = \sum_{k=1}^{\infty} 2x_k h_k e_k;$$

moreover, since $\|\omega(x, h)\| = \|h\|^2$, $F(x)$ is uniformly Fréchet-differentiable (see (7.1)).

The ball D_r (with center at zero and radius r) is of course not compact in c_0. The same is true of the intersection G_r of D_r and the half-space $x \geqslant 0$. Since the operator $F(x)$ maps any ball D_r onto G_{r^2}, it follows that $F(x)$ is not completely continuous. However, we can show that, for every fixed $x \in c_0$, the linear operator $DF(x, h)$ from c_0 to c_0 is completely continuous. Indeed, consider the ball $\|h\| \leqslant 1$. We have

$$\| R_n DF(x, h) \| \leqslant 2 \sup_{k \geqslant n} | x_k | \sup_{k \geqslant n} | h_k | \leqslant 2 \sup_{k \geqslant n} | x_k |.$$

But as the vector $x = \sum_{k=1}^{n} x_k e_k$ is fixed and $x_k \to 0$, we have that for any given $\varepsilon > 0$ there exists n_0 such that $|x_k| < \varepsilon/2$ for $k \geqslant n_0$, whence

$$\| R_n DF(x, h) \| < \varepsilon.$$

* Bonic /1/.

** See, for example, Lyusternik and Sobolev /1/, p. 247.

Hence, by the above criterion for compactness in c_0, $DF(x, h)$ is completely continuous.

Thus, complete continuity of the Fréchet derivative $F'(x)$ does not imply complete continuity of $F(x)$.

7.5. On the strong continuity of differentiable mappings

In this subsection we present results due to K. Palmer.*

*Theorem 7.8.** *Let E_x be a reflexive Banach space, E_y a Banach space, and $F(x)$ a mapping from E_x to E_y, which is uniformly Fréchet-differentiable on every ball D_r. Then $F(x)$ is strongly continuous if and only if conditions 1 and 2 of Theorem 7.6 are satisfied.*

Proof. Sufficiency follows from Theorem 7.6, by Remark 7.1. Necessity follows from the following proposition.

*Lemma 7.1.** *If a strongly continuous mapping $F(x)$ from a reflexive Banach space E_x to a Banach space E_y is uniformly Fréchet-differentiable in any ball D_r, then the derivative is strongly continuous as a mapping from E_x to E_{xy}.*

We prove this by contradiction. Suppose that there exists a sequence $x_n \longrightarrow x_0$ such that $F'(x_n) \nrightarrow F'(x_0)$. Then there exist $\varepsilon > 0$ and a sequence $\{x_m\}$ such that for all m

$$\| F'(x_m) - F'(x_0) \| > \varepsilon.$$

Hence, by standard properties of linear operators, there exists a sequence of vectors $\{h_m\} \subset E_x$ such that $\|h_m\| = 1$ and for all m

$$\| F'(x_m) h_m - F'(x_0) h_m \| \geqslant \varepsilon. \tag{7.2}$$

Let t be a real number with $|t| < 1$, and write

$$\begin{aligned} \| F'(x_m) th_m - F'(x_0) th_m \| &\leqslant \\ \leqslant \| F'(x_m) th_m - [F(x_m + th_m) - F(x_m)] \| &+ \\ + \| F'(x_0) th_m - [F(x_0 + th_m) - F(x_0)] \| + \Delta &= \\ = \| \omega(x_m, th_m) \| + \| \omega(x_0, th_m) \| + \Delta&, \end{aligned}$$

where

$$\Delta = \| F(x_m + th_m) - F(x_m) - F(x_0 + th_m) + F(x_0) \|.$$

* Palmer /1/.

By uniform differentiability, however, we have

$$\| \omega (x_m, th_m) \| < 3^{-1} \varepsilon | t |, \quad \| \omega (x_0, th_m) \| < 3^{-1} \varepsilon | t |,$$

since the reflexivity of the space E_x and the weak convergence of $\{x_m\}$ to x_0 entail the existence of $r > 0$ such that

$$\| x_m \| < r, \quad \| x_0 \| < r,$$
$$\| x_m + th_m \| \leqslant \| x_m \| + | t | < \| x_m \| + 1 < r,$$
$$\| x_0 + th_m \| < r.$$

Now, since the unit sphere in a reflexive space is weakly compact, it follows that we can choose a subsequence $\{h_{m_n}\}$ of $\{h_m\}$ such that $h_{m_n} \rightarrow h_0$ and so, by the strong continuity of $F(x)$,

$$F(x_{m_n}) \rightarrow F(x_0),$$
$$F(x_{m_n} + th_{m_n}) \rightarrow F(x_0 + th_0),$$
$$F(x_0 + th_{m_n}) \rightarrow F(x_0 + th_0)$$

for fixed t (as $n \rightarrow \infty$). It follows that

$$\| F(x_{m_n} + th_{m_n}) - F(x_{m_n}) - F(x_0 + th_{m_n}) + F(x_0) \| \leqslant$$
$$\leqslant \| F(x_{m_n} + th_{m_n}) - F(x_0 + th_0) \| +$$
$$+ \| F(x_0 + th_{m_n}) - F(x_0 + th_0) \| + \| F(x_{m_n}) - F(x_0) \|,$$

and thus, for sufficiently large n,

$$\| F(x_{m_n} + th_{m_n}) - F(x_{m_n}) - F(x_0 + th_{m_n}) + F(x_0) \| < 3^{-1} | t | \varepsilon.$$

By previous results, it follows that for sufficiently large n

$$\| F'(x_{m_n}) h_{m_n} - F'(x_0) h_{m_n} \| < \varepsilon.$$

This contradicts inequality (7.2), proving the lemma.

Since (see, for example, VM, Theorem 1.2) strongly continuous mappings in reflexive spaces are completely continuous, it follows from Lemma 7.1 that $F'(x)$ is compact as an operator from E_x to E_{xy}, so that the hypotheses of Theorem 7.8 imply condition 2 of Theorem 7.6. Finally, since a strongly continuous operator $F(x)$ is completely continuous, and since by Theorem 7.3 the Fréchet derivative $F'(x)$ of a completely continuous operator $F(x)$ is a completely continuous linear operator for each fixed x, it follows that the hypotheses of Theorem 7.8 imply condition 1 of Theorem 7.6. This completes the proof of Theorem 7.8.

Remark 7.2. In connection with Theorem 7.8, observe that compactness and uniform differentiability of $F(x)$ in every open ball do not guarantee strong continuity, as shown by the following example. Consider the functional

$$f(x)=(x,\ x)$$

in a real Hilbert space H. This functional is uniformly Fréchet-differentiable in every open ball, and it is compact; but it is not strongly continuous, since, if $\{e_n\}$ is an orthonormal sequence in H, then $e_n \rightharpoonup 0$ but $f(e_n)=1$ for each e_n.

Remark 7.3. The assumption of strong differentiability in Lemma 7.1 is essential. This can be seen from the following example. Consider the real function of a real variable

$$f(t)=t^2\sin(1/t), \quad t\neq 0; \quad f(0)=0.$$

Then

$$f'(t)=2t\sin(1/t)-\cos(1/t), \qquad t\neq 0,$$
$$f'(0)=\lim_{t\to 0}(1/t)f(t)=0,$$

i. e., this function is Fréchet-differentiable but not uniformly so. In finite-dimensional spaces, continuity is equivalent to strong continuity, so that strong continuity and Fréchet differentiability of $f(t)$ do not imply strong continuity of $f'(t)$.

Addendum to subsection 5.3. In the proof of Theorem 5.1 we used the most widely accepted definition of a convex functional. However, if $f(x)$ is Gâteaux-differentiable, inequality (8.5) may serve as an equivalent definition of convexity. This was our approach in /9/ (see VM, p. 102). This definition has several advantages; it is, in particular, useful in proving lower weak semicontinuity of convex functionals.

Let $\nabla f(x_0)$ be the subgradient (supporting functional) of $f(x)$ at the point x_0. At points where f is not differentiable, ∇f is a multivalued mapping. In this case, inequality (8.9) may be taken as a definition of the convexity of f. From (8.9) it follows that the graph of ∇f is a monotone set.

Chapter III

MINIMA OF NONLINEAR FUNCTIONALS

In this chapter we study real-valued finite functionals for critical points and minima, where our functionals are defined in real reflexive Banach spaces or in duals of normed separable spaces. It is characteristic for reflexive spaces that the unit sphere is weakly compact,* and an analogous statement holds for spaces which are duals of normed separable spaces.** As a matter of fact, some of the propositions will be valid for a broader class of linear spaces to be indicated in the text.

Our main objective in this chapter is to prove various propositions on the existence and uniqueness of minima for nonlinear functionals (§ 9) and on the convergence of minimizing sequences. To this end, in § 8 we present the basic propositions on weak semicontinuity of functionals needed in our investigation of minima.

§ 8. ON WEAK SEMICONTINUITY AND CONTINUITY OF FUNCTIONALS

8.1. Criteria for weak semicontinuity of functionals

Definition 8.1. A real functional $f(x)$ defined on a set U in a normed space is said to be lower (upper) semicontinuous at $x_0 \in U$ if, for any sequence $\{x_n\} \subset U$ converging weakly to x_0 ($x_n \rightharpoonup x_0$),

$$f(x_0) \leqslant \varliminf_{n\to\infty} f(x_n) \quad (f(x_0) \geqslant \varlimsup_{n\to\infty} f(x_n)).$$

As in the classical case (see, for example Natanson /1/), this definition implies that the sum of lower semicontinuous functionals is again lower semicontinuous.

Definition 8.2. A set σ in a normed space is said to be weakly closed (or sequentially weakly complete) if it is sequentially weakly closed, i. e., for any sequence $\{x_n\} \subset \sigma$ converging weakly to x_0, we have $x_0 \in \sigma$.

* See, for example, Dunford and Schwartz /1/, p. 425.

** See Hille, Phillips /1/, 2.10.

Remark 8.1. This definition of a weakly closed set in a normed space E differs from the conventional one. The usual definition is that a set $X \subset E$ is weakly closed if it is closed in the weak topology of E.

The neighborhoods in the weak topology are defined as follows. Let $x_0 \in E$; let ε be an arbitrary positive number, and $l_1, l_2, \ldots, l_n$ an arbitrary finite sequence of continuous linear functionals in E^*. A neighborhood is any set of the form

$$U(x_0; l_1, l_2, \ldots, l_n, \varepsilon) = \\ = \{x \in E: |l_k(x - x_0)| < \varepsilon, \ k = 1, 2, \ldots, n.\}$$

The family of all such neighborhoods determines the weak topology of the space E. The family of all neighborhoods $U(x_0; l_1, l_2, \ldots, l_n, \varepsilon)$ forms a basis about the point x_0. The weakly closed sets in this topology have the following property. There may exist a sequence having a unique limit point in this topology, while none of its subsequences converge weakly to the point.* Note, however, that if σ is the closure of $\mathscr{X} \subset E$ in the sense of Definition 8.2, and $\bar{\sigma}$ the closure of $\mathscr{X} \subset E$ in the weak topology of E, then $\bar{\sigma} \supset \sigma$. In view of this inclusion, we may invoke the various propositions on weak completeness of sets in E. For example, Mazur's theorem (strongly closed convex sets in Banach spaces are weakly closed) is valid if "weakly closed" is understood in the sense of Definition 8.2.

*Theorem 8.1.*** A functional $f(x)$ defined in a normed space is weakly lower semicontinuous if and only if, for any real c, the set*

$$E_c = \{x: f(x) \leqslant c\} \tag{8.1}$$

is weakly closed.

Proof. Sufficiency. Suppose condition (8.1) holds. We consider a sequence $x_n \rightharpoonup x_0$ and prove that

$$f(x_0) \leqslant \varliminf_{n \to \infty} f(x_n) = b. \tag{8.2}$$

If we suppose the contrary, $f(x_0) > b$, there exists $\varepsilon > 0$ such that

$$f(x_0) > b + \varepsilon. \tag{8.3}$$

Since

$$b = \varliminf_{n \to \infty} f(x_n),$$

* An example is due to von Neumann /1/. It exemplifies the inadequacy of sequentially weak completeness.

** Vainberg /16, 18/.

there exists a subsequence $\{x_{n_k}\}$ such that

$$f(x_{n_k}) \leqslant b + \varepsilon/2. \tag{8.4}$$

But, by assumption, the set

$$E_{c_1} = \{x\colon f(x) \leqslant b + \varepsilon/2\}, \quad c_1 = b + \varepsilon/2$$

is weakly closed and, since $x_{n_k} \rightharpoonup x_0$ as $k \to \infty$, it follows from (8.4) that $\{x_{n_k}\} \subset E_{c_1}$. Thus also $x_0 \in E_{c_1}$, i. e.,

$$f(x_0) \leqslant b + \varepsilon/2.$$

This contradicts (8.3), proving sufficiency for $b > -\infty$. If $b = -\infty$, then for any real α there exists a subsequence $\{x_{n_k}\}$ such that $f(x_{n_k}) < \alpha$, so that $x_0 \in E_\alpha$ and, as α is arbitrary, $f(x_0) = -\infty$.

Necessity. Suppose the functional $f(x)$ is lower semicontinuous. We must prove that E_c is weakly closed. Consider any weakly fundamental sequence $\{x_n\} \subset F_c$, converging weakly to an element x_0 of the space.* Since $f(x)$ is weakly lower semicontinuous at x_0, it follows that (8.2) holds. But $f(x_n) \leqslant c$ for any n, and thus also $b \leqslant c$. Consequently, $f(x_0) \leqslant b \leqslant c$, i. e., $x_0 \in E_c$. Q. E. D.

Note that this theorem remains valid for locally convex spaces.

8.2 Conditions for weak semicontinuity of convex functionals

Definition 8.3. A real functional $f(x)$ defined on a convex set ω in a linear space is said to be *convex* on ω if, for any $x_1, x_2 \in \omega$ and any $\lambda \in (0, 1)$,

$$f(\lambda x_1 + (1-\lambda) x_2) \leqslant \lambda f(x_1) + (1-\lambda) f(x_2);$$

it is said to be *strictly convex* if equality cannot hold when $x_1 \neq x_2$.

Definition 8.4. A real functional $f(x)$ defined on an open set U in a normed space is said to be *lower (upper) semicontinuous* at a point $x_0 \in U$ if, for any sequence $\{x_n\} \subset U$ converging to x_0 $(x_n \to x_0)$,

$$f(x_0) \leqslant \varliminf_{n\to\infty} f(x_n) \quad (f(x_0) \geqslant \varlimsup_{n\to\infty} f(x_n)).$$

* If the space is semicomplete (sequentially weakly complete) the condition $x_n \rightharpoonup x_0$ may be dropped.

Lemma 8.1. If $f(x)$ is a lower semicontinuous real functional, then, for any real c, the set

$$E_c = \{x: f(x) \leqslant c\}$$

is closed.

Proof. Consider a fundamental sequence $\{x_n\} \subset E_c$. Let $x_0 = \lim_{n\to\infty} x_n$. Suppose that $x_0 \overline{\in} E_c$, i. e., $f(x_0) > c$. Then there exists $\varepsilon > 0$ such that $f(x_0) > c + \varepsilon$. Since $f(x)$ is lower semicontinuous at x_0, there exists a positive integer $n \geqslant n_0$ such that for $n_0 = n_0(\varepsilon)$ we have $f(x_n) > f(x_0) - \varepsilon/2 > c + \varepsilon/2$. This contradicts the fact that $\{x_n\} \subset E_c$, completing the proof of the lemma.

Lemma 8.2. If $f(x)$ is a convex lower semicontinuous functional defined in a Banach space, the set

$$E_c = \{x: f(x) \leqslant c\},$$

is weakly closed for any real number c.

Proof. Since $f(x)$ is convex, E_c is a convex set; moreover, since $f(x)$ is lower semicontinuous it follows by Lemma 8.1 that E_c is closed. But any convex and closed set in a Banach space is weakly closed (Hille and Phillips /1/, p. 36).

By Theorem 8.1, this lemma yields

Theorem 8.2. Every convex lower semicontinuous functional in a Banach space E is weakly lower semicontinuous.

A stronger proposition will be proved later (Theorem 8.10).

8.3. Conditions for weak semicontinuity and convexity of differentiable functionals*

If a real nonlinear functional is Gâteaux-differentiable, there are simple sufficient conditions for its convexity and weak lower semicontinuity. We cite some of these.

Theorem 8.3. Let $f(x)$ be a real functional, defined in a normed linear space and Gâteaux-differentiable; let its Gâteaux differential $Df(x, h)$ be linear and continuous in x_0. If the inequality

$$f(x) - f(x_0) - Df(x_0, x - x_0) \geqslant 0 \qquad (8.5)$$

holds at a point x_0 for any $x \in U$, where U is a neighborhood of x_0, then $f(x)$ is weakly lower semicontinuous at x_0. If inequality (8.5) holds for any x, x_0 in an open convex set U, then $f(x)$ is convex and

* Vainberg /9, 10, 15, 23/; Kachurovskii /1, 4/.

weakly semicontinuous in U (strictly convex if equality cannot hold in (8.5) when $x \neq x_0$).

Proof. Consider an arbitrary sequence $\{x_n\} \subset U$ converging weakly to x_0. Since $Df(x_0, x_n - x_0)$ is linear and continuous in $x_n - x_0$ and $x_n \rightharpoonup x_0$, it follows that

$$\lim_{n \to \infty} Df(x_0, x_n - x_0) = 0.$$

Thus by (8.5),

$$\underline{\lim}_{n \to \infty} [f(x_n) - f(x_0)] \geqslant \underline{\lim}_{n \to \infty} Df(x_0, x_n - x_0) = \\ = \lim_{n \to \infty} Df(x_0, x_n - x_0) = 0,$$

i. e.,

$$f(x_0) \leqslant \underline{\lim}_{n \to \infty} f(x_n).$$

The first assertion (weak semicontinuity of $f(x)$ at x_0 or on U) is thus proved.

To prove the convexity of $f(x)$, we apply (8.5) to two arbitrary (but distinct) vectors $x_1, x_2 \in U$:

$$f(x_1) \geqslant f(x_0) + Df(x_0, x_1 - x_0),$$
$$f(x_2) \geqslant f(x_0) + Df(x_0, x_2 - x_0).$$

Multiplying these inequalities by t and $1 - t$, respectively, where $0 < t < 1$, and setting $x_0 = tx_1 + (1 - t)x_2$, we get

$$tf(x_1) + (1 - t)f(x_2) \geqslant f(x_0) + Df(x_0, tx_1 + (1 - t)x_2 - x_0) = \\ = f(x_0) + Df(x_0, 0) = f(x_0),$$

i. e.,

$$tf(x_1) + (1 - t)f(x_2) \geqslant f(tx_1 + (1 - t)x_2).$$

This shows that $f(x)$ is convex or strictly convex, as the case may be. Q. E. D.

This theorem remains valid for a broad class of linear topological spaces, provided the sequence in the first part of the proof is a transfinite sequence.

Various sufficient conditions for the validity of inequality (8.5) are known (see VM, p. 102). We cite some of these.

Lemma 8.3. Let $f(x)$ be a functional with linear Gâteaux-differential $Df(x, h)$ which is continuous in h and exists on a convex set σ in a normed space. Suppose, moreover, that

$$Df(x, x-x_0) - Df(x_0, x-x_0) \geqslant 0 \qquad (x, x_0 \in \sigma). \tag{8.6}$$

Then (8.5) holds.

Proof. By the Lagrange formula (2.2),

$$f(x)-f(x_0) = Df(x_0+\tau(x-x_0), x-x_0) = Df(x_0, x-x_0) + \\ + [Df(x_0+\tau(x-x_0), x-x_0) - Df(x_0, x-x_0)].$$

But, by (8.6),

$$Df(x_0+\tau(x-x_0), x-x_0) - Df(x_0, x-x_0) = \\ = \frac{1}{\tau}[Df(x_0+\tau(x-x_0), \tau(x-x_0)) - Df(x_0, \tau(x-x_0))] \geqslant 0.$$

Thus,

$$f(x)-f(x_0) \geqslant Df(x_0, x-x_0).$$

Q. E. D.

Inequality (8.6) may be written in another form. Indeed, since the differential $Df(x, h)$ is continuous in h, it follows that

$$Df(x, h) = \langle F(x), h\rangle; \quad F(x) = \operatorname{grad} f(x).$$

Thus inequality (8.6) becomes

$$\langle F(x) - F(x_0), x-x_0\rangle \geqslant 0.$$

This inequality means that $F(x)$ is a monotone operator. We thus have the following proposition from Lemma 8.3 and Theorem 8.3 of VM (cited without proof in Vainberg and Kachurovskii /1/):

*Theorem 8.4. If the gradient of a real differentiable functional is a monotone operator, the functional is weakly lower semicontinuous.**

Lemma 8.4. Let $f(x)$ be a twice Gâteaux-differentiable functional on a convex set σ in a normed space and let

$$D^2f(x; h, h) \geqslant 0 \qquad (x \in \sigma). \tag{8.8}$$

Then inequality (8.5) holds.

* This theorem remains valid for separable locally convex spaces.

Proof. Using the Lagrange formula and inequality (8.8), we have

$$Df(x_0+\tau(x-x_0),\ x-x_0)-Df(x_0,\ x-x_0)=$$
$$=\tau D^2f(x_0+\tau_1(x-x_0);\ x-x_0,\ x-x_0)\geqslant 0.$$

Using (8.7), we thus get (8.5).

This lemma and Theorem 8.3 yield

Theorem 8.5. If the gradient $F(x)$ *of a real functional* $f(x)$ *has Gâteaux derivative* $F'(x)$ *satisfying*

$$\langle F'(x)h,\ h\rangle\geqslant 0,$$

then $f(x)$ *is lower weakly semicontinuous.*

Indeed, under these assumptions the above inequality is equivalent to (8.8).

We prove one further proposition on the weak semicontinuity of functionals.

Theorem 8.6. If a convex functional $f(x)$ *defined on an open convex set* U *in a normed space is Gâteaux-differentiable and, for any fixed* $x\in U$, *the differential* $Df(x, h)$ *is continuous in* h, *then* $f(x)$ *is weakly semicontinuous on* U.

Proof. Let $F(x)=\operatorname{grad} f(x)$; let x, x_0 be any two distinct vectors in U and set $h=x-x_0$. Consider the real function

$$\varphi(t)=f(x_0+th),\qquad 0\leqslant t\leqslant 1.$$

This function is convex, since $f(x)$ is a convex functional, and its derivative (see the proof of Theorem 5.1)

$$\varphi'(t)=\langle F(x_0+th),\ h\rangle$$

is nondecreasing. Hence,

$$f(x)-f(x_0)=\varphi(1)-\varphi(0)=\varphi'(\tau)\geqslant\varphi'(0)=\langle F(x_0),\ h\rangle,$$

since $0<\tau<1$. Therefore,

$$f(x)-f(x_0)-Df(x_0,\ x-x_0)\geqslant 0,$$

so that inequality (8.5) holds. Thus, by Theorem 8.3, the functional $f(x)$ is weakly lower semicontinuous.

Note, in addition, that Theorem 8.6 follows immediately from Theorems 5.1 and 8.3. We shall see that a yet stronger proposition is valid (Theorem 8.10). From Theorems 8.3 and 8.6 we have

Corollary 8.1. If a real Gâteaux-differentiable functional $f(x)$ is defined on an open convex set U, its differential $Df(x, h)$ being continuous in h for each $x \in U$, then $f(x)$ is convex on U if and only if inequality (8.5) holds. If equality cannot hold in (8.5) when $x \neq x_0$, the validity of (8.5) is a necessary and sufficient condition for $f(x)$ to be strictly convex.

Theorem 8.3 and Lemma 8.4 imply

Corollary 8.2. Let $f(x)$ be a real functional, defined on an open convex set U in a normed space, and let its gradient $F(x)$ be Gâteaux-differentiable. Then $f(x)$ is convex if and only if for any $x, h \in U$

$$\langle F'(x) h, h \rangle \geqslant 0.$$

If this inequality is strict for $h \neq 0$, $f(x)$ is a strictly convex functional.

8.4. Connection with support functionals*

Let E be a normed space with dual E^*.

Definition 8.5. A linear functional $y_0 \in E^*$ is a support functional** of a functional $f(x)$ at a point x_0 if

$$f(x) - f(x_0) \geqslant \langle y_0, x - x_0 \rangle, \tag{8.9}$$

where $\langle y_0, h \rangle$ is the value of y_0 at the vector h.

Theorem 8.3 and its proof remain valid if we replace (8.5) by (8.9). Hence the following proposition.

Theorem 8.7. If a finite real functional $f(x)$, defined on an open convex set $U \subset E$, has a support functional at every point of U, then $f(x)$ is convex and weakly lower semicontinuous on U.

The existence of a support functional at each point of U is a criterion for the convexity of a real functional defined on U, as may be seen from the following

Theorem 8.8. If $f(x)$ is a finite convex functional defined on an open convex set $U \subset E$, it has a support functional at each point $x \in U$.

Proof. Consider arbitrary vectors $x, x + h \in U$ and define a real function by

$$\varphi(t) = f(x + th), \quad (-\varepsilon < t < 1 + \varepsilon, \ \varepsilon > 0).$$

* See Dubovitskii and Milyutin /1/, Levitin and Polyak /1/, Dem'yanov and Rubinov /1/, for an investigation of this problem.

** Some authors use the term subgradient.

For fixed x, h, this function is convex on the interval $(-\varepsilon, 1+\varepsilon)$. Indeed, let $\alpha \geqslant 0$, $\beta \geqslant 0$, $\alpha + \beta = 1$. Then, by the convexity of $f(x)$, we have

$$\varphi(\alpha t_1 + \beta t_2) = f[\alpha(x + t_1 h) + \beta(x + t_2 h)] \leqslant \alpha\varphi(t_1) + \beta\varphi(t_2).$$

But any convex function has left and right derivatives $\varphi'(t-0)$ and $\varphi'(t+0)$ (see subsection 2.7), and $\varphi'(t-0) \leqslant \varphi'(t+0)$. In particular, $\varphi'(-0) \leqslant \varphi'(+0)$, i. e. (see subsection 2.7),

$$V_- f(x, h) \leqslant V_+ f(x, h), \tag{8.10}$$

where

$$V_- f(x, h) = -V_+ f(x, -h).$$

The directional variation $V_+ f(x, h)$ (see subsection 2.7) is by definition a positive-homogeneous operator in h. Moreover, it is subadditive in h, since

$$V_+ f(x, h_1 + h_2) = \lim_{t\to+0} \frac{f\left(x + \frac{t}{2}(h_1 + h_2)\right) - f(x)}{t/2} =$$

$$= \lim_{t\to+0} \frac{2f\left(\frac{x + th_1}{2} + \frac{x + th_2}{2}\right) - 2f(x)}{t} \leqslant \lim_{t\to+0} \frac{f(x+th_1) - f(x)}{t} +$$

$$+ \lim_{t\to+0} \frac{f(x + th_2) - f(x)}{t} = V_+ f(x, h_1) + V_+ f(x, h_2).$$

Thus $V_+ f(x, h)$ is a sublinear functional, again implying (8.10) (see below, (8.10')).

We require the following

Lemma 8.5. If $f(x)$ is a convex finite functional on an open convex set U, then to every $x \in U$ there corresponds at least one $y \in E^*$ such that

$$V_- f(x, h) \leqslant \langle y, h\rangle \leqslant V_+ f(x, h), \tag{8.11}$$

where $V_+ f(x, h)$ is the variation in the direction h, and

$$V_- f(x, h) = -V_+ f(x, -h).$$

Proof.* Fix a nonzero vector h_0 and consider the straight line $z = \lambda h_0$ $(-\infty < \lambda < +\infty)$ in E. Let y_0 be a linear functional such that

$$y_0(h_0) = V_+ f(x, h_0).$$

* See Dem'yanov and Rubinov /1/.

Since $V_+f(x, h)$ is positive-homogeneous in h, it follows that

$$y_0(\lambda h_0)=V_+f(x, \lambda h_0), \qquad \lambda \geqslant 0.$$

By the sublinearity of $V_+f(x, h)$, we have

$$0=V_+f(x, h-h)\leqslant V_+f(x, h)+V_+f(x, -h)$$

or

$$-V_+f(x, h)\leqslant V_+f(x, -h), \tag{8.10'}$$

so that for $\lambda < 0$

$$y_0(\lambda h_0)=\lambda V_+f(x, h_0)=-|\lambda| V_+f(x, h_0)\leqslant$$
$$\leqslant |\lambda| V_+f(x, -h_0)=V_+f(x, -|\lambda| h_0)=V_+f(x, \lambda h_0).$$

We thus have for any $z=\lambda h_0$

$$y_0(z)\leqslant V_+f(x, z). \tag{8.12}$$

Hence, by the Hahn-Banach theorem,* the linear functional y_0, defined on the straight line and satisfying (8.12), can be extended to a functional $y\in E^*$ satisfying the conditions

$$\langle y, h\rangle\leqslant V_+f(x, h), \qquad \forall h\in E,$$
$$\langle y, h_0\rangle=V_+f(x, h_0).$$

Finally, the last inequality implies

$$\langle y, -h\rangle\leqslant V_+f(x, -h)$$

or

$$\langle y, h\rangle=-\langle y, -h\rangle\geqslant -V_+f(x, -h)=V_-f(x, h),$$

so that

$$V_-f(x, h)\leqslant\langle y, h\rangle\leqslant V_+f(x, h).$$

Q. E. D.

* See, for example, Dunford and Schwartz /1/.

To complete the proof of Theorem 8.8, we again consider the convex function $\varphi(t) = f(x + th)$ and make use of its property (see Bourbaki /2/, p. 47):

$$\varphi'(+0) \leqslant \frac{\varphi(1) - \varphi(0)}{1 - 0} \leqslant \varphi'(1 - 0).$$

Hence

$$\varphi(1) - \varphi(0) \geqslant \varphi'(+0)$$

or

$$f(x + h) - f(x) \geqslant V_{+}f(x, h).$$

Combining this with (8.11), we get

$$f(x + h) - f(x) \geqslant \langle y, h \rangle,$$

so that y is the support functional of f at x. This completes the proof.

Theorems 8.7 and 8.8 give rise to the following two propositions:*

Theorem 8.9. *A finite real functional defined on an open convex set $U \subset E$ is convex on U if and only if it has a support functional at every point $x \in U$.*

Theorem 8.10. *Any finite convex functional defined on an open convex set in E is weakly lower semicontinuous.*

8.5. On the weak semicontinuity of quasiconvex functionals

Definition 8.6. A real functional $f(x)$ defined on a convex set ω in a linear space is said to be quasiconvex on ω if, for any $x_1, x_2 \in \omega$ and any nonnegative $\lambda \in (0, 1)$,

$$f(\lambda x_1 + (1 - \lambda) x_2) \leqslant \max[f(x_1), f(x_2)];$$

it is strictly quasiconvex if equality cannot hold when $x_1 \neq x_2$.

Lemma 8.6. *If $f(x)$ is a quasiconvex lower semicontinuous functional in a Banach space, the set*

$$E_c = \{x: f(x) \leqslant c\},$$

where c is an arbitrary real number, is weakly closed.

* Lavrent'ev has remarked that Theorem 8.10 also follows from Theorem 8.1.

Proof. By Definition 8.6, the set E_c is convex; by the lower semicontinuity of $f(x)$ and Lemma 8.1, E_c is closed. But (see Hille and Phillips /1/, p. 36) a convex closed set in a Banach space is weakly closed.

By Theorem 8.1, this lemma implies

Theorem 8.11. Any quasiconvex lower semicontinuous functional in a Banach space E *is weakly lower semicontinuous.**

Note that if the functional is not lower semicontinuous, Theorem 8.11 fails to hold, as exemplified by the following function of a real variable:

$$f(x)=\begin{cases} e^x, & x<1, \\ e^2, & 1\leqslant x\leqslant 2, \\ e^x, & x>2. \end{cases}$$

This function is quasiconvex but not weakly lower semicontinuous.

8.6. Examples of weakly lower semicontinuous functionals

Example 8.1. The functional $f(x)=\|x\|$ in a normed space is convex, so by Theorem 8.1 it is weakly lower semicontinuous.

Example 8.2. Let A be a bounded positive selfadjoint operator in a real Hilbert space. Then

$$Ax=2^{-1}\operatorname{grad}(Ax,\ x)$$

and in view of the fact that

$$(Ax-Ay,\ x-y)\geqslant 0,$$

it follows that A is a monotone operator, whence, by Theorem 8.4, the functional $f(x)=(Ax,\ x)$ is weakly lower semicontinuous.

Example 8.3. Let T be a bounded selfadjoint operator in a Hilbert space such that the positive part of its spectrum is arbitrary and the negative part consists of a finite number of eigenvalues, each of finite multiplicity. Then (see VM, p. 101) the functional

$$f(x)=(Tx,\ x)$$

is weakly lower semicontinuous.

Example 8.4.** Let $g(u,\ x)$ be a nondecreasing function of u for almost every $x\in G$, where G is a measurable set in finite-dimensional euclidean space; suppose this function generates a

* Polyak /1/.

** Vainberg /10/.

Nemytskii operator from L^p to L^q $(p > 1, p^{-1} + q^{-1} = 1)$, i. e., *

$$|g(u, x)| \leqslant a(x) + b|u|^{p-1}, \quad a(x) \in L^q, \quad b > 0.$$

Then the potential of the Nemytskii operator (see Example 5.1)

$$f(u) = \int_G dy \int_0^{u(y)} g(v, y)\, dv$$

is weakly lower semicontinuous in $L^p(G)$. In fact (see VM, p. 90),

$$\operatorname{grad} f(u) = g(u(x), x) = hu,$$

and, since $g(u, x)$ is nondecreasing in u for almost every $x \in G$, it follows that

$$\langle hu - hv, u - v \rangle =$$
$$= \int_G [g(u(x), x) - g(v(x), x)]\,(u(x) - v(x))\, dx \geqslant 0,$$

i. e., the gradient of the functional $f(x)$ is a monotone operator and, by Theorem 8.4, the functional $f(u)$ is thus weakly lower semicontinuous in L^p.

Let T be a bounded linear operator from L^2 to L^p. Then it transforms any sequence $u_n \rightharpoonup u_0$ $(u_n, u_0 \in D^2)$ into a sequence $Tu_n \rightharpoonup Tu_0$ $(Tu_n, Tu_0 \in L^p)$ and thus, as we have shown, the functional

$$\varphi(u) = f(Tu) = \int_G dy \int_0^{Tu} g(v, y)\, dv$$

is weakly semicontinuous in $L^2(G)$.

Example 8.5. Suppose $g(u, x)$ is a convex function of u for almost every $x \in G$, generating a Nemytskii operator from $L(G)$ to itself, i. e. (VM, Theorem 19.1),

$$|g(u, x)| \leqslant a(x) + b|u|, \quad a(x) \in L, \quad b > 0.$$

Then the functional

$$\psi(u) = \int_G g(u(x), x)\, dx$$

is weakly lower semicontinuous in L, as a convex functional.

* VM, Theorem 19.1.

Moreover, the functional $\psi(u)$ is continuous in L because (see VM, Theorem 19.1) the Nemytskii operator is continuous.

Example 8.6. Let B be a bounded positive linear operator from a real reflexive Banach space E to E^*. Consider the quadratic functional

$$f(x) = \langle Bx, x\rangle.$$

For this functional (see Example 2.1) we have

$$\operatorname{grad} f(x) = Bx + B^*x.$$

But

$$\langle \operatorname{grad} f(x) - \operatorname{grad} f(y), x - y\rangle = \langle B(x-y), x-y\rangle + \\ + \langle B^*(x-y), x-y\rangle = 2\langle B(x-y), x-y\rangle \geqslant 0.$$

Thus grad $f(x)$ is a monotone operator, and so it follows from Theorem 8.4 that the quadratic functional $\langle Bx, x\rangle$ is weakly lower semicontinuous.

Furthermore, $B + B^*$ is a potential operator, and since it is also monotone, Theorem 5.1 implies that it has a convex potential, i. e., $\langle Bx, x\rangle$ is a convex functional.

Example 8.7. Let $F(x)$ be a monotone potential operator which is positive-homogeneous of degree $k > 0$ (see Example 6.3), defined on a linear space E. Then by Theorems 5.1 and 8.4, its potential

$$\varphi(x) = \frac{1}{k+1}\langle F(x), x\rangle$$

is a convex weakly lower semicontinuous functional. If $F(x)$ is dissipative, i. e., $(-1)F(x)$ is a monotone operator, then $\varphi(x)$ is a weakly upper semicontinuous functional. Finally, if $F(x)$ is a compact operator defined in a normed space (see Theorem 8.12), then $\varphi(x)$ is weakly continuous.

8.7. On the weak continuity of functionals

Definition 8.7. A real functional $f(x)$ defined on a set in a normed space E is said to be weakly continuous at a point $x_0 \in U$ if it is both weakly lower continuous and weakly upper continuous there (relative to U). In other words (see Definition 8.1), for any sequence $\{x_n\} \subset U$ converging weakly to $x_0 \in U$ $(x_n \rightharpoonup x_0)$,

$$\lim_{n \to \infty} f(x_n) = f(x_0).$$

Theorem 7.1 immediately yields a necessary and sufficient condition for a functional defined in a reflexive Banach space with countable biorthogonal basis to be weakly continuous. We cite a simple sufficient condition for weak semicontinuity of a functional (see VM, Theorem 8.2).

Theorem 8.12. Let $f(x)$ be a Gâteaux-differentiable functional defined on an open convex set ω in a normed space, with compact gradient. Then $f(x)$ is weakly continuous on ω.

In view of the appropriate analogous condition for weak upper semicontinuity, Theorem 8.1 implies

Lemma 8.7. Given that a functional $f(x)$ defined in a normed space is weakly continuous, the set

$$E_c = \{x : f(x) = c\},$$

where c is an arbitrary real number, is weakly closed.

8.8. Semiconvex functionals

Let ω be a convex set in a normed space.

Definition 8.8. A real functional $f(x)$ defined on ω is said to be semiconvex if there exists a weakly continuous functional $g(x)$ on ω such that $f(x) + g(x)$ is convex on ω.

An example is the functional in Example 8.3. Every convex functional is of course semiconvex, but Example 8.3 shows that the converse does not always hold. If $f(x)$ is a semiconvex functional and $g(x)$ is weakly semicontinuous, then $f(x) + g(x)$ is semiconvex. Finally, any finite semiconvex functional $f(x)$ is weakly lower semicontinuous. In fact, there exists by definition a weakly continuous functional $g(x)$ such that $f(x) + g(x)$ is convex, and therefore (see Theorem 8.10) weakly lower semicontinuous. Hence, the functional

$$f(x) = [f(x) + g(x)] + [-g(x)]$$

is weakly lower semicontinuous, being a sum of such functionals.

§ 9. EXISTENCE AND UNIQUENESS THEOREMS FOR MINIMA

In this section we consider mainly real functionals defined in real reflexive Banach spaces. Recall that in such spaces any ball is weakly compact.

9.1. Extremum points of functionals and the generalized Weierstrass theorem

Let $f(x)$ be a real functional defined in a normed space E. A point $x_0 \in E$ is an extremum point of $f(x)$ if in some neighborhood $U(x_0)$ of x_0 the functional satisfies one of the inequalities

$$1)\ f(x) \leqslant f(x_0),\quad 2)\ f(x) \geqslant f(x_0) \quad (\forall x \in U(x_0)).$$

If the second inequality holds for all $x \in E$, then x_0 is called an absolute minimum point of $f(x)$. If $f(x)$ is Gâteaux-differentiable at x_0 and moreover

$$Df(x_0, h) = 0,$$

then x_0 is called a critical point of $f(x)$. Since the vanishing of the differential implies its continuity, the last equality becomes

$$\langle \operatorname{grad} f(x_0), h \rangle = 0,$$

and since this equality holds for any $h \in E$, we may say that x_0 is a critical point of $f(x)$ if

$$\operatorname{grad} f(x_0) = 0.$$

Theorem 9.1.* *Let $f(x)$ be a functional defined in a domain in a normed space E, and x_0 an interior point of ω at which the linear Gâteaux differential exists. Then the following assertions hold:*

1. *If x_0 is an extremum point, it is a critical point:*

$$\operatorname{grad} f(x_0) = 0. \tag{9.1}$$

2. *Suppose, in addition, that $f(x)$ is convex in some convex neighborhood $U(x_0) \subset \omega$ (or that $\operatorname{grad} f(x)$ is a monotone operator). Then x_0 is a minimum point of $f(x)$ if and only if (9.1) holds.*

Proof. Let h be any fixed vector in E. Then $f(x_0 + th)$ is a real function, defined in some neighborhood of the point $t = 0$ and differentiable at $t = 0$. Thus, if x_0 is an extremum point of the

* See Vainberg /9, 15/.

functional $f(x)$, then $t=0$ is an extremum point of the function $f(x_0+th)$, and so

$$\frac{d}{dt} f(x_0+th)\Big|_{t=0}=0$$

or

$$Df(x_0, h)=0,$$

or

$$\langle \operatorname{grad} f(x_0), h\rangle=0.$$

In view of the arbitrary choice of h, it then follows that $\operatorname{grad} f(x_0)=0$, proving part 1.

To prove part 2, we consider the functional

$$\varphi(z)=f(z+x_0)-f(x_0).$$

By hypothesis, it is convex, Gâteaux-differentiable at $z=0$ and $\varphi(0)=0$. If x_0 is a minimum point of $f(x)$, then, as we have shown, $\operatorname{grad}\varphi(0)=0$. Conversely, assuming this last equality (which is simply (9.1)) to hold, let us show that $z=0$ is a minimum point of $\varphi(z)$; this will imply that x_0 is a minimum point of $f(x)$. We prove this by contradiction. Suppose that for some vector $z_1 \in (U(x_0)\setminus x_0)$

$$\varphi(z_1)=\alpha<0.$$

Then, by the convexity of $\varphi(z)$,

$$\varphi(tz_1+(1-t)0)\leqslant t\varphi(z_1)=t\alpha \qquad (0<t<1).$$

Thus

$$[\varphi(0+tz_1)-\varphi(0)]/t\leqslant\alpha,$$

and so

$$\langle \operatorname{grad}\varphi(0), z_1\rangle=\lim_{t\to 0}\frac{\varphi(0+tz_1)-\varphi(0)}{t}\leqslant\alpha<0,$$

i. e.,

$$\operatorname{grad}\varphi(0)\neq 0.$$

This contradiction proves the theorem.

We quote yet another assertion (see VM, Theorem 9.2).

Theorem 9.2 *(First generalized Weierstrass theorem). A weakly lower semicontinuous functional defined on a bounded weakly closed subset σ of a reflexive Banach space E is bounded below and attains its infimum on σ.*

The assumption that the space in Theorem 9.2 is reflexive is essential. We claim, however, that it may be dropped if E is the dual of a separable normed space. To do this, we first review some standard material from functional analysis.*

Let E be a normed space and E^* its dual. A sequence of functionals $\{\varphi_n\} \subset E^*$ is said to be *E-weakly convergent* to $\varphi \in E^*$ (notation: $\varphi_n \xrightarrow[E]{} \varphi$) if for any $x \in E$

$$\lim_{n\to\infty} \varphi_n(x) = \varphi(x).$$

It turns out that E^* is weakly complete with respect to weak convergence of this kind, and if E is separable every bounded set in E^* is weakly compact in this sense.

A nonlinear real functional $f(\varphi)$ defined on E^* is *E-weakly lower semicontinuous* at a point φ_0 if, for any sequence $\varphi_n \xrightarrow[E]{} \varphi_0$,

$$f(\varphi_0) \leqslant \varliminf_{n\to\infty} f(\varphi_n).$$

A set $\sigma \subset E^*$ is *E-weakly closed* if, for any sequence $\{\varphi_n\} \subset E^*$ converging *E-weakly* to φ_0, we have $\varphi_0 \in \sigma$. An example of an E-weakly closed set is the ball $\|\varphi\| \leqslant a$ in E^*.

Observe that if a functional is E-weakly lower semicontinuous, then it is weakly lower semicontinuous, since weak convergence of a sequence of elements implies its E-weak convergence.

Theorem 9.3 *(Second generalized Weierstrass theorem). Let E be a separable normed space and σ a bounded E-weakly closed subset of the dual E^*. Then any functional $f(\varphi)$ defined and E-weakly lower semicontinuous on σ is bounded below on σ and attains its infimum there.*

Proof. If $f(\varphi)$ is not bounded below on σ, there exists a sequence $\{\varphi_n\} \subset \sigma$ such that

$$f(\varphi_n) < -n.$$

Since $\|\varphi_n\| \leqslant C = \text{const}$ by assumption and every bounded set in E^* is E-weakly compact, it follows that there exists a subsequence

* See Kantorovich and Akilov /1/, pp. 248–249, and also Hille and Phillips /1/, 2.10.

$\varphi_{n_k} \xrightarrow[E]{} \varphi_0$, and $\varphi_0 \in \sigma$ because is E-weakly closed. Hence, since $f(\varphi)$ is E-weakly lower semicontinuous, we have

$$f(\varphi_0) \leqslant \varliminf_{k \to \infty} f(\varphi_{n_k}) = -\infty,$$

which is impossible. Consequently,

$$d = \inf_{\varphi \in \sigma} f(\varphi) \tag{9.2}$$

exists [and is finite].

Let $\{\varphi_n\} \subset \sigma$ be a minimizing sequence, i. e.,

$$\lim_{n \to \infty} f(\varphi_n) = d.$$

Since σ is E-weakly compact and E-weakly closed, we can choose a subsequence of $\{\varphi_n\}$, say $\{\varphi_{m_n}\}$, such that

$$\varphi_{m_n} \xrightarrow[E]{} \varphi_0 \in \sigma.$$

Hence, as $f(\varphi)$ is E-weakly continuous on σ,

$$f(\varphi_0) \leqslant \varliminf_{n \to \infty} f(\varphi_{m_n}) = \lim_{n \to \infty} f(\varphi_n) = d,$$

so that (9.2) gives $f(\varphi_0) = d$. Q. E. D.

Remark 9.1. The assumption in Theorems 9.2 and 9.3 that σ is bounded may be dropped if f satisfies the condition*

$$\lim_{R \to \infty} f(z) = +\infty \qquad (R = \| z \|).$$

This follows immediately from the proofs of Theorems 9.2 and 9.3.

9.2. On the m-property of functionals

Let ω be an open convex set in a Banach space E; let ω' be its boundary and $\bar{\omega} = \omega \cup \omega'$. The set $\bar{\omega}$ is weakly closed. Consider all possible sets $\bar{\omega}$ on which the functional $\varphi(x)$ is weakly lower semicontinuous.

* See VM, subsection 13.4.

Definition 9.1. A nonlinear functional $\varphi(x)$ is said to possess the m-property (minimum property) if the following condition holds: among the various weakly closed and weakly compact sets $\bar{\omega}$ on which $\varphi(x)$ is weakly lower semicontinuous, there is at least one, $\bar{\omega}_0$ say, such that $\varphi(x) > \varphi(x_0)$ on ω_0', where x_0 is any point of ω_0.

By the generalized Weierstrass Theorem, a functional $\varphi(x)$ possessing the m-property has a minimum on $\bar{\omega}_0$ and the minimum point cannot belong to ω_0', so that it must be in ω_0. Hence, if the functional is Gâteaux-differentiable, we have the following

Theorem 9.4. Let $\varphi(x)$ be a Gâteaux-differentiable functional defined in a reflexive Banach space and possessing the m-property. Then the functional has at least one critical point x_0, i.e., a point such that* $\operatorname{grad} \varphi(x_0) = 0$.

This theorem remains valid if we assume that E is not necessarily reflexive but that $E = X^*$, where X is a normed separable space.

Corollary 9.1. If the functional $\varphi(x)$ in Theorem 9.4 is assumed to be not merely weakly lower semicontinuous but also strictly convex, the minimum point x_0 is unique.

Indeed, if $\varphi(x)$ is convex, then by Theorem 8.10 it is weakly lower semicontinuous, so that it has at least one minimum point. Moreover, since $\varphi(x)$ is strictly convex, it cannot have more than one minimum point. Indeed, if there are two distinct minimum points x_1 and x_2, set $d_1 = \varphi(x_1)$, $d_2 = \varphi(x_2)$ and suppose that $d_1 \leqslant d_2$; then, by the strict convexity of $\varphi(x)$,

$$\varphi(\lambda x_1 + (1-\lambda) x_2) < \lambda d_1 + (1-\lambda) d_2 \leqslant d_2 \qquad (0 < \lambda < 1).$$

However, for any neighborhood $U(x_2)$ of x_2, there exists a sufficiently small λ such that $\lambda x_1 + (1-\lambda) x_2 = z \in U(x_2)$. Hence, using previous results, we have the inequality

$$\varphi(z) < \varphi(x_2) = d_2, \quad z \in U(x_2),$$

which contradicts our assumption that x_2 is a minimum point of $\varphi(x)$.

In the next subsection, we shall present sufficient conditions for functionals to have the m-property;** these conditions will yield various theorems on the existence and uniqueness of minima of functionals.

* See Vainberg /5, 9/.

** Vainberg /5, 9/.

9.3. Criteria for unconditional minima

We begin with a simple proposition.

Lemma 9.1. If a strictly convex functional $f(x)$ *defined in a linear space* E *has a minimum at a point* x_0, *then* x_0 *is an absolute minimum point, and there are no other minimum points; moreover* $f(x_0 + tx)$ *is an increasing function of* $t\,(t > 0)$ *for any* $x \neq 0$.

Proof. Set

$$\varphi(z) = f(x_0 + z) - f(x_0).$$

The functional $\varphi(z)$ is convex and by assumption has a minimum at the zero of the space. We claim that $\varphi(z) > 0$ at any point $z \neq 0$. It will then follow that $f(x) > f(x_0)$ for $x \neq x_0$. We first observe that in proving Corollary 9.1 we in fact showed that if a strictly convex functional has a minimum, then the latter is strict and there are no other minimum points. Hence there exists a neighborhood U of zero such that if $x \in U$, $x \neq 0$, then $\varphi(x) > \varphi(0) = 0$. Now consider any nonzero point $x \in E$. For sufficiently small λ, we have $\lambda x \in U$. But $\varphi(x)$ is strictly convex, so that

$$\varphi(\lambda x + (1-\lambda)0) < \lambda\varphi(x) + (1-\lambda)\varphi(0) = \lambda\varphi(x) \quad (0 < \lambda < 1).$$

Since $\lambda x \in U$,

$$0 = \varphi(0) < \varphi(\lambda x) < \lambda\varphi(x) < \varphi(x).$$

This establishes the first part of the lemma.

In order to prove the second part, we consider the real function of a real variable

$$\psi(t) = f(x_0 + tx) - f(x_0), \quad x \neq 0, \quad t \geqslant 0.$$

This function is strictly convex and so (see Bourbaki /2/, p. 46, Proposition 5) $[\psi(t) - \psi(0)]/t$ is an increasing function of t, i. e.,

$$\frac{\psi(t_2)}{t_2} > \frac{\psi(t_1)}{t_1}, \quad t_1 < t_2.$$

Hence, since $\psi(t) > 0$ for $t > 0$, it follows that $\psi(t_2) > \psi(t_1)$ and thus $f(x_0 + tx)$ is an increasing function for nonnegative t. Q. E. D.

Theorem 9.5. Let $f(x)$ *be a finite weakly lower semicontinuous functional defined in a reflexive Banach space* E, *satisfying the condition*

$$\overline{\lim_{R\to+\infty}} f(x) = +\infty \qquad (R = \|x\|). \tag{9.3}$$

Then $f(x)$ *has an unconditional minimum point.*

Proof. Condition (9.3) implies that, for any fixed vector x_0, there exists a sphere $S_r = \{x \in E\colon \|x\| = r > \|x_0\|\}$ on which $f(x) > f(x_0)$. Since the ball $D_r = \{x \in E\colon \|x\| \leqslant r\}$ is weakly closed and $f(x)$ is weakly lower semicontinuous, it follows that $f(x)$ has the m-property, i. e., it has a local minimum at some interior period of D_r. Q. E. D.

Note that if we replace the upper limit in (9.3) by a limit, a local minimum at a point x^* becomes an absolute minimum of $f(x)$, i. e., $f(x^*) \leqslant f(x)$ for all $x \in E$, since in this case $f(x) > f(x_0)$ not only when $\|x\| = r$, but also when $\|x\| > r$. Moreover, instead of requiring that E be reflexive, we can stipulate that $E = X^*$, where X is a separable space.

Theorem 9.6.* *Let* $f(x)$ *be a real functional defined in a real reflexive Banach space* E; *suppose* $f(x)$ *is Gâteaux-differentiable and that* $\operatorname{grad} f(x) = F(x)$ *satisfies the following conditions:*

1) *the function* $\langle F(tx), x\rangle$ *is continuous in* t *on* $[0, 1]$ *for any* $x \in E$,
2) $\langle F(x+h) - F(x), h\rangle \geqslant 0 \qquad (\forall x, h \in E)$,
3) $\lim\limits_{R\to\infty} \dfrac{\langle F(x), x\rangle}{R} = +\infty \qquad (R = \|x\|)$.

Then $f(x)$ *has a minimum point* x_0 *and* $\operatorname{grad} f(x_0) = 0$. *If condition 2 is an equality only when* $h = 0$, *i.e.,* $F(x)$ *is a strictly monotone operator, then the minimum point of the functional is unique and* $f(x)$ *has an absolute minimum there.*

Proof. By Theorem 8.4, condition 2 implies that the functional $f(x)$ is weakly lower semicontinuous. Further, by (6.4),

$$f(x) = f(0) + \int_0^1 \langle F(tx), x\rangle\, dt,$$

where the existence of the integral is guaranteed by condition 1. Condition 2 thus yields

$$f(x) - f(0) = \int_0^1 \langle F(tx), x\rangle\, dt = \langle F(0), x\rangle + $$

$$+ \int_0^1 \langle F(tx) - F(0), x\rangle\, dt \geqslant \langle F(0), x\rangle + $$

$$+ \int_{1/2}^1 \langle F(tx) - F(0), x\rangle\, dt = \frac{1}{2}\langle F(0), x\rangle + \frac{1}{2}\langle F(t_1 x), x\rangle,$$

* Vainberg /9, 15, 17, 19/.

where $0.5 \leqslant t_1 \leqslant 1$. Hence,

$$f(x) - f(0) \geqslant 2^{-1} \| x \| \left(\frac{\langle F(t_1 x), t_1 x \rangle}{\| t_1 x \|} - \| F(0) \| \right),$$

so that, by condition 3, we obtain (9.3). The hypotheses of Theorem 9.5 thus hold, so that $f(x)$ has a local minimum at an interior point x_0 of the ball D_r. At this point,

$$\operatorname{grad} f(x_0) = 0,$$

because $f(x)$ is Gâteaux-differentiable. Finally, if $F(x)$ is strictly monotone, then (Theorem 5.1) $f(x)$ is strictly convex, so Corollary 9.1 and Lemma 9.1 entail that x_0 is the unique minimum point and that $f(x)$ has an absolute minimum there. Q. E. D.

Remark 9.2. If $\operatorname{grad} f(x) = F(x)$ is strictly monotone, the uniqueness of the minimum for $f(x)$ may also be proved directly. In fact, if $f(x)$ has minima at two distinct points $x_1, x_2 \in E$, then $F(x_1) = 0$ and $F(x_2) = 0$, whence

$$\langle F(x_1) - F(x_2), x_1 - x_2 \rangle = 0.$$

But this contradicts the strict monotonicity of $F(x)$.

Theorem 9.7. Let $F(x)$ be a monotone potential operator, defined in a reflexive Banach space E and satisfying the following conditions:

1. $\langle F(tx), x \rangle$ is integrable with respect to t on $(0, 1)$ for any $x \in E$,

2. $\langle F(x), x \rangle \geqslant \| x \| \gamma(\| x \|)$,

where $\gamma(t)$ is integrable on $[0, R]$ for any $R > 0$,

3. $$\overline{\lim_{R \to \infty}} \int_0^R \gamma(t)\, dt = c > 0.$$

Then the potential $f(x)$ of $F(x)$ has a minimum point. The minimum point is unique and $f(x)$ has an absolute minimum there if F is strictly monotone.

Proof. As before, we find that $f(x)$ is a weakly lower semicontinuous functional satisfying the condition

$$f(x) - f(0) = \int_0^1 \langle F(tx), x \rangle\, dt \geqslant$$

$$\geqslant \int_0^1 \| x \| \gamma(t \| x \|)\, dt = \int_0^{\| x \|} \gamma(z)\, dz.$$

Hence, setting $\|x\|=R$, we obtain

$$\overline{\lim_{R\to\infty}} f(x) \geqslant f(0) + \overline{\lim_{R\to\infty}} \int_0^R \gamma(z)\,dz = f(0)+c,$$

so that for some $r=R$, all points on the sphere $\|x\|=r$ satisfy

$$f(x) > f(0),$$

i. e., $f(x)$ has the m-property. Consequently, $f(x)$ has a minimum at some interior point x_0 of the ball $\|x\| \leqslant r$, and so grad $f(x_0)=0$. The uniqueness of the minimum and the fact that x_0 is an absolute minimum point are proved as in Theorem 9.6.

We shall need the following

Lemma 9.2. Let $f(x)$ be a real Gâteaux-differentiable functional defined in a real normed space E. If

$$\operatorname{grad} f(x) = F(x)$$

satisfies the condition

$$\langle F(x), x\rangle > 0, \tag{9.4}$$

on the sphere $S=\{x\in E: \|x\| = R>0\}$, then there are no vectors z on S such that

$$\min_{\|x\|\leqslant R} f(x) = f(z),$$

when the minimum on the left exists.

The proof follows easily by contradiction. Suppose that

$$\min_{\|x\|\leqslant R} f(x) = f(x_0), \quad \|x_0\| = R.$$

Then the real function $\varphi(t) = f(x_0 + t(-x_0))$ satisfies the inequality $\varphi(t) \geqslant \varphi(0)$ for sufficiently small $t>0$, and so $\varphi'(+0)\geqslant 0$. But

$$\varphi'(+0) = \lim_{t\to+0} \frac{f(x_0+t(-x_0)) - f(x_0)}{t} = \langle F(x_0), -x_0\rangle,$$

so that

$$\varphi'(+0) = -\langle F(x_0), x_0\rangle < 0.$$

This contradiction proves the lemma.

Lemma 9.3. Let $f(x)$ *be a real Gâteaux-differentiable functional defined in a real normed space* E *and let* $F(x) = \operatorname{grad} f(x)$. *If for all points on the sphere*

$$S_a = \{x \in E : \| x - a \| = R > 0\}$$

we have

$$\langle F(x), x - a \rangle > 0,$$

then there are no points z *on* S_a *for which*

$$\min_{\| x - a \| \leqslant R} f(x) = f(z),$$

when the minimum on the left exists.

To prove this, we need only consider the functional

$$\psi(t) = f(z + th), \quad h = a - z$$

and repeat the argument of the previous lemma.

Theorem 9.8. Let $f(x)$ *be a real Gâteaux-differentiable functional defined in a real reflexive Banach space, which is weakly lower semicontinuous and satisfies the condition*

$$\langle F(x), x \rangle > 0 \tag{9.4}$$

for any vector $x \in E$, $\|x\| = R > 0$, *and* $F(x) = \operatorname{grad} f(x)$. *Then there exists an interior point* x_0 *of the ball* $\|x\| \leqslant R$ *at which* $f(x)$ *has a local minimum, so that* $\operatorname{grad} f(x_0) = 0$.

Proof. Consider the ball

$$D_R = \{x \in E : \| x \| \leqslant R;\ R > 0\}.$$

Since E is reflexive, this ball is weakly closed and the functional $f(x)$ is weakly lower semicontinuous there. Hence, by Theorem 9.2, $f(x)$ has a minimum on D_R and Lemma 9.2 entails that the minimum occurs at an interior point of D_R. Q. E. D.

Remark 9.3. In the hypotheses of Theorem 9.8, the reflexivity of E may be replaced by the requirement that $E = X^*$, where X is a separable normed space. In this case, one uses Theorem 9.2 instead of Theorem 9.3. We also note that by Lemma 9.3 inequality (9.4) may be replaced by

$$\langle F(x), x - a \rangle > 0 \tag{9.4'}$$

on the sphere $\|x-a\| = R > 0$. Then there exists an interior point of the ball $\|x-a\| \leqslant R$, at which $f(x)$ takes on a local minimum and $F(x_0) = 0$.

Theorem 9.9. Let $F(x) = \operatorname{grad} f(x)$ *be a strictly monotone operator defined in a reflexive Banach space* E, *which satisfies the following conditions: for any* $x \in E$, *the function* $\langle F(tx), x\rangle$ *is integrable with respect to* t *on* $[0, 1]$,

$$\langle F(x) - F(0), x\rangle \geqslant \|x\|\gamma(\|x\|),$$

where $\gamma(z)$ *is integrable on* $[0, R]$ *for any* $R > 0$, *and the function*

$$c(R) = \int_0^R \gamma(z)\,dz$$

satisfies the inequality $c(R) > R\|F(0)\|$ *for some* R. *Then the functional* $f(x)$ *has a unique minimum point, at which it attains an absolute minimum.*

Proof. Since $F(x)$ is strictly monotone, it follows (Theorems 5.1 and 8.4) that $f(x)$ is strictly convex and weakly lower semicontinuous. From the hypotheses of the theorem and formula (6.4) it follows that

$$\begin{aligned} f(x) - f(0) &= \int_0^1 \langle F(tx), x\rangle\,dt = \\ &= \langle F(0), x\rangle + \int_0^1 \langle F(tx) - F(0), tx\rangle \frac{dt}{t} \geqslant \\ &\geqslant \langle F(0), x\rangle + \int_0^1 \|x\|\gamma(t\|x\|)\,dt = \\ &= \langle F(0), x\rangle + \int_0^R \gamma(z)\,dz \geqslant R\left(\frac{1}{R}\int_0^R \gamma(z)\,dz - \|F(0)\|\right), \end{aligned}$$

where $R = \|x\|$. Hence, for some R_0,

$$f(x) > f(0) \quad (\|x\| = R_0). \tag{9.5}$$

Since by hypothesis the ball $\|x\| \leqslant R_0$ is weakly closed and $f(x)$ is weakly lower semicontinuous there, it follows from Theorem 9.2, that $f(x)$ has a minimum in the ball $\|x\| \leqslant R_0$, occurring by inequality (9.5) at an interior point. Finally, since $f(x)$ is strictly convex the minimum point is unique (Corollary 9.1 and Lemma 9.1), and $f(x)$ has an absolute minimum there. Q. E. D.

Theorem 9.10. Let $f(x)$ be a twice Gâteaux-differentiable functional defined in a reflexive Banach space and suppose it satisfies the conditions:

1. $D^2f(y+tx; x, x)$ is integrable with respect to t on $[0, 1]$ for any $x, y \in E$.

2. $D^2f(x; h, h) \geqslant \gamma(\|x\|)\|h\|^2$ for any $x, h \in E$, where $\gamma(z)$ is a positive decreasing function, integrable on $[0, R]$ for any $R > 0$, and for some $R > 0$

$$\frac{1}{R}\alpha(R) \equiv \frac{1}{R}\int_0^R z\gamma(z)\,dz > \|F(0)\|, \tag{9.6}$$

where $F(x) = \operatorname{grad} f(x)$.

Then $f(x)$ has a unique minimum point, at which it takes on an absolute minimum.

Proof. Since $F(x) = \operatorname{grad} f(x)$, it follows that $Df(x, h) = \langle F(x), h\rangle$; $D^2f(x; h, h) = \langle DF(x, h), h\rangle$, and, proceeding as on p. 110 of VM, we obtain

$$Df(x, h) = Df(0, h) + \int_0^1 D^2f(tx; h, x)\,dt,$$

or

$$\langle F(x), h\rangle = \langle F(0), h\rangle + \int_0^1 \langle DF(tx, x), h\rangle\,dt.$$

Hence

$$\langle F(x), x\rangle = \langle F(0), x\rangle + \int_0^1 D^2f(tx; x, x)\,dt \geqslant$$
$$\geqslant \langle F(0), x\rangle + \int_0^1 \gamma(u\|x\|)\|x\|^2\,du$$

and

$$f(x) - f(0) = \int_0^1 \langle F(tx), tx\rangle \frac{dt}{t} \geqslant$$
$$\geqslant \int_0^1 \left[\langle F(0), tx\rangle + \int_0^1 \gamma(ut\|x\|)\,t^2\|x\|^2\,du\right]\frac{dt}{t} =$$
$$= \langle F(0), x\rangle + \int_0^1\int_0^1 t\gamma(ut\|x\|)\|x\|^2\,du\,dt \geqslant$$
$$\geqslant \langle F(0), x\rangle + \int_0^1\int_0^1 t\gamma(t\|x\|)\|x\|^2\,du\,dt =$$

$$= \langle F(0), x\rangle + \int_0^1 t\gamma(t\|x\|)\|x\|^2\,dt =$$

$$= \langle F(0), x\rangle + \int_0^{\|x\|} z\gamma(z)\,dz \geqslant \int_0^{\|x\|} z\gamma(z)\,dz - \|x\|\|F(0)\|.$$

Setting $R = \|x\|$, we have

$$f(x) - f(0) \geqslant R\left(\frac{1}{R}\int_0^R z\gamma(z)\,dz - \|F(0)\|\right). \tag{9.7}$$

Hence, by the hypotheses of the theorem, we have that for some $R > 0$

$$f(x) - f(0) > 0,$$

i. e., $f(x)$ has the m-property and thus has a minimum point x_0. By our hypotheses, we have that for $x \neq y$

$$\langle F(x) - F(y), x - y\rangle = \langle DF(y + t(x-y), x-y), x-y\rangle =$$
$$= D^2 f(y + t(x-y); x-y, x-y) \geqslant$$
$$\geqslant \gamma(\|y + t(x-y)\|)\|x-y\|^2 > 0,$$

so that $F(x)$ is a strictly monotone operator and x_0 is hence the unique absolute minimum point of $f(x)$. Q. E. D.

Note that conditions (9.6) holds if

$$\overline{\lim_{R\to+\infty}} \frac{1}{R}\alpha(R) = +\infty.$$

9.4. Conditional extrema

If a real weakly lower semi-continuous functional $f(x)$, defined on a reflexive Banach space E, does not have the m-property, one cannot state for any ball

$$D_r = \{x \in E : \|x\| < r, r > 0\}$$

that the functional $f(x)$ has a minimum on D_r at an interior point x_0, i. e., that $\|x_0\| < r$. In fact, $f(x)$ may have a minimum on the sphere $\|x\| = r$. For example, if $F(x)$ is a compact positive-homogeneous potential operator (see Examples 6.3 and 8.7), its potential

$$\varphi(x) = \frac{1}{k+1}\langle F(x), x\rangle, \qquad F(tx) = t^k F(x)$$

takes on its minimum and maximum on D_r. If, however, $\varphi(x)$ takes on negative values, then the homogeneity entails that $\varphi(x)$ takes on its maximum and minimum on the sphere $\|x\| = r$ for any $r > 0$.

If $f(x)$ attains its minimum on D_r at a point x_0 of the sphere $S_r = \{x \in E : \|x\| = r\}$, there exists a neighborhood $U(x_0)$ of x_0 such that

$$f(x_0) \leqslant f(x), \quad \forall x \in S_r \cap U(x_0). \tag{9.8}$$

If (9.8) holds, we say that $f(x)$ has a conditional minimum at x_0 relative to the sphere S_r. This definition also applies in a more general case.

Let E be a real linear topological space on which real functionals $f(x)$ and $\varphi(x)$ are defined. Let $V_c(\varphi)$ denote the (linear) manifold $\varphi(x) = c =$ const.

Definition 9.2. A point $x_0 \in V_c(\varphi)$ is said to be a *conditional extremum point* of $f(x)$ relative to the manifold $V_c(\varphi)$ if there exists a neighborhood $U(x_0)$ of x_0 such that, for all $x \in U(x_0) \cap V_c(\varphi)$, the difference $f(x) - f(x_0)$ has the same sign.

Let the functionals $\varphi(x)$ and $f(x)$ have continuous linear Gâteaux differentials at x_0, so that grad $f(x_0)$ and grad $\varphi(x_0)$ exist.

Definition 9.3. A point $x_0 \in V_c(\varphi)$ is an *ordinary point* of the manifold $V_c(\varphi)$ if $\operatorname{grad} \varphi(x_0) \neq 0$.

Definition 9.4. A point $x_0 \in V_c(\varphi)$ is a *conditionally critical point* of the functional $f(x)$ relative to the manifold $V_c(\varphi)$ if

$$\operatorname{grad} f(x_0) = \mu \operatorname{grad} \varphi(x_0),$$

where μ is some number.

9.5. Fundamental theorem on conditional extrema*

We shall assume that E is a real linear topological space with "enough" continuous linear functionals (for example, a locally convex space). We denote the variations of functionals $f(x)$ and $\varphi(x)$, which are continuous in h and x, by $df(x, h)$ and $d\varphi(x, h)$, respectively, and refer to them as Fréchet differentials, since in a normed space a Gâteaux differential which is continuous in x is a Fréchet differential (Theorem 2.1). Finally, we let E' denote the strong dual of E.

We shall need the following proposition (Lavrent'ev and Lyusternik /1/, p. 162).

* See Vainberg and Engel'son /2/.

Lemma 9.4. Let $y_1, y_2 \in E'$, $y_1 \neq 0$. If the null space X of the functional $y_1(x) = \langle y_1, x\rangle$ is contained in the null space of the functional $y_2(x)$, then there exists a real number λ such that $y_2 = \lambda y_1$.

Let x_0 be an ordinary point of the manifold $V_c(\varphi)$ and X the null space of the differential $d\varphi(x_0, h)$, known as the tangent hyperplane to the manifold $V_c(\varphi)$ at the point x_0. By our choice of x_0, there exists $h_1 \in E$ such that $d\varphi(x_0, h) \neq 0$, i. e., $h_1 \notin X$. Now let h_0 be a fixed vector in X, so that $d\varphi(x_0, h_0) = 0$. We then have

Lemma 9.5. For any number t of sufficiently small absolute value, there exists $s(t)$ such that

1) $x_0 + th_0 + sh_1 \in V_c(\varphi)$,

2) $\lim\limits_{t\to 0} s/t = 0$.

Proof. By the definition of the Gâteaux differential at the point $x_0 + sh_1$, we may write

$$\varphi(x_0 + sh_1 + th_0) - \varphi(x_0 + sh_1) = \\ = t[d\varphi(x_0 + sh_1, h_0) + \alpha(s, t)], \tag{9.9}$$

where $\alpha(s,t) \to 0$ as $t \to 0$. Applying the Lagrange formula to the left-hand side of this equality, we get

$$d\varphi(x_0 + sh_1 + \tau th_0, th_0) = t[d\varphi(x_0 + sh_1, h_0) + \alpha(s, t)], \\ 0 < \tau < 1,$$

whence

$$\alpha(s, t) = d\varphi(x_0 + sh_1 + \tau th_0, h_0) - d\varphi(x_0 + sh_1, h_0). \tag{9.10}$$

Since h_0 and h_1 are fixed, we have by the continuity of $d\varphi(x, h)$ in x that for any $\varepsilon > 0$ there exists $\rho > 0$ such that, whenever $|t| < \rho$ and $|s| < \rho$,

$$|\alpha(s, t)| < \varepsilon. \tag{9.11}$$

Now, by (9.10), the function $\alpha(s,t)$ is jointly continuous in both variables. Moreover, by the definition of the Gâteaux differential, the left-hand side of (9.9) may be transformed as follows:

$$\varphi(x_0 + th_0 + sh_1) - \varphi(x_0 + sh_1) = \\ = \varphi(x_0 + th_0 + sh_1) - \varphi(x_0) - [\varphi(x_0 + sh_1) - \varphi(x_0)] = \\ = \varphi(x_0 + th_0 + sh_1) - \varphi(x_0) - s[d\varphi(x_0, h_1) + \beta(s)],$$

where $\lim\limits_{s\to 0} \beta(s) = 0$. The continuity of $\beta(s)$ may be proved in the same manner as for $\alpha(s,t)$. Replacing the left-hand side of (9.9) by the transformed expression, we get

$$\varphi(x_0 + th_0 + sh_1) - \varphi(x_0) - s[d\varphi(x_0, h_1) + \beta(s)] = \\ = t[d\varphi(x_0 + sh_1, h_0) + \alpha(s, t)]. \qquad (9.12)$$

By what we have just proved, the function

$$T(s, t) = -t\frac{d\varphi(x_0 + sh_1, h_0) + \alpha(s, t)}{d\varphi(x_0, h_1) + \beta(s)} \qquad (9.13)$$

is jointly continuous in both variables for sufficiently small $|s|$, since $d\varphi(x_0, h_1) \neq 0$ and $\lim_{s\to 0}\beta(s) = 0$. By (9.11) and the continuity of $d\varphi(x, h)$, we have

$$\lim_{s\to 0,\, t\to 0}\alpha(s, t) = 0, \quad \lim_{s\to 0} d\varphi(x_0 + sh_1, h_0) = d\varphi(x_0, h_0) = 0,$$

so that

$$\lim_{s\to 0,\, t\to 0}\frac{T(s, t)}{t} = 0. \qquad (9.14)$$

This implies that, if $0 < \eta < 1/2$, there exists $\delta(\eta) > 0$ such that, if $|s| \leqslant \delta$ and $|t| \leqslant \delta$, then

$$|T(s, t)| < |t|\eta. \qquad (9.15)$$

Now, for any fixed t with $|t| \leqslant \delta$ the function $T(s, t)$ defined by (9.13) maps the interval $[-\delta, +\delta]$ into a proper subset of itself, since, by (9.15), if $|t_0| \leqslant \delta$, then for any $s \in [-\delta, +\delta]$

$$|T(s, t)| < \delta\eta < \delta/2.$$

Therefore, by Brouwer's theorem,* the mapping $T(s, t_0)$ has a fixed point, i. e., there exists s_0 such that

$$s_0 = -t_0\frac{d\varphi(x_0 + s_0h_1, h_0) + \alpha(s, t)}{d\varphi(x_0, h_1) + \beta(s_0)}. \qquad (9.16)$$

By (9.16), it follows from (9.12) that for every t_0 of sufficiently small absolute value there exists s_0 such that

$$\varphi(x_0 + t_0h_0 + s_0h_1) = \varphi(x_0) = c.$$

* See for example, Nemytskii /1/, p. 153, or Lyusternik and Sobolev /1/.

The first part of the lemma is thus proved. Finally, (9.16) entails that $s_0 \to 0$ as $t_0 \to 0$, so that by (9.14), the second part of the lemma also holds.

Theorem 9.11 (Fundamental theorem). Let x_0 be an ordinary point of $V_c(\varphi)$ which is a conditional extremum point of a functional $f(x)$ relative to $V_c(\varphi)$. Then*

$$\operatorname{grad} f(x_0) = \lambda \operatorname{grad} \varphi(x_0). \tag{9.17}$$

Proof. By Lemma 9.4, it suffices to prove that the tangent hyperplane X to $V_c(\varphi)$ at the point x_0 is contained in the null space of the linear functional $df(x_0, h)$. We prove this by reductio ad absurdum. Suppose there exists a vector $h' \in X$ such that $df(x_0, h') = a \neq 0$. Since x_0 is an ordinary point of $V_c(\varphi)$, there exists a vector $h_1 \notin X$. Consider the difference

$$f(x_0 + th' + sh_1) - f(x_0 + sh_1) = t[df(x_0 + sh_1, h') + \alpha(s, t)].$$

As in the proof of Lemma 9.5, we get

$$f(x_0 + th' + sh_1) - f(x_0) = s[df(x_0, h_1) + \beta_1(s)] + \\ + t[df(x_0 + sh_1, h') + \alpha_1(s, t)].$$

Moreover,

$$df(x_0 + sh_1, h') = df(x_0, h') + \gamma(s) = a + \gamma(s),$$

where by the continuity of $df(x, h)$ in x,

$$\lim_{s \to 0} \gamma(s) = 0.$$

The previous equality thus yields

$$f(x_0 + th' + sh_1) - f(x_0) = \\ = t\left[a + \gamma(s) + \alpha_1(s, t) + \frac{s}{t}(df(x_0, h_1) + \beta_1(s))\right]. \tag{9.18}$$

By Lemma 9.5, for any t of sufficiently small absolute value, there exists s such that

$$x_0 + th' + sh_1 \in V_c(\varphi),$$

and $s/t \to 0$ as $t \to 0$. Hence, since $\alpha_1(s, t)$, $\gamma(s)$ and $\beta_1(s)$ tend to zero as $t \to 0$, it follows that

* See Vainberg and Engel'son /2/; compare Lavrent'ev and Lyusternik /1/, p. 165.

$$\left|\gamma(s)+\alpha_1(s,\ t)+\frac{s}{t}(df(x_0,\ h_1)+\beta_1(s))\right|<\frac{1}{2}|a|,$$

provided $|t|$ is sufficiently small. For such t, therefore, the sign of the expression on the right of (9.18) is determined by the sign of t; in other words, the sign of the difference

$$f(x_0+th'+sh_1)-f(x_0) \tag{9.19}$$

is determined by that of t. But addition of elements and multiplication by a scalar are continuous operations; and so, for any neighborhood $U(x_0)$, there exists a sufficiently small positive Δ such that whenever $|t|<\Delta$ the point $x_0+th'+sh_1$ belongs to $U(x_0)$. Hence, since the difference (9.19) changes sign, x_0 is not an extremum point. This contradiction proves the theorem.

Remark 9.4. In regard to Theorem 9.11, if E is a normed space, it is sufficient that the functionals $f(x)$ and $\varphi(x)$ be Fréchet differentiable. In that case the proof does not require the continuity of $df(x,h)$ and $d\varphi(x,h)$ in x. For normed spaces, this theorem is due to Lyusternik.*

Theorem 9.11 formed the basis for the authors' investigation of fixed directions and eigenfunctions of nonlinear operators in normed spaces. Some of these results are presented in VM.

9.6. Examples

We now present a few applications of (9.17).

Example 9.1. Let $f(x)$ be a Fréchet-differentiable weakly lower semicontinuous functional and $\varphi(x)$ a nonnegative strictly convex Fréchet-differentiable functional such that the set

$$D_c(\varphi)=\{x\in E:\varphi(x)\leqslant c,\ c>0\}$$

is bounded for any finite c, $\varphi(x)=0$ only if $x=0$, and both functionals are defined in some real reflexive Banach space E. We may take $\varphi(x)$, for example, to be $\|x\|$, if the norm in E is Fréchet-differentiable and the unit sphere is strictly convex. Theorem 9.1 and Lemma 9.1 imply that grad $\varphi(x)\neq 0$ if $x\neq 0$.

Since $\varphi(x)$ is strictly convex and continuous, the set $D_c(\varphi)$ is strictly convex and closed. Thus $D_c(\varphi)$, as a convex closed set in a reflexive space, is weakly closed. Thus, by Theorem 9.2, there exists a vector $x_0\in D_c(\varphi)$ such that

$$f(x_0)=\min_{D_c(\varphi)} f(x)=m(c).$$

* See Lyusternik /1/ for an equivalent formulation.

We claim that if $m(c) \neq f(0)$ the equation

$$\operatorname{grad} f(x) = \mu \operatorname{grad} \varphi(x) \tag{9.20}$$

has a continuum of nontrivial solutions.

Indeed, if for any $c > 0$ the functional $f(x)$ assumes its minimum on $D_c(\varphi)$ at a point x_0 of the manifold

$$V_c(\varphi) = \{x \in E : \varphi(x) = c\},$$

then by Theorem 9.11 (Remark 9.4) equation (9.20) has a solution $x_0 \neq 0$ for some real μ, and thus equation (9.20) has a continuum of solutions. If for some $c = r > 0$ the functional $f(x)$ has no conditionally critical points relative to $V_r(\varphi)$, then, proceeding as in the proof of Theorem 13.9 in VM, we consider the functional

$$\psi_\alpha(x) = \alpha\varphi(x) + f(x), \tag{9.21}$$

where α lies in the interval $\left(0, \frac{f(0) - m(r)}{\varphi(x_0)}\right)$. This functional is weakly lower semicontinuous and Fréchet-differentiable, as a sum of two functionals having these properties. By Theorem 8.1 and 8.10, the functional $\varphi(x)$ is weakly lower semicontinuous. By Theorem 9.2, therefore, there exists a point $x_\alpha \in D_r(\varphi)$ such that

$$\min_{D_r(\varphi)} \psi_\alpha(x) = \psi_\alpha(x_\alpha).$$

We claim that $x_\alpha \neq 0$. In fact, $\psi_\alpha(0) = f(0)$ and since by assumption

$$0 < \alpha < [\varphi(x_0)]^{-1}[f(0) - m(r)],$$

it follows that

$$\psi_\alpha(0) = f(0) > \alpha\varphi(x_0) + m(r) = \alpha\varphi(x_0) + f(x_0) = \\ = \psi_\alpha(x_0) \geqslant \psi_\alpha(x_\alpha),$$

i. e., $x_\alpha \neq 0$. We show, moreover, that $x_\alpha \notin V_r(\varphi)$. In fact, if we suppose that $x_\alpha \in V_r(\varphi)$, Theorem 9.11 implies that $\operatorname{grad} \psi_\alpha(x_\alpha) = \lambda \operatorname{grad}\varphi(x_\alpha)$. But (9.21) entails that

$$\operatorname{grad} \psi_\alpha(x_\alpha) = \alpha \operatorname{grad} \varphi(x_\alpha) + \operatorname{grad} f(x_\alpha),$$

whence

$$\operatorname{grad} f(x_\alpha) = (\lambda - \alpha) \operatorname{grad} \varphi(x_\alpha),$$

where $x_\alpha \in V_r(\varphi)$, i. e., the functional $f(x)$ has conditionally critical points relative to $V_r(\varphi)$, contradicting our assumption. Thus $x_\alpha \in (D_r(\varphi) \setminus V_r(\varphi))$ and by Theorem 9.1 we have that $\operatorname{grad} \psi_\alpha(x_\alpha) = 0$, or

$$\operatorname{grad} f(x_\alpha) = -\alpha \operatorname{grad} \varphi(x_\alpha)$$

for any $\alpha \in (0, [f(0) - m(r)][\varphi(x_0)]^{-1})$.

Equation (9.20) thus has a continuum of solutions, corresponding to real values of μ. This assertion clearly remains valid if $f(x)$ is weakly upper semicontinuous and $f(0) \neq \max_{D_c(\varphi)} f(x) = M(c)$. Finally, if $f(x)$ is weakly continuous, the equation has a continuum of solutions without any restrictions on the value of $f(x)$ at zero, since $m(c) = M(c)$ implies $f(x) = \text{const}$. Observe that instead of a reflexive E, we may take $E = X$, with X a separable normed space.

Example 9.2. Let $f(x)$ be a real Fréchet-differentiable weakly lower semicontinuous functional such that

$$\lim_{R \to +\infty} f(x) = +\infty \quad (R = \|x\|),$$

and $\varphi(x)$ a nonnegative strictly convex Fréchet-differentiable functional which vanishes only at zero, when both functionals are defined in some real reflexive Banach space E (or in the dual E of some real separable normed space). Here, as opposed to the previous example, the sets $D_c(\varphi)$ may be unbounded, though they are strictly convex and closed, therefore also weakly closed. Since the sets $D_c(\varphi)$ are weakly closed, it follows from Theorems 9.2 and 9.3 (together with Remark 9.1) that for any $c > 0$ there exists a vector $x_0 \in D_c(\varphi)$ such that

$$f(x_0) = \min_{D_c(\varphi)} f(x) = m(c).$$

As in the previous example, we conclude that, if $f(0) \neq m(c)$, equation (9.20) has a continuum of nontrivial solutions for certain real values of μ.

Example 9.3. Let E be a real reflexive Banach space with strictly convex unit ball and Fréchet-differentiable norm. Let $F(x)$ be a strongly potential compact operator, positive-homogeneous of degree $k > 0$, and consider also the functional $\varphi(x) = \|x\|^{\alpha+1}$, $\alpha > 0$ As we have seen, the potential of $F(x)$ (see Example 8.7) is weakly continuous and has the form

$$f(x) = \frac{1}{k+1} \langle F(x), x \rangle,$$

while $\varphi(x)$ is convex (see Example 5.7 and Theorem 5.1). Moreover, every point of the manifold

$$V_c(\varphi) = \{x \in E:\ \varphi(x) = c,\ c > 0\}$$

is ordinary, since for $x \neq 0$

$$\langle \operatorname{grad} \varphi(x),\ x \rangle = (\alpha + 1) \| x \|^{\alpha+1} > 0.$$

Suppose that $f(x)$ takes on negative values.* Then its homogeneity entails that its minimum on the ball $\varphi(x) \leqslant c$ occurs on the boundary $V_c(\varphi)$, so that, by Theorem 9.11,

$$\operatorname{grad} f(x_0) = \mu_0 \operatorname{grad} \varphi(x_0), \qquad (x_0 \in V_c(\varphi))$$

or

$$F(x_0) = \lambda_0 \| x_0 \|^{\alpha} \operatorname{grad} \| x_0 \|, \quad \lambda_0 = (\alpha + 1) \mu_0. \tag{9.22}$$

Hence, we find that

$$\lambda_0 = (1 + k) \| x_0 \|^{-\alpha-1} f(x_0) < 0.$$

It follows from (9.22) that if $\alpha = k$, the vector $x = tx_0$ is a solution of the equation

$$F(x) = \lambda \| x \|^{\alpha} \operatorname{grad} \| x \| \tag{9.23}$$

for $\lambda = \lambda_0$ and any $t > 0$, i. e., every vector of the ray tx is a solution of equation (9.23) for $\lambda = \lambda_0$. For $\alpha \neq k$, every vector of the ray $x = t^{-1}x_0$ $(t > 0)$ is also a solution of equation (9.23), corresponding to $\lambda = \lambda_0 t^{\alpha-k}$. Varying t, we see that for every negative λ there is a solution of equation (9.23) collinear with x_0. If $f(x)$ takes on positive values, all assertions regarding equation (9.23) remain valid if we replace λ by $(-\lambda)$. Note that if E is a real Hilbert space and $\alpha = 1$, equation (9.23) becomes

$$F(x) = \lambda x. \tag{9.24}$$

Nontrivial solutions of such an equation ($F(0) = 0$) are *eigenfunctions* or *eigenvectors* of the operator F, and the corresponding values of λ are the *eigenvalues* of the operator F.

* This is the case, for instance, if $F(x) \not\equiv 0$ and $F(-x) = F(x)$.

Consequently, if x_0 is an eigenvector of F corresponding to an eigenvalue λ_0, i. e., $F(x_0) = \lambda_0 x_0$, then the homogeneity of F entails that every vector on the ray tx_0 is also an eigenvector. Thus, if $(F(x), x)$ takes negative (positive) values, then $\lambda_0 < 0$, ($\lambda_0 > 0$). If, furthermore, $k = 1$, the "eigenray" tx_0 corresponds to λ_0, and if $k \neq 1$, every negative (positive) number is an eigenvalue of F.

§ 10. CONVERGENCE OF MINIMIZING SEQUENCES

10.1. Preliminaries

Let E be a linear space, $f(x)$ a real functional defined on E, σ a subset of E and

$$d = \inf_{x \in \sigma} f(x).$$

Definition 10.1. Any sequence $\{x_n\} \subset E$ such that

$$\lim_{n \to \infty} f(x_n) = d$$

is called a *minimizing sequence*.

This definition remains in force if $d = f(x_0)$, where x_0 is a conditional minimum point of $f(x)$ (relative to σ) or if it is an unconditional minimum point, i. e., a local or absolute minimum point of $f(x)$ relative to E. We are interested in the convergence of minimizing sequences to an unconditional minimum point of the functional.*

In investigating this problem, we need convex functionals with certain special properties.

Of course, not every convex functional has an absolute minimum. Even strictly convex functionals need not have an absolute minimum. For example, the real function $f(t) = e^t$ (where t is real) is strictly convex but has no unconditional minimum. If, moreover, a convex (or even strictly convex) functional has a minimum, not every minimizing sequence converges to a minimum point. Before presenting examples, we note that if a strictly convex functional $f(x)$ has a minimum at a point x_0, then it increases along every ray issuing from x_0. Indeed, the functional

$$\varphi(x) = f(x_0 + x) - f(x_0)$$

* The problem of the convergence of minimizing sequences to unconditional minimum points has been investigated by many authors; see, for example, Dem'yanov and Rubinov /1/, Kachurovskii /5/, Levitin and Polyak /1/.

is also strictly convex and has a minimum at zero. This minimum is strict and unique (§ 9). Now consider a vector $x \neq 0$. By the strict convexity,

$$\varphi(\lambda x + (1-\lambda)\,0) < \lambda\varphi(x) + (1-\lambda)\,\varphi(0) = \lambda\varphi(x).$$

Hence, by Lemma 9.1,

$$\varphi(0) < \varphi(\lambda x) < \lambda\varphi(x) < \varphi(x), \qquad 0 < \lambda < 1,$$

and the functional $\varphi(x)$ increases along the ray connecting the zero vector to any vector $x \in E$, i.e., $f(x)$ increases along any ray with initial point x_0.

Here we remark that if E is a finite-dimensional space, then any ball is compact. Consequently, if a real function $f(x)$ increases along any ray issuing from some point x_0, there exists a monotone minorant $c(t)$ for $t \geqslant 0$, $c(0) = 0$, which may be assumed to be continuous, strictly increasing and satisfying the condition

$$f(x) - f(x_0) \geqslant c(\|x - x_0\|). \tag{10.1}$$

Of course, if inequality (10.1) holds for a functional on an infinite-dimensional space, then x_0 is a minimum point of $f(x)$ and, moreover, any minimizing sequence converges to x_0.

However, there exist strictly convex functionals on infinite-dimensional spaces, assuming a minimum at a point x_0 and therefore increasing along any ray issuing from x_0, which nevertheless do not satisfy (10.1). This possibility is illustrated by the following examples.

Example 10.1. Let $x = (\xi_1, \xi_2, \xi_3, \ldots)$ be a vector in the real infinite-dimensional space l^2, which we write as

$$x = \sum_{k=1}^{\infty} \xi_k e_k,$$

where $e_1 = (1, 0, 0, \ldots)$, $e_2 = (0, 1, 0, \ldots)$, $e_3 = (0, 0, 1, 0, 0, \ldots), \ldots$ Consider the nonlinear functional

$$f(x) = \sum_{k=1}^{\infty} \frac{\xi_k^2}{k^2}$$

in l^2. This functional is strictly convex and has a unique minimum at zero. The sequence $x_n = \sqrt{n}\, e_n$ $(n = 1, 2, 3, \ldots)$ is a minimizing sequence, since

$$f(x_n) = \frac{1}{n} \to 0.$$

as $n \to \infty$. However, $\|x_n\| = \sqrt{n}$ as $n \to \infty$. This example is due to Z. Paritskaya.

Example 10.2. Consider the quadratic functional

$$f(u) = \int_0^1 u^2(t)\,dt$$

in $C[0,1]$. This functional is strictly convex, since for any $u(t), v(t) \in C[0,1]$, $u \neq v$, the inequality

$$\lambda u^2 + (1-\lambda)v^2 - (\lambda u + (1-\lambda)v)^2 = \lambda(1-\lambda)(u-v)^2 > 0$$

implies that, for $0 < \lambda < 1$,

$$f(\lambda u + (1-\lambda)v) < \lambda f(u) + (1-\lambda)f(v).$$

Now $u_n = t^{n/2}$ $(0 \leqslant t \leqslant 1)$ is a minimizing sequence, for $f(u_n) = 1/(n+1) \to 0$ as $n \to \infty$, but this sequence diverges in $C[0,1]$. This example is due to Yu. S. Ektov.

Example 10.3. Let $\{e_k\}$ be a complete orthonormal system in some real Hilbert space H. Consider the strictly convex functional

$$f(x) = \sum_{k=1}^{\infty} \frac{(x, e_k)^2}{k^2}, \quad x \in H,$$

which vanishes only at $x = 0$, and $f(x) > 0$ for $x \neq 0$. The sequence $x_n = \sqrt{n}\, e_n$ $(n = 1, 2, 3, \ldots)$ is a minimizing sequence for this functional, since

$$f(x_n) = \frac{1}{n} \to 0$$

but $\|x_n\| = \sqrt{n} \to \infty$.

10.2. Boundedness of minimizing sequences

The minimizing sequences in the examples just considered were unbounded. The question naturally arises of when every minimizing sequence is bounded. We shall quote an appropriate condition, which uses

Definition 10.2. A real finite functional $f(x)$ defined in a normed space E, is said to be increasing if, for any vector $y \in E$, there exists $r > 0$ such that $f(x) > f(y)$ whenever $\|x\| > r$.

Note that if

$$\lim_{R \to \infty} f(x) = +\infty \qquad (R = \| x \|),$$

$f(x)$ is an increasing functional.

Lemma 10.1. Suppose $f(x)$ is a real increasing functional defined and bounded below in a normed space E, and let

$$d = \inf_{x \in E} f(x).$$

Alternatively, let x_0 be a (possibly local) minimum point of $f(x)$ and let $f(x_0) = d$. Then every minimizing sequence $\{x_n\}$,

$$\lim_{n \to \infty} f(x_n) = d,$$

is bounded.

Proof. Supposing the contrary, we have that there exists a subsequence $\{y_m\} \subset \{x_n\}$ such that $\|y_m\| > m \quad (m = 1, 2, 3, \ldots)$. Let $m_0 > d$. Since $f(x)$ is increasing, there exists $r \geqslant m_0$ such that

$$f(y_m) > f(y_{m_0}) \geqslant m_0 \quad (m = m_1, \ m_1 + 1, \ \ldots; \ m_1 > m_0).$$

Hence

$$d = \lim_{m \to \infty} f(y_m) \geqslant m_0 > d.$$

This contradiction establishes the lemma.

10.3. Weak convergence of minimizing sequences

Theorem 10.1. An increasing strictly convex functional $f(x)$, defined in a reflexive Banach space E, has an absolute minimum at some point x_0; it has no other minimum points and every minimizing sequence converges weakly to the minimum point.

Proof. Since $f(x)$ is increasing, there exists a ball $D_r = \{x \in E : \|x\| \leqslant r\}$ on whose surface $f(x) > f(0)$. By Theorem 8.10, the convexity of $f(x)$ implies that it is weakly lower semicontinuous. Now, in a reflexive Banach space the ball D_r is weakly closed and weakly compact, so that $f(x)$ has the m-property, i. e., there exists an interior point x_0 of D_r at which $f(x)$ has a minimum. This minimum point is unique and $f(x)$ attains its absolute minimum there (Lemma 9.1). In order to prove the second part of the theorem, consider a minimizing sequence $\{x_n\}$; it is bounded by Lemma 10.1. Since the space E is reflexive, there exists a subsequence $y_m \rightharpoonup y_0$ and by the weak lower-semicontinuity of $f(x)$

$$f(y_0) \leqslant \underline{\lim}_{m\to\infty} f(y_m) = \lim_{m\to\infty} f(y_m) = f(x_0).$$

Hence $y_0 = x_0$, for if $y_0 \neq x_0$ Lemma 9.1 would imply that $f(y_0) > f(x_0)$.

It remains to show that $\{x_n\}$ converges weakly to x_0. Indeed, were this not the case, there would be a subsequence of $\{x_n\} \setminus \{y_m\}$, say $\{z_k\}$, converging weakly to $z_0 \neq x_0$. By the above arguments, we would have

$$f(z_0) \leqslant \underline{\lim}_{k\to\infty} f(z_k) = \lim_{k\to\infty} f(z_k) = f(x_0),$$

contradicting Lemma 9.1. Q. E. D.

Note that the theorem remains valid if $f(x)$ is not necessarily strictly convex, but only strictly quasiconvex and lower semi-continuous.

Theorem 10.2. Let $f(x)$ be a finite weakly lower semi-continuous functional defined in a reflexive Banach space E, and satisfying

$$\lim_{R\to+\infty} f(x) = +\infty \qquad (R = \| x \|). \tag{10.2}$$

Then $f(x)$ has at least one absolute minimum point and every minimizing sequence has a subsequence converging weakly to such a point.

Proof. The existence of an absolute minimum point follows from Theorem 9.5. Denote this point by x_0. Since $f(x)$ is increasing (see (10.2)), Lemma 10.1 guarantees that any minimizing sequence $\{x_n\}$ is bounded, i. e., contains a subsequence $y_m \rightharpoonup y_0$. The fact that $f(x)$ is weakly lower semicontinuous implies that

$$f(y_0) \leqslant \underline{\lim}_{m\to\infty} f(y_m) = \lim_{m\to\infty} f(y_m) = f(x_0).$$

Since

$$f(x_0) = \inf_{x\in E} f(x),$$

it follows that $f(y_0) = f(x_0)$, i. e., y_0 is an absolute minimum point of $f(x)$. Q. E. D.

Note that if x_0 is the unique absolute minimum point, one can prove as in Theorem 10.1 that $x_n \to x_0$.

Theorem 10.3. Under the hypotheses of Theorem 9.6, the assertion of Theorem 10.2 is valid; moreover, if condition 2 is an equality only when $h = 0$, *every minimizing sequence converges weakly to the unique absolute minimum point of the functional.*

Proof. As in the proof of Theorem 9.6, the hypotheses imply those of Theorem 10.2, so that the assertion of the latter is true. If equality can hold in condition 2 of Theorem 9.6 only when $h = 0$, $F(x)$ is a strongly monotone operator, and thus its potential $f(x)$ is a strictly convex functional. Since $f(x)$ is also increasing, every minimizing sequence converges weakly to a minimum point, as guaranteed by Theorem 10.1. Q. E. D.

10.4. Well-posedness of the minimization problem

The following definition is due to Tikhonov /1/.

Definition 10.3. The minimization problem for a real functional defined on a subset of a vector space is well-posed if it is solvable and has a unique solution, and if every minimizing sequence converges to this solution in the strong topology of the space.

In this subsection we consider the unconditional minimum problem for Gâteaux-differentiable functionals defined in real Banach spaces, and give sufficient conditions for this problem to be well-posed in the above sense.

These sufficient conditions essentially lead to inequalities of type (10.1), which, as we have seen, imply that the minimization problem for functionals is well-posed.

Theorem 10.4. Let* $F(x) = \operatorname{grad} f(x)$ *be an operator defined on a reflexive Banach space* E *such that the following conditions hold: for any* $h, y \in E$, *the function* $\langle F(y+th), h\rangle$ *is integrable with respect to* t *on* $[0, 1]$ *and*

$$\langle F(y+h) - F(y), h\rangle \geqslant \|h\|\gamma(\|h\|), \tag{10.3}$$

where $\gamma(t)$ *is a nonnegative function, integrable on* $[0, R]$ *for any* $R > 0$, *such that*

$$c(R) = \int_0^R \gamma(t)\,dt \tag{10.4}$$

* See Vainberg /19/, Theorem 2.1, and the corresponding footnote.

is increasing, and for some R

$$c(R) > R\|F(0)\|. \qquad (10.5)$$

Then the minimization problem for $f(x)$ *is well-posed.*

Proof. Condition (10.3) implies, first, that $F(x)$ is monotone, so that $f(x)$ is weakly lower semicontinuous, and, second, that (see (6.4))

$$f(x) - f(0) = \int_0^1 \langle F(tx), x\rangle\, dt =$$

$$= \int_0^1 \langle F(tx) - F(0), x\rangle\, dt + \langle F(0), x\rangle \geqslant$$

$$\geqslant \int_0^1 \|x\|\gamma(t\|x\|)\, dt - \|x\|\|F(0)\| = \int_0^R \gamma(z)\, dz - R\|F(0)\|,$$

where $R = \|x\|$. By (10.5), $f(x)$ therefore has the m-property, i. e., $f(x)$ has a minimum point x_0, and so $F(x_0) = 0$. It follows that (see VM, p. 81)

$$f(x) - f(x_0) = \int_0^1 \langle F(x_0 + t(x - x_0)) - F(x_0), x - x_0\rangle\, dt \geqslant$$

$$\geqslant \int_0^1 \|x - x_0\|\gamma(t\|x - x_0\|)\, dt = c(\|x - x_0\|).$$

The minimum is hence unique and every minimizing sequence converges to it.

Note that (10.5) holds if, say,

$$\overline{\lim_{R\to\infty}} \frac{c(R)}{R} = +\infty.$$

Theorem 10.5. *Let* $f(x)$ *be a twice Gâteaux-differentiable functional defined in a reflexive Banach space* E *and satisfying the following conditions:*

1. *The differential* $D^2 f(y + tx; x, x)$ *is integrable with respect to* $t \in [0, 1]$ *for any* $x, y \in E$.

2. $D^2 f(x; h, h) \geqslant \|h\|\gamma(\|h\|)$, $(\forall x, h \in E)$, *where* $\gamma(t)$ *is a non-negative function, integrable on* $[0, R]$ *for any* $R > 0$, *the function*

$$c(R) = \int_0^R \gamma(z)\, dz$$

is increasing, and for some R

$$\frac{1}{R}c(R) > \|F(0)\|, \qquad F(x) = \operatorname{grad} f(x). \tag{10.6}$$

Then the minimization problem for the functional $f(x)$ *is well-posed.*

Proof. By (6.4), we may write

$$f(x) - f(0) = \langle F(0), x\rangle + \int_0^1 \langle F(tx) - F(0), x\rangle\, dt.$$

But

$$\langle F(0), x\rangle = Df(0, x), \qquad \langle F(tx), x\rangle = Df(tx, x)$$

and, by the Lagrange formula (Lemma 2.1),

$$Df(tx, x) - Df(0, x) = D^2 f(\tau tx; tx, x),$$

where $0 < \tau < 1$. Thus

$$f(x) - f(0) = \langle F(0), x\rangle + \int_0^1 D^2 f(\tau tx; tx, x)\, dt.$$

Hence

$$f(x) - f(0) \geqslant -\|x\|\|F(0)\| + \int_0^1 \gamma(t\|x\|)\|x\|\, dt =$$

$$= \int_0^{\|x\|} \gamma(z)\, dz - \|x\|\|F(0)\| = R\left(\frac{1}{R}c(R) - \|F(0)\|\right),$$

where $R = \|x\|$. Condition (10.6) now yields

$$f(x) > f(0) \quad \text{for} \quad \|x\| = R.$$

Since the assumptions imply the monotonicity* of $F(x)$, it follows that $f(x)$ has the m-property, i. e., there exists a minimum point of $f(x)$. Now since $F(x_0) = 0$, we have

$$f(x) - f(x_0) = \int_0^1 \langle F(x_0 + t(x - x_0)) - F(x_0), x - x_0\rangle\, dt =$$

* By Theorem 8.4, if F is monotone, $f(x)$ is weakly lower semicontinuous.

$$=\int_0^1 \langle DF(x_0+\tau t(x-x_0),\ t(x-x_0)),\ x-x_0\rangle\, dt \geqslant$$

$$\geqslant \int_0^1 \gamma(t\|x-x_0\|)\|x-x_0\|\, dt = \int_0^R \gamma(z)\, dz = c(R),$$

where $R=\|x-x_0\|$. Thus

$$f(x)-f(x_0)\geqslant c(\|x-x_0\|).$$

The minimum is therefore unique and any minimizing sequence converges to x_0. Q. E. D.

Theorem 10.6. Under the hypotheses of Theorem 9.10, let $\gamma(z)$ be continuous for $z\geqslant 0$. Then the minimization problem for $f(x)$ is well-posed.

Proof. Let x_0 be the minimum point of $f(x)$; its existence and uniqueness were established in Theorem 9.10. Since $F(x) = \operatorname{grad} f(x)$, it follows from Theorem 9.1 that $F(x_0) = 0$. Using (6.4), we obtain

$$f(x)-f(x_0)=\int_0^1 \langle F(x_0+t(x-x_0)),\ x-x_0\rangle\, dt =$$

$$=\int_0^1 \langle F(x_0+t(x-x_0))-F(x_0),\ x-x_0\rangle\, dt.$$

Hence, by the Lagrange formula (Lemma 2.1), we have

$$f(x)-f(x_0)=\int_0^1 \langle DF(x_0+\tau t(x-x_0),\ t(x-x_0)),\ x-x_0\rangle\, dt=$$

$$=\int_0^1 D^2 f(x_0+\tau t(x-x_0);\ t(x-x_0),\ x-x_0)\, dt,$$

where $0<\tau<1$. Condition 2 of Theorem 9.10 therefore gives

$$f(x)-f(x_0)\geqslant \int_0^1 \gamma(\|x_0+\tau t(x-x_0)\|)\, tR^2\, dt \geqslant$$

$$\geqslant \int_0^1 \gamma(\|x_0\|+tR)\, tR^2\, dt = \int_0^R \gamma(\|x_0\|+z)\, z\, dz = c(R),$$

where $R=\|x-x_0\|$. But

$$c'(R)=\gamma(\|x_0\|+R)\, R>0,$$

and so $c(R)$ is an increasing continuous function and $c(0) = 0$. Thus

$$f(x) - f(x_0) \geqslant c(\|x - x_0\|),$$

and this inequality of type (10.1) implies the truth of our assertion.

10.5. Well-posedness of the minimization problem for uniformly convex functionals

Definition 10.4. A real functional $f(x)$ defined in a normed space E is said to be uniformly convex if

$$\Delta_\lambda \equiv \lambda f(x) + (1-\lambda) f(y) - f(\lambda x + (1-\lambda) y) \geqslant \\ \geqslant a(\lambda)\gamma(\|x - y\|), \qquad (10.7)$$

where $\gamma(t)$ is an increasing function, $\gamma(0) = 0$,

$$0 < a(\lambda) \leqslant 1 \quad \text{for} \quad 0 < \lambda < 1, \quad a(0) = a(1) = 0, \\ a(1/2) = 1.$$

This definition remains in force if the functional $f(x)$ is considered on a convex set $\omega \subset E$. A uniformly convex functional is clearly strictly convex.

If $f(x)$ is Gâteaux-differentiable and $F(x) = \operatorname{grad} f(x)$, then we have

*Lemma 10.2.** *If $f(x)$ is a uniformly convex functional and*

$$\lim_{\lambda \to +0} \frac{a(\lambda)}{\lambda} = b > 0, \qquad (10.8)$$

then $\operatorname{grad} f(x) = F(x)$ *is a uniformly monotone operator. If $F(x)$ is a uniformly monotone operator such that $\gamma(z) \geqslant cz$, $c > 0$, then $f(x)$ is a uniformly convex functional satisfying (10.8).*

The proof is roughly similar to that of Theorem 5.1.

Theorem 10.7. Let $f(x)$ be a finite uniformly convex functional, defined in a Banach space E, bounded above on some ball $D_r = \{x \in E\colon \|x\| \leqslant r;\ r > 0\}$, and suppose that outside this ball the values of $f(x)$ exceed its values on D_r. Then the minimization problem for $f(x)$ is well-posed.

Proof. We first show that $f(x)$ is also bounded below on D_r. In fact, we have by hypothesis that for any $x \in D_r$

$$f(x) \leqslant M = \text{const}.$$

* Levitin and Polyak /1/.

Setting $\lambda = 1/2$ in the definition of uniform convexity, we see that

$$2^{-1}[f(x)+f(-x)]-f(0) \geqslant \gamma(2\|x\|),$$

whence, as $f(-x) \leqslant M$, we arrive at the inequality

$$f(x) \geqslant -M + 2f(0) + 2\gamma(2\|x\|).$$

Thus, $f(x)$ is indeed bounded below on D_r, and since outside D_r the values of $f(x)$ exceed its values on D_r, it follows that the infimum d exists:

$$d = \inf_{x \in E} f(x).$$

Let $\{x_n\} \subset E$ be a minimizing sequence:

$$\lim_{n \to \infty} f(x_n) = d.$$

Then for any $\varepsilon > 0$, there exists n_0 such that for any $n \geqslant n_0$

$$d \leqslant f(x_n) < d + \varepsilon.$$

Hence, for any m, $n \geqslant n_0$, we have

$$\Delta_\lambda = \lambda f(x_n) + (1-\lambda) f(x_m) - f(\lambda x_n + (1-\lambda) x_m) < \\ < \lambda(d+\varepsilon) + (1-\lambda)(d+\varepsilon) - d = \varepsilon,$$

since $f(\lambda x_n + (1-\lambda)x_m) \geqslant d$. Since $f(x)$ is uniformly convex, this implies that $\gamma(\|x_n - x_m\|) < \varepsilon$ for $\lambda = 1/2$. Since γ is an increasing function and $\gamma(0) = 0$, this last inequality and the arbitrary choice of $\varepsilon > 0$ entail that $\{x_n\}$ is a fundamental sequence, so that it converges to some vector $x_0 \in E$. But since $f(x)$ is bounded on D_r it is continuous,* so that the convergence of $\{x_n\}$ to x_0 implies

$$\lim_{n \to \infty} f(x_n) = f(x_0) = d.$$

Thus x_0 is an absolute minimum point for $f(x)$. Being a strictly convex functional, $f(x)$ has no other minimum points (Lemma 9.1). Thus any minimizing sequence converges to the absolute minimum point and there can be no other minimum points. Q. E. D.

Remark 10.1. The assertion of Theorem 10.7 remains valid if $f(x)$ is a continuous uniformly convex functional bounded below.

* Bourbaki /1/, p. 92.

Remark 10.2. An even uniformly convex functional is increasing, because the inequality

$$2^{-1}[f(x)+f(-x)]-f(0)\geqslant\gamma(2\|x\|)$$

then implies

$$f(x)\geqslant f(0)+\gamma(2\|x\|).$$

10.6. Well-posedness of the minimization problem for some special functionals

We now consider some nonlinear functionals which are closely tied up with the theory of nonlinear equations with monotone (not necessarily potential) operators (see §§ 19, 20).

Let E be a real reflexive Banach space, contained in a real Hilbert space H, H in turn being dense in the dual E^*. We assume that the topologies of E and H and of H and E^* are compatible, i. e., $\|x\|_H\leqslant a\|x\|_E$ for $x\in E$; $\|y\|_{E^*}\leqslant b\|x\|_H$ for $y\in H$, where $a>0$, $b>0$, and the value of the linear functional $y\in E^*$ at a vector $x\in E$, i. e., $\langle y, x\rangle$, coincides with the scalar product on H if $y\in H$.

Let F be a hemicontinuous nonlinear operator from E to E^* and B a bounded operator from E^* to E, whose restriction B_H to H is a positive selfadjoint operator in H, $A=B_H^{1/2}$. Then (§§ 15, 16) A is a bounded operator from H to E and the adjoint A^* is bounded from E^* to H. Consider the functional

$$\varphi(y)=\|y-A^*F(Ay)\|_H,$$

defined in H; the following proposition is valid.

Theorem 10.8. If the hemicontinuous operator $F(x)$ satisfies the condition

$$\langle F(x_1)-F(x_2),\ x_1-x_2\rangle\leqslant\alpha(x_1-x_2,\ x_1-x_2),$$

where $\alpha>0$, $\alpha\|B_H\|<1$, $x_1,\ x_2\in E$, then the functional $\varphi(y)$ has a unique absolute minimum point and its minimization problem is well-posed.

Proof. The existence in H of a unique solution to the equation $y-A^*F(Ay)=0$ will follow from Theorem 19.6 (see the proof of that theorem).

We prove the second part of the theorem. Let y_0 be the unique zero of the functional $\varphi(y)$. Then

$$\varphi(y)-\varphi(y_0)=\|(y-y_0)-A^*[F(Ay)-F(Ay_0)]\|\frac{\|y-y_0\|}{\|y-y_0\|}\geqslant$$
$$\geqslant(y-y_0-A^*[F(Ay)-F(Ay_0)],\ y-y_0)\frac{1}{\|y-y_0\|}=$$
$$=\|y-y_0\|-\langle F(Ay)-F(Ay_0),\ Ay-Ay_0\rangle\frac{1}{\|y-y_0\|}\geqslant$$
$$\geqslant\|y-y_0\|-\alpha(Ay-Ay_0,\ Ay-Ay_0)\|y-y_0\|^{-1}\geqslant$$
$$\geqslant\|y-y_0\|(1-\alpha\|B_H\|),$$

i. e.,

$$\varphi(y)-\varphi(y_0)\geqslant(1-\alpha\|B_H\|)\|y-y_0\|.$$

This shows that every minimizing sequence converges to the minimum point. Q. E. D.

We now consider a more general case. Let B_H be an indefinite operator in H (see subsection 6.5), i. e.,

$$\alpha=\inf(B_Hx,\ x)<0,\quad \beta=\sup(B_Hx,\ x)>0,\quad \|x\|=1,$$

where E_t is a resolution of the identity for the operator B_H, $P_1=(E_\beta-E_0)$ is the projection from H onto $H_1\subset H$ and P_2 the projection from H onto $H_2=H\ominus H_1$. Set

$$B_H^+=2^{-1}(|B_H|+B_H),\quad \overline{B}_H=2^{-1}(|B_H|-B_H),$$
$$A=(B_H^+)^{1/2}-(B_H^-)^{1/2},\quad T=(B_H^+)^{1/2}+(B_H^-)^{1/2}.$$

Then (see §§ 15, 16) the operator A from E to H is bounded, as is the operator T^* from E^* to H, $B=AT^*$ and $T(P_1-P_2)=A$.

Theorem 10.9. Suppose that:

1. F is a hemicontinuous operator from a reflexive Banach space E to E^*, such that

$$\langle F(x)-F(y),\ x-y\rangle\geqslant\frac{1+\alpha}{m}\|x-y\|_H^2,$$

where $\alpha>0$, $m>0$.

2. The interval $(0,m)$ does not contain the spectrum of the indefinite operator B_H.

Then there exists a unique point x_0 at which the functional

$$f(x)=\|A^*F(Ay)-(P_1-P_2)x\|_H$$

vanishes, and the minimization problem for $f(x)$ is well-posed.

Proof. The existence of a unique solution in H to the equation $A^*F(Ax) = (P_1 - P_2)x$ follows from Theorem 19.7.

We prove the second part of the theorem. Let x_0 be the unique absolute minimum point of $f(x)$, i. e., $f(x_0) = 0$. Then

$$\begin{aligned} f(x) - f(x_0) &= \| A^*[F(Ax) - F(Ax_0)] - (P_1 - P_2)(x - x_0) \| \frac{\|x - x_0\|}{\|x - x_0\|} \geqslant \\ &\geqslant \{(A^*[F(Ax) - F(Ax_0)], x - x_0) - \\ &- ((P_1 - P_2)(x - x_0), x - x_0)\} \frac{1}{\|x - x_0\|} = \\ &= \langle F(Ax) - F(Ax_0), A(x - x_0)\rangle \frac{1}{\|x - x_0\|} - (\| P_1(x - x_0)\|^2 - \\ &- \| P_2(x - x_0)\|^2) \frac{1}{\|x - x_0\|}. \end{aligned}$$

But

$$\langle F(Ax) - E(Ax_0), Ax - Ax_0\rangle \geqslant \frac{1+\alpha}{m} \|A(x - x_0)\|^2,$$

and since the projections P_1 and P_2 reduce the operator A, it follows that

$$\begin{aligned} \| Ax \|^2 &= \| AP_1 x \|^2 + \| AP_2 x \|^2 \geqslant \| AP_1 x \|^2 = \\ &= \int_{\sqrt{m}}^{\sqrt{\beta}} t^2 \, d(E_t P_1 x, P_1 x) \geqslant m \|P_1 x\|^2. \end{aligned}$$

Consequently,

$$\begin{aligned} f(x) - f(x_0) \geqslant ((1+\alpha) \|P_1(x - x_0)\|^2 - \\ - \|P_1(x - x_0)\|^2 + \|P_2(x - x_0)\|^2) \frac{1}{\|x - x_0\|}. \end{aligned}$$

Finally, setting $b = \min(1, \alpha)$, we have

$$f(x) - f(x_0) \geqslant b \|x - x_0\|.$$

Hence every minimizing sequence converges to x_0. Q. E. D.

Let E be a real reflexive Banach space whose dual is uniformly convex. For such spaces (see § 22) the norm is uniformly Fréchet-differentiable. Consider the duality mapping

$$U(x) = \| x \| \operatorname{grad} \| x \|, \quad U(0) = 0.$$

By Lemma 2.5,

$$\langle U(x), x\rangle = \|x\|^2, \quad \| U(x) \| = \|x\|,$$

where $\langle z, x\rangle$ is the value of a linear functional $z \in E^*$ at the vector $x \in E$. Let $F(x)$ be a continuous and bounded operator from E to E with bounded inverse, and v a fixed vector in E. Consider the absolute minimum problem for the functional

$$f_v(x) = \|F(x) - v\|, \qquad x \in E,$$

i. e., we wish to find a vector x_v such that $f_v(x_v) = 0$. Under the above assumptions, we have

Theorem 10.10. Let

$$\langle U(x-y),\ F(x) - F(y)\rangle \geqslant \gamma(\|x-y\|)\,\|x-y\|,$$

where $\gamma(t)$ is an increasing continuous function, $\gamma(0) = 0$. Then the absolute minimum problem for $f_v(x)$ is well-posed.

Proof. The existence of a unique point x_v such that $f_v(x_v) = 0$ follows from Theorem 22.4. We now write

$$f_v(x) - f_v(x_v) = \|F(x) - F(x_v)\| = \frac{\|F(x) - F(x_v)\|\ \|U(x - x_v\|}{\|x - x_v\|} \geqslant$$
$$\geqslant \langle U(x - x_v),\ F(x) - F(x_v)\rangle \frac{1}{\|x - x_v\|} \geqslant \gamma(\|x - x_v\|),$$

i. e.,

$$f_v(x) - f_v(x_v) \geqslant \gamma(\|x - x_v\|).$$

This shows that every minimizing sequence converges to x_v. Q. E. D.

A similar proposition concerning the well-posedness of the absolute minimum problem for $f_v(x)$ also holds under the hypotheses of Theorem 21.1.

Chapter IV

MINIMIZATION METHODS FOR NONLINEAR FUNCTIONALS

In the last section of the previous chapter we studied the convergence of minimizing sequences. In this chapter we consider various methods for the construction of such sequences. We consider only three of the known minimization techniques — the method of steepest descent, the Ritz method and Newton's method.

§ 11. METHOD OF STEEPEST DESCENT

11.1. Idea of the method

We first illustrate the underlying idea for the special case of a Hilbert space.

Let $f(x)$ be a real Gâteaux-differentiable nonlinear functional, defined and bounded below in a real Hilbert space H. Set $d = \inf_{x \in H} f(x)$ and $F(x) = \operatorname{grad} f(x)$. Let x_1 be some vector in H such that $F(x_1) \neq 0$. If $f(x)$ is strictly convex, we know (see Lemma 9.1, Corollary 9.1 and Theorem 9.1) that it may have only one minimum point, where it assumes the value d; thus, if $F(x_1) \neq 0$, x_1 is not a minimum point of $f(x)$. Choose a vector $h \in H$ of length $\|h\| = \|F(x_1)\|$; we wish to fix the direction of h so that the derivative

$$\frac{d}{dt} f(x_1 + th) = (F(x_1 + th), h)$$

has its minimum value when $t = 0$, i. e., we wish h to be a direction of steepest decrease for $f(x)$ at x_1. To this end we first find h such that $(F(x_1), h)$ has a maximum value, and then change the sign of h, so that $(F(x_1), -h)$ has a minimum value. Since

$$(F(x_1), h) \leqslant \|F(x_1)\| \, \|h\| = \|F(x_1)\|^2,$$

it follows that $(F(x_1), h)$ has a maximum only when $h = F(x_1)$ and a minimum when $h = -F(x_1)$, i.e., when the direction of h is that of the antigradient of $f(x)$. Set $h_1 = -F(x_1)$ and consider the real function

$$\varphi(t) = f(x_1 + th_1), \quad t \geqslant 0.$$

By construction, $\varphi(t)$ decreases in some right half-neighborhood of $t = 0$. Let t_1 be the smallest positive t for which $\varphi(t_1) = \min \varphi(t)$. Set

$$x_2 = x_1 + t_1 h_1 = x_1 - t_1 F(x_1).$$

By construction

$$f(x_2) = f(x_1 + t_1 h_1) < f(x_1).$$

Supposing that $F(x_2) \neq 0$, we can repeat the above argument. Thus, in general, if $F(x_k) \neq 0$, we have the following process:

$$x_{n+1} = x_n - t_n F(x_n) \qquad (n = 1, 2, 3, \ldots), \tag{11.1}$$

known as a steepest descent process or gradient descent process.

Although the sequence (11.1) has the property

$$f(x_{n+1}) < f(x_n) \qquad (n = 1, 2, 3, \ldots), \tag{11.2}$$

it may not be a minimizing sequence.

Definition 11.1. A sequence $\{x_n\}$, satisfying (11.2) is known as a *relaxation sequence* and the numbers t_n in (11.1) are known as *relaxation multipliers*.

Even in the case of a Hilbert space difficulties arise when one tries to determine the relaxation multipliers. Only in the simplest cases can one succeed in evaluating them effectively.* The t_n are therefore replaced by positive numbers ε_n, which are either given a priori or allowed to vary freely within prescribed limits. When the t_n are replaced by ε_n the process (11.1) is called a descent process (of gradient type).

Now consider the more general case that a real Gâteaux-differentiable functional $f(x)$ is defined and bounded below in a real Banach space E. Set $F(x) = \operatorname{grad} f(x)$ and let x_1 be any vector in E such that $F(x_1) \neq 0$. Choose $h \in E$ such that $\|h\| = \|F(x_1)\|$. The direction of h is to be such that the derivative

* See, for example, Yakovlev /1/.

$$\frac{d}{dt} f(x_1 + th) = \langle F(x_1 + th),\ h\rangle$$

has a minimum at $t = 0$.

We first select h such that $\langle F(x_1), h\rangle$ has a maximum value, and then take the vector $(-h)$. Then $\langle F(x_1), -h\rangle$ has a minimum value. Since

$$\langle F(x_1),\ h\rangle \leqslant \|F(x_1)\|\ \|h\| = \|F(x_1)\|^2, \tag{11.3}$$

h must be so chosen that $\langle F(x_1), h\rangle = \|F(x_1)\|^2$. To do this, consider an operator U from E^* to E, such that

$$\|Uy\| = \|y\|, \qquad \langle y,\ Uy\rangle = \|y\|^2.$$

If E is reflexive and the norm of E^* is Gâteaux-differentiable, a suitable operator U is (Lemma 2.5)

$$Uy = \|y\| \operatorname{grad} \|y\|.$$

Setting

$$h = UF(x_1),$$

we obtain

$$\langle F(x_1),\ UF(x_1)\rangle = \|F(x_1)\|^2$$

and by (11.3) the supremum of $\langle F(x_1), h\rangle$ is attained. It remains only to set

$$h = h_1 = -UF(x_1).$$

Now select t such that the function

$$\varphi(t) = f(x_1 + th_1), \qquad t \geqslant 0$$

has a minimum (usually the first minimum is chosen) at this point. Let $\varphi(t_1) = \min \varphi(t)$. Then

$$f(x_2) = f(x_1 + t_1 h_1) < f(x_1),$$

and

$$x_2 = x_1 - t_1 UF(x_1).$$

Continuing this process, we get

$$x_{n+1} = x_n - t_n UF(x_n), \qquad n = 1, 2, 3, \ldots \tag{11.4}$$

This is a process of descent. The t_n are usually not evaluated, and instead one considers the process

$$x_{n+1} = x_n - \varepsilon_n UF(x_n). \tag{11.5}$$

The choice of $\{\varepsilon_n\}$ is restricted in such a way that the process of steepest descent will converge.

The operator U in (11.5) may be replaced by any operator A (not necessarily linear) from E^* to E such that $\|A_n y\| \leqslant a_n \|y\|$, where $a_n > 0$, $y \in E^*$.

11.2. Convergence of the gradient descent method for Gâteaux-differentiable functionals

We begin with a few auxiliary propositions.

Lemma 11.1. Let $F(x)$ be a real Gâteaux-differentiable functional defined on a real reflexive Banach space E, whose gradient $F(x)$ satisfies the condition

$$\begin{gathered} \langle F(x+h) - F(x), h \rangle \leqslant M(r) \| h \|^2; \\ x, x+h \in D_r = \{x \leqslant E: \| x \| \leqslant r\}, \end{gathered} \tag{11.6}$$

where $M(r)$ is an arbitrary increasing function for $r \geqslant 0$, and E^* is a space with Gâteaux-differentiable norm. If $\varepsilon_n M_n \leqslant 1/2$, where

$$M_n = \max[1, M(R_n)], \; R_n \geqslant \| x_n \| + \| F(x_n) \|,$$

*then (11.5) is a relaxation process.**

Proof. Using the Lagrange formula, (11.5), and the properties of the operator U, we have

$$\begin{aligned} f(x_n) - f(x_{n+1}) &= \langle F(x_{n+1} + \tau(x_n - x_{n+1})), x_n - x_{n+1} \rangle = \\ &= \langle F(x_n), x_n - x_{n+1} \rangle - \langle F(x_{n+1} + \tau(x_n - x_{n+1})) - \\ &- F(x_n), x_{n+1} - x_n \rangle \geqslant \varepsilon_n \| F(x_n) \|^2 - M_n \| x_{n+1} - x_n \|^2 = \\ &= \varepsilon_n \| F(x_n) \|^2 - \varepsilon_n^2 M_n \| F(x_n) \|^2 \end{aligned}$$

or

$$f(x_n) - f(x_{n+1}) \geqslant 2^{-1} \varepsilon_n \| F(x_n) \|^2 > 0. \tag{11.7}$$

Q. E. D.

* Relaxation processes are the subject of an extensive article by Lyubich and Maistrovskii /2/, where a bibliography is also provided.

A sufficient condition for (11.6) to hold is that the operator $F(x)$ satisfy a Lipschitz condition

$$\| F(x+h) - F(x) \| \leqslant M(r) \| h \|, \qquad (11.8)$$

where the Lipschitz constant may also be an increasing function.

Let the hypotheses of Lemma 11.1 hold, and suppose $\inf_{x \in E} f(x) = d > -\infty$. Set

$$r_n = f(x_n) - d.$$

Inequality (11.7) implies that

$$r_n - r_{n+1} = f(x_n) - f(x_{n+1}) > 0,$$

i. e., this is a decreasing sequence of nonnegative numbers and so we may speak of

$$\lim_{n \to \infty} r_n = r_0 \geqslant 0.$$

If we assume that the set

$$E_0 = \{x \in E \colon f(x) \leqslant f(x_0)\} \qquad (11.9)$$

is bounded, Lemma 11.1 implies that the sequence $\{x_n\}$ is bounded. Of course, the set E_0 is bounded if, say, $f(x)$ is an increasing functional (see Definition 10.2) or if

$$\lim_{R \to \infty} f(x) = +\infty \quad (R = \| x \|).$$

If $F(x)$ satisfies (11.8), the boundedness of $\{x_n\}$ implies the boundedness of $\|F(x_n)\|$ and hence the boundedness of M_n, so that $1 \leqslant M_n \leqslant M_0$. If we now suppose that $\varepsilon_n \geqslant {}^1/_4 M_n$, we have $\varepsilon_n \geqslant {}^1/_4 M_0 > 0$ and thus it follows from (11.7) that

$$\lim_{n \to \infty} F(x_n) = 0. \qquad (11.10)$$

We have thus proved

Lemma 11.2. Suppose that:

1. E is a real reflexive Banach space and E^* is a space with Gâteaux-differentiable norm.

2. $f(x)$ is a real Gâteaux-differentiable functional bounded below and increasing (or the set E_0 is bounded), and its gradient satisfies a Lipschitz condition (11.8).

3. *The relaxation multipliers* ε_n *satisfy the inequalities* $1/4 \leqslant \varepsilon_n M_n \leqslant 1/2$.

Then (11.5) is a relaxation process and $\lim_{n\to\infty} F(x_n) = 0$.

Remark 11.1. If we assume in addition that

$$f(x) - d \leqslant \| F(x) \|^{\alpha}, \quad \alpha > 0,$$

where $d = \inf f(x)$, then x_n is clearly a minimizing sequence.

Theorem 11.1. *Suppose that:*

1. E *is a real reflexive Banach space and* E^* *is a space with Gâteaux-differentiable norm.*

2. $f(x)$ *is a real Gâteaux-differentiable functional defined on* E *and its gradient* $F(x)$ *is such that the function* $\langle F(tx), x\rangle$ *is integrable with respect to* $t \in (0, 1)$; *furthermore, the gradient satisfies the following inequalities*

$$\| F(x+h) - F(x) \| \leqslant M(r) \| h \|; \tag{11.8}$$
$$r, \; r+h \in D(r) = \{x \in E : \| x \| \leqslant r\},$$
$$\langle F(x+h) - F(x), h\rangle \geqslant \| h \| \gamma(\| h \|), \tag{11.11}$$

where $M(r)$ *is a nonnegative increasing continuous function defined for* $r \geqslant 0$, $\gamma(t)$ *is an increasing function, continuous for* $t \geqslant 0$, *such that* $\gamma(0) = 0$,

$$c(R) = \frac{1}{R} \int_0^R \gamma(z)\, dz$$

is an increasing function such that $c(R) > \|F(0)\|$ *for some* R.

3. $1/4 \leqslant \varepsilon_n M_n \leqslant 1/2$, $\quad M_n = \max[1, M(R_n)]$, $R_n \geqslant \|x_n\| + \|F(x_n)\|$.

Then

$$x_{n+1} = x_n - \varepsilon_n U F(x_n) \tag{11.5}$$

is a relaxation sequence and a minimizing sequence, converging to an absolute minimum point of $f(x)$.

Proof. From (6.4) it follows that

$$f(x) - f(0) = \int_0^1 \langle F(tx), x\rangle\, dt =$$
$$= \langle F(0), x\rangle + \int_0^1 \langle F(tx) - F(0), x\rangle\, dt.$$

Using (11.11), we have

$$f(x)-f(0)\geqslant\langle F(0),\ x\rangle+\|x\|\int_0^1\gamma(t\|x\|)\,dt\geqslant$$
$$\geqslant-\|x\|\|F(0)\|+\int_0^{\|x\|}\gamma(z)\,dz=R(-\|F(0)\|+c(R)),$$

where $R=\|x\|$. Condition 2 of the theorem then yields

$$\lim_{R\to\infty}f(x)=+\infty\quad(R=\|x\|),$$

i. e., $f(x)$ is an increasing functional.

It follows from (11.11) (see Theorem 5.1) that $f(x)$ is strictly convex, and so by Theorem 10.1 there exists a unique point $x_0\in E$ such that

$$f(x_0)=\inf_{x\in E}f(x)=d$$

and $f(x)$ has no other minimum points. By Lemma 1.2, (11.5) is a relaxation sequence and

$$\lim_{n\to\infty}F(x_n)=0.$$

But it follows from (11.11) that

$$\|F(x_n)\|\geqslant\left\langle F(x_n),\ \frac{x_n-x_0}{\|x_n-x_0\|}\right\rangle=$$
$$=\left\langle F(x_n)-F(x_0),\ \frac{x_n-x_0}{\|x_n-x_0\|}\right\rangle\geqslant\gamma(\|x_n-x_0\|),$$

since $F(x_0)=0$ (Theorem 9.1). Hence

$$\|x_n-x_0\|\leqslant\gamma^{-1}(\|F(x_n)\|),$$

where γ^{-1} is the inverse function to γ, whose existence is guaranteed by condition 2 of the theorem. The last inequality implies

$$\lim_{n\to\infty}x_n=x_0.$$

Q. E. D.

11.3. Convergence of the gradient descent method for twice Gâteaux-differentiable functionals

In the previous subsection we needed neither the differentiability nor even the continuity of $F(x)$. Here we shall assume that the gradient $F(x)$ of the functional $f(x)$ is Gâteaux-differentiable.

Lemma 11.3. Suppose that:

1. E is a real reflexive Banach space and E^* has Gâteaux-differentiable norm.

2. $f(x)$ is a real Gâteaux-differentiable functional, defined in E, which is increasing and bounded below, with Gâteaux-differentiable gradient $F(x)$ satisfying the inequality

$$\langle F'(x)h_1, h_2\rangle \leqslant M(\|x\|), \|h_1\| \|h_2\| \qquad (11.12)$$

for any $x, h_1, h_2 \in E$; here $M(r)$ is a positive nondecreasing function, defined for $r \geqslant 0$ and bounded on any closed interval, and the function $\langle F'(x+th_1)h_1, h_1\rangle$ is continuous in $t \in [0, 1]$.

3. The multipliers ε_n in (11.5) satisfy the inequalities $1/2 \leqslant \varepsilon_n M_n \leqslant 1$ and $\varepsilon_n \leqslant 1$, where

$$M_n = \max[1, M(R_n)], \quad R_n = \|x_n\| + \|F(x_n)\|.$$

Then (11.5) is a relaxation process for $f(x)$ and $\lim_{n\to\infty} F(x_n) = 0$.

Proof. By the Taylor formula (Theorem 4.7) we have

$$-[f(x_{n+1}) - f(x_n)] = -\langle F(x_n), x_{n+1} - x_n\rangle -$$
$$-\int_0^1 \langle F'(x_n + t(x_{n+1} - x_n))(x_{n+1} - x_n), x_{n+1} - x_n\rangle (1-t)\,dt =$$
$$= -\langle F(x_n), x_n - x_{n+1}\rangle -$$
$$-\frac{1}{2}\langle F'(x_n + \theta(x_{n+1} - x_n))(x_{n+1} - x_n), x_{n+1} - x_n\rangle,$$

where x_n and x_{n+1} are as in (11.5) and $0 < \theta < 1$. Using (11.12), we have

$$f(x_n) - f(x_{n+1}) \geqslant -\langle F(x_n), x_{n+1} - x_n\rangle -$$
$$-\frac{1}{2} M(\|x_n + \theta(x_{n+1} - x_n)\|)\|x_{n-1} - x_n\|^2$$

or, since

$$\|x_n + \theta(x_{n+1} - x_n)\| \leqslant \theta\|x_{n+1}\| + (1-\theta)\|x_n\| \leqslant$$
$$\leqslant \theta(\|x_n\| + \varepsilon_n\|F(x_n)\|) + (1-\theta)\|x_n\| \leqslant \theta(\|x_n\| + \|F(x_n)\|) +$$
$$+ (1-\theta)\|x_n\| \leqslant \theta R_n + (1-\theta)R_n = R_n,$$

it follows that

$$f(x_n)-f(x_{n+1})\geqslant-\langle F(x_n),\ x_{n+1}-x_n\rangle-\frac{1}{2}M(R_n)\|x_{n+1}-x_n\|^2.$$

Now, using (11.5) and the properties of the operator U, we get

$$f(x_n)-f(x_{n+1})\geqslant\varepsilon_n\|F(x_n)\|^2-\frac{1}{2}\varepsilon_n^2M_n\|F(x_n)\|^2=$$
$$=\varepsilon_n\|F(x_n)\|^2\left(1-\frac{1}{2}\varepsilon_nM_n\right)\geqslant\frac{1}{2}\varepsilon_n\|F(x_n)\|^2,$$

i. e.,

$$f(x_n)-f(x_{n+1})\geqslant\frac{1}{2}\varepsilon_n\|F(x_n)\|^2>0. \tag{11.7}$$

This proves the first part of the lemma.

To prove the second part, we set $d=\inf\limits_{x\in E}f(x)$ and consider the sequence

$$r_n=f(x_n)-d.$$

This is a decreasing sequence, because

$$r_n-r_{n+1}=f(x_n)-f(x_{n+1})\geqslant\frac{1}{2}\varepsilon_n\|F(x_n)\|^2>0,$$

so that the limit

$$\lim_{n\to\infty}r_n=r_0\geqslant0 \tag{11.13}$$

exists. Since $f(x)$ is increasing, it follows from (11.7) that the sequence $\{x_n\}$ is bounded. We claim that the sequence $\{\|F(x_n)\|\}$ is bounded. In fact, a corollary of the Hahn-Banach theorem* and the reflexivity of E entail that for any fixed $x\neq0$ there exists a unit vector $h\in E$ such that

$$\langle F(x)-F(0),\ h\rangle=\|F(x)-F(0)\|.$$

By the generalized Lagrange formula,

$$\|F(x)-F(0)\|<\langle F'(\tau x)x,\ h\rangle,\quad 0<\tau<1,$$

whence

$$\|F(x)\|\leqslant\|F(0)\|+\|F(x)-F(0)\|=\|F(0)\|+\langle F'(\tau x)x,\ h\rangle$$

* See, for example, Dunford and Schwartz /1/, p. 65.

and by (11.12),

$$\|F(x)\| \leqslant \|F(0)\| + M(\tau\|x\|)\|x\| \leqslant \|F(0)\| + M(\|x\|)\|x\|,$$

i. e., $\{\|F(x_n)\|\}$ is a bounded sequence. Since $\|x_n\|$ and $\|F(x_n)\|$ are bounded, so are $R_n = \|x_n\| + \|F(x_n)\|$ and M_n, and thus

$$1 \leqslant M_n \leqslant M_0 < +\infty.$$

Hence, by the hypothesis of the lemma, we have

$$0 < \frac{1}{2M_0} \leqslant \varepsilon_n \leqslant 1.$$

This inequality together with (11.7) and (11.13) imply

$$\lim_{n\to\infty} F(x_n) = 0.$$

Q. E. D.

Theorem 11.2. Suppose that:

1. E is a real reflexive Banach space and E^* has Gâteaux-differentiable norm.

2. $f(x)$ is a twice Gâteaux-differentiable real functional defined on E; the function $\langle F'(y+tx)x, x\rangle$ is continuous in $t \in [0, 1]$ for any $x, y \in E$; the inequalities

$$\langle F'(x)h_1, h_2\rangle \leqslant M(\|x\|)\|h_1\|\|h_2\| \tag{11.12}$$

hold for any $x, h_1, h_2 \in E$, where $M(r)$ is a positive nondecreasing function defined for $r \geqslant 0$ and bounded on any closed interval; furthermore, for any $x, y \in E$,

$$\langle F'(y+tx)x, x\rangle \geqslant \|x\|\gamma(\|x\|), \tag{11.14}$$

where $\gamma(t)$ is a continuous increasing function for $t \geqslant 0$, $\gamma(0) = 0$, and for some R the increasing function

$$c(R) = \frac{1}{R}\int_0^R \gamma(z)\,dz$$

satisfies the inequality $c(R) > \|F(0)\|$.

3. $0 < \varepsilon_n \leqslant 1$, $1/2 \leqslant \varepsilon_n M_0 \leqslant 1$, $M_n = \max[1, M(R_n)]$, $R_n = \|x_n\| + \|F(x_n)\|$.

Then

$$x_{n+1} = x_n - \varepsilon_n UF(x_n) \qquad (n = 1, 2, 3, \ldots) \tag{11.5}$$

is a relaxation process for $f(x)$, $\{x_n\}$ *is a minimizing sequence and converges to an absolute minimum point of* $f(x)$.

Proof. By Corollary 8.2, it follows from (11.14) that $f(x)$ is strictly convex. We claim that $f(x)$ is an increasing functional. Indeed, in view of (6.4), we have

$$f(x)-f(0)=\int_0^1 \langle F(tx),\ x\rangle\,dt=$$
$$=\langle F(0),\ x\rangle+\int_0^1 \langle F(tx)-F(0),\ x\rangle\,dt,$$

so that the generalized Lagrange formula gives

$$f(x)-f(0)=\langle F(0),\ x\rangle+\int_0^1 \langle F'(\tau tx)\,tx,\ x\rangle\,dt.$$

Hence, by (11.14),

$$f(x)-f(0)\geqslant\langle F(0),\ x\rangle+\int_0^1 \|x\|\gamma(t\|x\|)\,dt=$$
$$=\langle F(0),\ x\rangle+\int_0^R \gamma(z)\,dz\geqslant-R\|F(0)\|+\int_0^R \gamma(z)\,dz,$$

where $R=\|x\|$. Thus

$$f(x)\geqslant f(0)+R(-\|F(0)+c(R)).$$

By the assumptions of the theorem, this implies

$$\lim_{R\to\infty} f(x)=+\infty \quad (R=\|x\|).$$

Since $f(x)$ is a strictly convex increasing functional, it follows from Theorem 10.1, that there is a unique point $x_0\in E$ such that

$$f(x_0)=\inf_{x\in E} f(x)=d,$$

and $f(x)$ has no other minimum points.

Now, by Theorem 9.1 we have $F(x_0)=0$, and, since all the assumptions of Lemma 11.3 are valid, it follows that

$$\lim_{n\to\infty} F(x_n)=0,$$

where $\{x_n\}$ is the sequence defined by (11.5). Finally, since $F(x_0)=0$, it follows from (11.14) that

$$\|F(x_n)\| \geqslant \left\langle F(x_n), \frac{x_n - x_0}{\|x_n - x_0\|}\right\rangle = \left\langle F(x_0) - F(x_0), \frac{x_n - x_0}{\|x_n - x_0\|}\right\rangle =$$
$$= \left\langle F'(x_0 + \tau(x_n - x_0))(x_n - x_0), \frac{x_n - x_0}{\|x_n - x_0\|}\right\rangle \geqslant \gamma(\|x_n - x_0\|),$$

or

$$\|F(x_n)\| \geqslant \gamma(\|x_n - x_0\|). \tag{11.15}$$

It is assumed, of course, that $x_n \neq x_0$ in (11.15), since $F(x_n) \neq 0$ for any n. But, by assumption, $\gamma(x)$ is an increasing continuous function, so that it has an inverse γ^{-1}, which is also continuous and increasing and $\gamma^{-1}(0)=0$. We therefore have by (11.15) that

$$\|x_n - x_0\| \leqslant \gamma^{-1}(\|F(x_n)\|),$$

and, since $F(x_n) \to 0$ as $n \to \infty$, $\lim_{n\to\infty} \|x_n - x_0\| = 0$. Q. E. D.

§ 12. RITZ METHOD*

In this section, E is a real separable normed space and $f(x)$ a finite real functional defined on E. Only in subsections 12.5—12.7 shall we need the completeness of E. Assuming that $f(x)$ is bounded below on E, we shall minimize $f(x)$ by the Ritz method. Ritz himself applied the method to a certain specific problem. The method was later extended by Mikhlin** and other authors.†

The proofs of certain propositions concerning the Ritz method, will require

Definition 12.1. A functional $f(x)$ is said to be upper (lower) semicontinuous at a point $x_0 \in E$ if, for any ε, there exists $\delta > 0$ such that, whenever $\|x - x_0\| < \delta$,

$$f(x_0) - f(x) > -\varepsilon \quad (f(x_0) - f(x) < \varepsilon).$$

A functional $f(x)$ is said to be upper (lower) semicontinuous on a set $M \subset E$ if it is upper (lower) semicontinuous at each point $x \in M$.

* Ritz /1—2/.

** Mikhlin /1—3/.

† See, for example, Zeragiya /1, 4/ and Kachurovskii /5/.

12.1. Ritz approximations and Ritz systems

Let $f(x)$ be a real functional, defined and bounded below on a normed space E. The Ritz method for minimization of $f(x)$ is as follows. We first select a "coordinate system," i. e., a linearly independent system of vectors

$$\varphi_1, \varphi_2, \ldots, \varphi_n, \ldots, \tag{12.1}$$

with the following properties: 1) $\varphi_n \in E$ for any positive integer n; 2) the set of all linear combinations of vectors of (12.1) is dense in E. We now construct a sequence of finite-dimensional subspaces $\{E_n\}$, where E_n is the n-dimensional space spanned by $\varphi_1, \varphi_2, \ldots, \varphi_n$. Since $f(x)$ is bounded below on E, it is also bounded below on E_n. Let

$$d_n = \inf_{x \in E_n} f(x) \qquad (n = 1, 2, 3, \ldots).$$

By construction, $d_1 \geqslant d_2 \geqslant \ldots \geqslant d_n \geqslant \ldots$. Suppose that for any n there is a point $x_n \in E_n$ such that

$$f(x_n) = d_n.$$

Since $x_n \in E_n$, we have

$$x_n = \sum_{k=1}^{n} a_k \varphi_k, \tag{12.2}$$

where $a_k = a_k(n)$. The vectors x_n are called Ritz approximations. In this and in following subsections we shall see how to find them.

Let $f(x)$ be Gâteaux-differentiable on E, $F(x) = \operatorname{grad} f(x)$. Then $f(x)$ is Gâteaux-differentiable on E_n too, and for any $x, h \in E_n$, i. e.,

$$x = \sum_{k=1}^{n} \alpha_k \varphi_k, \quad h = \sum_{k=1}^{n} \beta_k \varphi_k,$$

where α_k and β_k are arbitrary, we have

$$\frac{d}{dt} f(x + th)\Big|_{t=0} = \langle F(x), h \rangle = \sum_{i=1}^{n} \beta_i \left\langle F\left(\sum_{k=1}^{n} \alpha_k \varphi_k\right), \varphi_i \right\rangle.$$

It thus follows by Theorem 9.1 that, if x_n is an absolute minimum point of $f(x)$ on E_n,

$$\left\langle F\left(\sum_{k=1}^{n} a_k \varphi_k\right), \varphi_i \right\rangle = 0 \qquad (i = 1, 2, \ldots, n), \tag{12.3}$$

and if $f(x)$ is a convex functional, system (12.3) furnishes a necessary and sufficient condition for the vectors x_n defined by (12.2) to be Ritz approximations. System (12.3) for the coefficients a_k is known as a Ritz system.

We claim that if $f(x)$ is convex, every solution of (12.3) substituted into (12.2) yields a Ritz approximation, thus giving an absolute minimum point of $f(x)$ on E_n. This assertion follows from the following

Lemma 12.1. If $f(x)$ is a convex functional defined in a linear space (not necessarily normed) which has two distinct minimum points, then its values at these points must coincide.

Proof. Let x_1 and x_2 be minimum points of $f(x)$, $d_1 = f(x_1)$ and $d_2 = f(x_2)$, say $d_1 < d_2$. Without loss of generality, we may assume that $x_2 = 0$ and $d_2 = 0$; otherwise (see the proof of Lemma 9.1) we simply consider the functional

$$\varphi(z) = f(x_2 + z) - f(x_2).$$

By assumption, therefore, $d_1 < 0 = d_2 = f(x_2)$. Let U be a neighborhood of the zero vector such that $f(x) \geqslant 0$ for any $x \in U$. Then, for sufficiently small positive λ,

$$0 = f(0) \leqslant f(\lambda x_1)$$

and, as $f(x)$ is convex,

$$f(\lambda x_1 + (1 - \lambda) 0) \leqslant \lambda f(x_1) + (1 - \lambda) f(0) = \lambda f(x_1) = \lambda d_1.$$

Consequently,

$$0 = f(0) \leqslant \lambda d_1,$$

i. e., $d_1 \geqslant 0$, contradicting the assumption $d_1 < d_2 = 0$. Q. E. D.

12.2. Solvability of Ritz systems

As we have seen, the vectors defined by (12.2) are Ritz approximations for a convex Gâteaux-differentiable functional $f(x)$ defined on E if and only if the coefficients a_k satisfy the Ritz system (12.3). Of course, if $f(x)$ is strictly convex on E, it is strictly convex on $E_n \subset E$, and so, by Corollary 9.1, system (12.3) can have no more than one solution.

Here we investigate the solvability of system (12.3). Now, a solution $(a_1, a_2, \ldots, a_n)$ of system (12.3) may furnish (via (12.2)) either a critical point of $f(x)$ (i. e., a point at which the gradient of $f(x)$ vanishes), a local minimum point of $f(x)$ on E_n, or an absolute minimum point of $f(x)$ on E_n. Only in the last case does system (12.3) define a Ritz approximation. If $f(x)$ is a convex functional, then (Theorem 9.1) each of its critical points is a minimum point, and by Lemma 12.1 even an absolute minimum point. We present a few propositions on the solvability of system (12.3).

Lemma 12.2. *If a real increasing* Gâteaux-differentiable functional $f(x)$, defined on E, is lower semicontinuous on every subspace $E_n \subset E$, then the Ritz system (12.3) is solvable for any n.*

Proof. Since $f(x)$ is increasing, there exists $r > 0$ such that $f(x) > f(0)$ for $\|x\| \geqslant r$. Consider the ball

$$D_r^{(n)} = \{x \in E_n \colon \| x \| \leqslant r\}$$

in E_n. By a standard theorem of classical analysis, the lower semicontinuity of $f(x)$ on $D_r^{(n)}$ entails that there exists a point $x_0 \in D_r^{(n)}$ at which $f(x)$ assumes its smallest value:

$$f(x_0) = \inf_{x \in D_r^{(n)}} f(x), \quad x_0 = \sum_{k=1}^{n} a_k^{(0)} \varphi_k.$$

This cannot be a boundary point of $D_r^{(n)}$, because $f(x) > f(0)$ on the surface of the ball. Outside the ball, however, we have

$$f(x) > f(0) \geqslant f(x_0).$$

Therefore,

$$f(x_0) = \inf_{x \in E_n} f(x)$$

and, as we have seen, $a_1^{(0)}, a_2^{(0)}, \ldots, a_n^{(0)}$ satisfy system 12.3. Q. E. D.

This lemma differs only slightly from a similar proposition of Mikhlin (/3/, p. 336).

Lemma 12.3. *If $f(x)$ is a real Gâteaux-differentiable functional defined on E and lower semicontinuous on every subspace $E_n \subset E$, and for some $r > 0$*

* See Definition 10.2.

$$\langle F(x),\ x\rangle > 0, \quad \|x\| = r, \quad F(x) = \operatorname{grad} f(x), \tag{12.4}$$

then the Ritz system (12.3) is solvable for any n.*

Proof. Consider the ball $D_r = \{x \in E_n : \|x\| \leqslant r\}$ in E_n. There is an absolute minimum point of $f(x)$ on this ball, for $f(x)$ is semicontinuous on D_r. Denote this point by x_0. By Lemma 9.2, x_0 is an interior point of D_r, so that for any $h \in E_n$ we have

$$\frac{d}{dt} f(x_0 + th)\Big|_{t=0} = 0.$$

Since $h = \sum_{k=1}^{n} \beta_k \varphi_k$ and the β_k are arbitrary, it follows that

$$\langle F(x_0),\ \varphi_i\rangle = 0 \qquad (i = 1,\ 2,\ \ldots,\ n),$$

where x_0, as a vector in E_n, is of the form

$$x_0 = \sum_{k=1}^{n} a_k^{(0)} \varphi_k, \tag{12.5}$$

so that system (12.3) is solvable and $a_k = a_k^{(0)}$ $(k=1,\ 2,\ \ldots,\ n)$. Q. E. D.

Remark 12.1. Inequality (12.4) holds if, for example,

$$\overline{\lim_{R\to\infty}}\ \langle F(x),\ x\rangle = +\infty \quad (R = \|x\|).$$

Lemma 12.4. If $f(x)$ *is an increasing convex Gateaux-differentiable functional, defined on* E, *the Ritz system (12.3) is solvable for any* n.

Proof. Any finite convex functional $f(x)$ is weakly lower semicontinuous on E (Theorem 8.10), hence also on E_n for any n. But in finite-dimensional spaces weak semicontinuity is equivalent to semicontinuity, so that $f(x)$ is lower semicontinuous on every $E_n \subset E$. The assumptions of Lemma 12.2 then hold, so that the proof is complete. One can, of course consider weakly lower-semicontinuous functionals instead of convex functionals.

Remark 12.2. Under the assumptions of Lemma 12.4, Ritz systems may have more than one solution, but according to Lemma 12.1 every solution of (12.3) yields, via (12.2), a vector at which $f(x)$ assumes an absolute minimum on E_n. If $f(x)$ is not only convex but strictly convex, then every Ritz system has a unique solution. It is evident that instead of assuming that $f(x)$

* Compare Kachurovskii /5/.

is convex (strictly convex) we may assume that the operator $F(x) = \operatorname{grad} f(x)$ is monotone (strictly monotone).

12.3. Minimization of functionals by Ritz approximations

Lemma 12.5. Suppose that the Ritz approximations (12.2) exist for every n, where $f(x)$ is defined and bounded below on E. Then, if $f(x)$ is upper semicontinuous, the Ritz approximations form a minimizing sequence.*

Proof. Let $d = \inf_{x \in E} f(x)$ and let $\{u^{(n)}\} \subset E$ be a minimizing sequence such that

$$f(u^{(n)}) \leqslant d + 1/n \qquad (n = 1, 2, 3, \ldots).$$

In view of the completeness of the coordinate system (12.1), it follows that for any vector $u^{(n)}$ and any positive δ_n there exists a vector $v^{(m)} \in E_m$,

$$v^{(m)} = \sum_{k=1}^{m} a_k^{(m)} \varphi_k \qquad (m = m(n) \geqslant n),$$

such that $\| u^{(n)} - v^{(m)} \| < \delta_n$. Since $f(x)$ is upper semicontinuous, δ_n can be taken sufficiently small so that, for any $u \in E$ with $\|u^{(n)} - u\| < \delta_n$, we have

$$f(u^{(n)}) - f(u) \geqslant -1/n.$$

Setting $u = v^{(m)}$, we find

$$f(v^{(m)}) \leqslant f(u^{(n)}) + 1/n \leqslant d + 2/n.$$

This inequality means that $\{v^{(m)}\}$ is a minimizing sequence. But the Ritz approximations (12.2) satisfy

$$f(x_m) = d_m = \inf_{x \leqslant E_m} f(x),$$

so that

$$f(x_m) \leqslant f(v^{(m)}) \leqslant d + 2/n.$$

* Mikhlin /3/, pp. 336–337; Gel'fand and Fomin /1/, p. 197.

Since $f(x_m) \geqslant d$, we have

$$\lim_{n\to\infty} f(x_m) = d.$$

Q. E. D.

To guarantee the existence of Ritz approximations, it suffices (see the proof of Lemma 12.2)) that $f(x)$ be increasing and lower semicontinuous on every subspace $E_n \subset E$.

From Lemmas 12.2 and 12.5 we have

*Theorem 12.1.** *If* $f(x)$ *is a real increasing Gâteaux-differentiable functional, defined on* E, *continuous and bounded below, then the Ritz system (12.3) is solvable for any* n *and the Ritz approximations form a minimizing sequence.*

This theorem remains valid if we drop the assumption that the functional is continuous, requiring it only to be upper semicontinuous on E and lower semicontinuous on each subspace E_n spanned by the vectors $\varphi_1, \varphi_2, \ldots, \varphi_n$.

From Lemmas 12.3 and 12.5 we obtain

Theorem 12.2. Let $f(x)$ *be a real Gâteaux-differentiable functional, defined and bounded below on* E, *satisfying the following conditions: it is upper semicontinuous on* E, *lower semicontinuous on each subspace* E_n, *and for some* $r > 0$

$$\langle F(x), x\rangle > 0, \quad \|x\| = r, \quad F(x) = \operatorname{grad} f(x). \tag{12.4}$$

Then the Ritz system is (12.3) solvable for any n *and the Ritz approximations form a minimizing sequence.*

A sufficient condition for the validity of (12.4) was indicated in Remark 12.1.

Theorem 12.3. Let $f(x)$ *be a real Gâteaux-differentiable functional* $f(x)$, *defined and bounded below on* E, *convex and continuous. Then, if either condition (12.4) holds or* $f(x)$ *is increasing,*** *the Ritz system (12.3) is solvable for any* n *and the Ritz approximations form a minimizing sequence.*

The proof differs only slightly from those of Theorems 12.1 and 12.2. The theorem remains valid if $f(x)$ is assumed to be weakly lower semicontinuous instead of convex and upper semicontinuous instead of continuous.

Remark 12.3. If we drop the assumption of Theorems 12.1 – 12.3 that the functional is Gâteaux-differentiable then, naturally, it is meaningless to speak of the solvability of Ritz systems, but

* Compare Mikhlin /3/, Theorem 10.1.

** See Definition 10.2.

the other assertions of these theorems remain valid; namely, the Ritz approximations (12.2) exist for any n and form a minimizing sequence.

The assumptions of Theorems 12.1—12.3 do not guarantee that the functional has a minimum or that (if it does) the Ritz approximations converge to the minimum point. The main import of these theorems is the existence of the infima

$$d = \inf_{x \in E} f(x), \quad d_n = \inf_{x \in E_n} f(x),$$

of vectors $x_n \in E_n$, such that $f(x_n) = d_n$, and the fact that $d_n \to d$ as $n \to \infty$. Conditions for the existence of a point x_0 such that $f(x_0) = d$ and for the convergence of the Ritz approximations to x_0 are studied in the following subsection.

12.4. Weak convergence of Ritz approximations

In this subsection E is a real reflexive Banach space. Combining the propositions of the previous subsection with those of subsection 10.3 we obtain various theorems, of which we cite the following.

Theorem 12.4. If $f(x)$ is a real strictly convex functional, continuous and increasing on E, then:

1. There exists a unique minimum point x_0 at which $f(x)$ assumes its infimum.

2. The Ritz approximations (12.2) exist for any n and converge weakly to x_0.

Proof. Let E_n be the subspace spanned by $\varphi_1, \varphi_2, \ldots, \varphi_n$ (see subsection 12.1). Since $f(x)$ is strictly convex and increasing, it follows (Theorem 10.1) that there is a unique point $x_n \in E_n$ such that

$$f(x_n) = d_n = \inf_{x \in E_n} f(x).$$

By Lemma 12.5, these Ritz approximations will form a minimizing sequence provided

$$\inf_{x \in E} f(x) = d > -\infty.$$

But, by Theorem 10.1, there is a unique point $x_0 \in E$ such that $f(x_0) = d$ and every minimizing sequence converges weakly to x_0, so that $x_n \rightharpoonup x_0$. Q. E. D.

The assumption of this theorem that $f(x)$ is strictly convex may be replaced by the following assumptions: a) $f(x)$ is weakly lower semicontinuous; b) $f(x)$ has a unique minimum point, coinciding with its absolute minimum point.

Note that the remark made after the proof of Theorem 10.1 is also relevant here.

Theorem 12.5. Let $f(x)$ be a finite increasing upper semicontinuous and weakly lower semicontinuous functional, defined on E. Then $f(x)$ has at least one absolute minimum point, and a suitable subsequence of Ritz approximations is weakly convergent to some absolute minimum point.

Proof. Since the functional $f(x)$ is increasing, there exists $r > 0$ such that $f(x) > f(0)$ for $\|x\| \geqslant r$. As we know, in a reflexive space E the ball

$$D_r = \{x \in E : \|x\| \leqslant r\}$$

is weakly compact, weakly complete and weakly closed; hence, since $f(x)$ is weakly lower semicontinuous (see Theorem 9.2), there exists a vector $x_0 \in D_r$ such that

$$f(x_0) = \inf_{x \in E} f(x).$$

This vector is an interior point of D_r, since $f(x) > f(0)$ on the surface of the ball. Consequently,

$$f(x_0) = d = \inf_{x \in E} f(x).$$

Applying the same reasoning to every subspace $E_n \subset E$, we see that there exists a vector $x_n \in E_n$ such that

$$f(x_n) = d_n = \inf_{x \in E_n} f(x).$$

Thus the Ritz approximations exist for any n and, by Lemma 12.5, they form a minimizing sequence. Hence, by Lemma 10.1, the sequence $\{x_n\}$ is bounded, and so it contains a subsequence $y_n \rightharpoonup y_0 \in E$. In view of the fact that $f(x)$ is weakly lower semicontinuous, it follows that

$$f(y_0) \leqslant \varliminf_{n \to \infty} f(y),$$

but, as we have proved, $f(x_n) \to d = f(x_0)$ as $n \to \infty$, so that

$$f(y_0) \leqslant \varliminf_{n \to \infty} f(y_n) = \lim_{n \to \infty} f(x_n) = f(x_0) = d.$$

Since d is the infimum, it follows that $f(y_0) = d$, i. e., $f(x)$ has an absolute minimum at y_0 and $y_n \longrightarrow y_0$. Q. E. D.

Theorem 12.6. Let $f(x)$ be a real functional, continuous and Gâteaux-differentiable, with gradient $F(x)$ satisfying the following conditions:

1. $\langle F(tx), x\rangle$ *is a continuous function of* t *in* $[0,1]$ *for every* $x \in E$.
2. $\langle F(x+h) - F(x), h\rangle \geqslant 0$ *for any* $x, h \in E$.
3. $\lim\limits_{R\to\infty} \frac{\langle F(x), x\rangle}{R} = +\infty$ $(R = \|x\|)$.

Then the assertion of Theorem 12.5 holds. If equality can hold in condition 2 only when $h = 0$, the assertion of Theorem 12.4 holds.

Proof. By Theorem 8.4, condition 2 implies that $f(x)$ is weakly lower semicontinuous, while conditions 1, 2 and 3 imply (see the proof of Theorem 9.6) that

$$\lim_{R\to\infty} f(x) = +\infty \quad (R = \|x\|),$$

which guarantees that $f(x)$ is an increasing functional. The assumptions of Theorem 12.5 therefore hold, so that the first part of our theorem is proved. If equality is possible in condition 2 only when $h = 0$, then $f(x)$ is strictly convex so that all the assumptions of Theorem 12.4 hold. Q. E. D.

The hypotheses of Theorem 12.6 guarantee solvability of the Ritz system (12.3) for any n.

12.5. Strong convergence of Ritz approximations

Theorem 12.7. Let $f(x)$ be a real functional defined on E, bounded and Gâteaux-differentiable, with gradient $F(x)$ satisfying the conditions of Theorem 10.4, where $\gamma(t) > 0$ for $t > 0$. Then the Ritz system (12.3) is solvable for any n and the Ritz approximations converge to the unique absolute minimum point of $f(x)$.

Proof. From (10.3) it follows that $F(x)$ is strictly monotone, so that (Theorem 5.1) $f(x)$ is strictly convex. Inequality (10.5) then implies (see the proof of Theorem 10.4) that $f(x)$ is an increasing functional, but it is known* that a convex functional bounded on a ball is continuous on E. Thus all assumptions of Theorems 12.1 and 12.4 hold. Hence the following conclusions: there exists a unique minimum point x_0 such that

$$f(x_0) = \inf_{x \in E} f(x);$$

* See Bourbaki /1/, p. 92.

the solution of the Ritz system is unique for any n; and the Ritz approximations form a minimizing sequence. By Theorem 12.4, this minimizing sequence converges to x_0. Q. E. D.

Theorem 12.8. Let $f(x)$ be a functional defined on E, bounded on some ball and satisfying the assumptions of Theorem 10.5. Then the conclusions of Theorem 12.7 hold.

Proof. Condition 2 of Theorem 10.5 implies (see the proof of Theorem 10.5) that $f(x)$ is an increasing and convex functional with unique minimum point x_0; since $f(x)$ is also bounded, all the assumptions of Theorems 12.1 and 12.4 are valid, whence solvability of the Ritz system for any n and the fact that the Ritz approximations comprise a minimizing sequence. The convergence of this sequence to x_0 follows from Theorem 10.5. Q. E. D.

Theorem 12.9. Let $f(x)$ be a twice Gâteaux-differentiable functional, defined in a reflexive Banach space E and satisfying the following conditions:

1. $D^2 f(y+tx; x, x)$ is integrable with respect to t on $[0,1]$ for any $x, y \in E$.

2. $D^2 f(x; h, h) \geqslant \gamma(\|x\|)\|h\|^2$ for any $x, h \in E$, where $\gamma(z)$ is a positive continuous decreasing function on $[0, +\infty)$ such that for some $R > 0$

$$\frac{1}{R}\int_0^R z\gamma(z)\,dz > \|F(0)\|,$$

where $F(x) = \operatorname{grad} f(x)$.

3. $f(x)$ is bounded on some ball.

Then the conclusions of Theorem 12.7 are valid.

Proof. Conditions 1—2 imply (see the proofs of Theorems 9.10 and 10.6) that $f(x)$ is an increasing strictly convex functional, while condition 3 entails that $f(x)$ is continuous, as a convex functional bounded on a bounded set. Thus, all the assumptions of Theorems 12.1 and 12.4 hold, implying the existence of a unique absolute minimum point x_0 of $f(x)$, the solvability of system (12.3) for any n, and the fact that the approximations (12.2) form a minimizing sequence. The convergence of this sequence to x_0 follows from Theorems 10.6. Q. E. D.

Theorem 12.10. Under the assumptions of Theorem 10.7, there exists a unique point x_0 such that

$$f(x_0) = \inf_{x \in E} f(x),$$

the Ritz approximations exist for any n and converge to x_0.

Proof. Let $\{E_n\}$ be a sequence of finite-dimensional subspaces of E (see subsection 12.1). As in the proof of Theorem 10.7, we see that there exist a unique vector $x_0 \in E$ and, for every n, a unique vector $x_n \in E_n$ such that

$$f(x_0) = \inf_{x \in E} f(x), \quad f(x_n) = \inf_{x \in E_n} f(x).$$

Now, as a convex functional bounded on a ball, $f(x)$ is continuous. Consequently, by Lemma 12.5, the Ritz approximations $\{x_n\}$ form a minimizing sequence, and they converge to x_0, by Theorem 10.7.

§ 13. NEWTON—KANTOROVICH METHOD

Newton's method for solution of the equation $\varphi(t) = 0$, where $\varphi(t)$ is a real function of a real variable, has been generalized to the abstract case* and extended by Kantorovich and others** to the solution of nonlinear equations in various spaces. We shall apply the method to minimization of functionals.

13.1. Construction of Newton approximations

Let $f(x)$ be a real thrice Gâteaux-differentiable functional, defined and bounded below in a normed space E. Set $F(x) = \operatorname{grad} f(x)$ and suppose that x_0 is an absolute minimum point of $f(x)$. If x_1 is an initial approximation to x_0, Taylor's formula (see subsection 4.5) yields†

$$f(x) = f(x_1) + f'(x_1)(x - x_1) + 2^{-1} f''(x_1)(x - x_1)^2 + r_3(x_1, x),$$

or

$$f(x) = f(x_1) + \langle F(x_1), x - x_1 \rangle + \\ + 2^{-1} \langle F'(x_1)(x - x_1), x - x_1 \rangle + r_3(x_1, x).$$

Dropping $r_3(x_1, x)$, we obtain a quadratic functional

$$\varphi_1(x) = f(x_1) + \langle F(x_1), x - x_1 \rangle + 2^{-1} \langle F'(x_1)(x - x_1), x - x_1 \rangle.$$

* Bennett /1/, Kantorovich /1/.

** See Kantorovich and Akilov /1/, which includes bibliographic references.

† We are using the integral form of the remainder in Taylor's formula. If one uses the Peano form of the remainder (see subsection 4.6), it suffices to assume that $f(x)$ is twice differentiable or that the operator $F(x)$ is differentiable.

The idea behind the Newton– Kantorovich method is to find the next approximation x_2 as an absolute minimum point of the quadratic functional $\varphi_1(x)$. To this end, we first find grad $\varphi_1(x_1)$. We write

$$\frac{d}{dt}\varphi_1(x+th)\Big|_{t=0} =$$
$$=\langle F(x_1), h\rangle + 2^{-1}\langle F'(x_1)h, x-x_1\rangle + 2^{-1}\langle F'(x_1)(x-x_1), h\rangle.$$

Since $F(x)$ is a potential operator, it follows from Theorem 6.3 that

$$\langle F'(x_1)h, x-x_1\rangle = \langle F'(x_1)(x-x_1), h\rangle,$$

whence

$$\frac{d}{dt}\varphi_1(x+th)\Big|_{t=0} = \langle F(x_1) + F'(x_1)(x-x_1), h\rangle,$$

and so

$$\Phi_1(x) \equiv \text{grad}\,\varphi_1(x) = F(x_1) + F'(x_1)(x-x_1).$$

If x_2 is an absolute minimum point for $\varphi_1(x)$, then, by Theorem 9.1, $\Phi_1(x_2) = 0$, or

$$F(x_1) + F'(x_1)(x_2 - x_1) = 0,$$

whence, provided the inverse operator

$$\Gamma(x_1) = [F'(x_1)]^{-1}$$

exists, we have

$$x_2 = x_1 - \Gamma(x_1)F(x_1).$$

The next approximation is found similarly. That is, the functional $f(x)$ is approximated by the quadratic functional

$$\varphi_2(x) = f(x_2) + \langle F(x_2), x-x_2\rangle + 2^{-1}\langle F'(x_2)(x-x_2), x-x_2\rangle,$$

and, repeating the same reasoning, we find that

$$x_3 = x_2 - \Gamma(x_2)F(x_2).$$

Continuing this process, we get

$$x_{n+1} = x_n - \Gamma(x_n)F(x_n) \qquad (n = 1, 2, 3, \ldots), \tag{13.1}$$

where

$$\Gamma(x) = [F'(x)]^{-1}.$$

The process (13.1) is known as the Newton *iterative process* (or *Newton method*), and the vectors x_n given by this formula are *Newton approximations*. Each approximation x_{n+1} is an absolute minimum point of the functional

$$\varphi_n(x) = f(x_n) + \langle F(x_n), x - x_n \rangle + \\ + 2^{-1} \langle F'(x_n)(x - x_n), x - x_n \rangle. \qquad (13.2)$$

The vector x_{n+1}, then, is to be found from the equality

$$\Phi_n(x) \equiv \operatorname{grad} \varphi_n(x) = F(x_n) + F'(x_n)(x - x_n) = 0. \qquad (13.3)$$

We have obtained this equality as a necessary condition for $\varphi_n(x)$ to have a minimum. If, however, $f(x)$ is a convex functional, it is also a sufficient condition for the existence of an absolute minimum of $\varphi_n(x)$. This follows from

Lemma 13.1. If $f(x)$ is a convex functional, the quadratic functional $\varphi_n(x)$ defined by (13.2) is also convex.

Proof. By Corollary 8.2, if $f(x)$ is convex, then for any $x, h \in E$

$$\langle F'(x)h, h \rangle \geqslant 0,$$

and so

$$\langle \Phi_n(x) - \Phi_n(y), x - y \rangle = \langle F'(x_n)(x - y), x - y \rangle \geqslant 0,$$

i. e., $\Phi_n(x)$ is a monotone operator, so that (Theorem 5.1) $\varphi_n(x)$ is a convex functional. Q. E. D.

Since $\varphi_n(x)$ is convex, equation (13.3) is, by Theorem 9.1, a necessary and sufficient condition for the existence of an absolute minimum point (Lemma 12.1). If, furthermore, the operator

$$\Gamma(x) = [F'(x)]^{-1}$$

exists for any $x \in E$, equation (13.3) is solvable for any n.

The Newton process (13.1) arose here from the need to minimize the functional $f(x)$. Since the gradient $F(x)$ of $f(x)$ vanishes at any minimum point of the latter, the same process may be obtained (see Kantorovitch and Akilov /1/) by approximate solution of the equation

$$F(x) = 0. \qquad (13.4)$$

Let $F(x)$ be twice Gâteaux-differentiable. Taylor's formula (see subsection 4.5) then gives

$$F(x)=F(x_1)+F'(x_1)(x-x_1)+r_2(x_1, x),$$

where x_1 is an initial approximation for the root of equation (13.4). Dropping $r_2(x_1, x)$, we replace equation (13.4) by the equation

$$F(x_1)+F'(x_1)(x-x_1)=0,$$

whose solution x_2 is taken as the second approximation for the root of equation (13.4). If the operator $\Gamma(x_1)=[F'(x_1)]^{-1}$ exists, we can write

$$x_2=x_1-\Gamma(x_1)F(x_1).$$

Repeating these arguments, we again arrive at the iterative process (13.1). The process clearly breaks off if $F(x_n)=0$ for some n. In that case, x_n is a root of equation (13.4) or a minimum point of the functional $f(x)$, provided the latter is convex.

13.2. Relaxation property of the process

We now present a simple sufficient condition for the Newton approximations (13.1) to form a relaxation process, i. e., to satisfy the inequalities $f(x_{n+1})\leqslant f(x_n)$ $(n=1,2,3,\ldots)$.

Lemma 13.2. Let $f(x)$ *be a real functional, defined and thrice Gâteaux-differentiable on a normed space* E, $F(x)=\operatorname{grad} f(x)$. *Let* $\langle F''(x+th)h^2, h\rangle$ *be continuous in* $t\in[0,1]$ *for any* $x, h\in E$. *If*

$$\langle F'(x)h, h\rangle\geqslant m\|h\|^2, \quad m>0, \tag{13.5}$$

$$\langle F''(x)h^2, h\rangle\leqslant N\|h\|^3 \tag{13.6}$$

and the approximations (13.1) satisfy the inequalities

$$N\|x_{n+1}-x_n\|\equiv N\|\Gamma(x_n)F(x_n)\|\leqslant 3m, \tag{13.7}$$

then

$$f(x_{n+1})\leqslant f(x_n).$$

Proof. By Taylor's formula (see subsection 4.5), we may write

$$f(x_{n+1}) - f(x_n) = \langle F(x_n), x_{n+1} - x_n \rangle + \\ + 2^{-1} \langle F'(x_n)(x_{n+1} - x_n), x_{n+1} - x_n \rangle + r_3(x_n, x_{n+1}),$$

where

$$r_3(x_n, x_{n+1}) = 2^{-1} \int_0^1 \langle F''(x_n + t(x_{n+1} - x_n))(x_{n+1} - x_n)^2, \\ x_{n+1} - x_n \rangle (1 - t)^2 \, dt,$$

or

$$f(x_{n+1}) - f(x_n) = \langle F(x_n) + F'(x_n)(x_{n+1} - x_n), x_{n+1} - x_n \rangle - \\ - 2^{-1} \langle F'(x_n)(x_{n+1} - x_n), x_{n+1} - x_n \rangle + r_3(x_n, x_{n+1}).$$

By (13.1), we thus have

$$f(x_{n+1}) - f(x_n) = \\ = - 2^{-1} \langle F'(x_n)(x_{n+1} - x_n), x_{n+1} - x_n \rangle + r_3(x_n, x_{n+1}).$$

But if $x_{n+1} \neq x_n$ then, by inequality (13.5),

$$A = 2^{-1} \langle F'(x_n)(x_{n+1} - x_n), x_{n+1} - x_n \rangle \geqslant \\ \geqslant \frac{m}{2} \| x_{n+1} - x_n \|^2 > 0$$

and, in view of (13.6),

$$| r_3(x_n, x_{n+1}) | \leqslant \frac{1}{6} N \| x_{n+1} - x_n \|^3,$$

whence, by (13.7), we have

$$A^{-1} | r_3(x_n, x_{n+1}) | \leqslant \frac{1}{3} \frac{N}{m} \| x_{n+1} - x_n \| \leqslant 1,$$

i. e.,

$$| r_3(x_n, x_{n+1}) | \leqslant A,$$

and so

$$f(x_{n+1}) - f(x_n) = - A + r_3(x_n, x_{n+1}) \leqslant 0.$$

Q. E. D.

13.3. Convergence of Newton approximations

Theorem 13.1. *Let $f(x)$ be a real functional, defined and thrice Gâteaux-differentiable in a reflexive Banach space E, $F(x) = \operatorname{grad} f(x)$. Let the function $\langle F''(x+th)h^2, h\rangle$ be integrable with respect to t on $[0,1]$ for any $x, h \in E$. If*

$$\langle F'(x)h, h\rangle \geqslant m\|h\|^2, \quad (m>0), \tag{13.5}$$

$$\langle F''(x)h^2, h_1\rangle \leqslant N\|h\|^2\|h_1\|, \tag{13.6'}$$

$$d = \frac{N}{m}\|x_2 - x_1\| < 1 \tag{13.8}$$

and $\Gamma(x)F(x) \equiv [F'(x)]^{-1}F(x)$ is a continuous operator, then the functional $f(x)$ has a unique minimum point x_0, the Newton approximations (13.1) converge to this point, and the rate of convergence is given by

$$\|x_n - x_0\| \leqslant \frac{m}{N}\sum_{i=n-1}^{\infty} d^{2^i}.$$

Proof. From (13.5) it follows, by Corollary 8.2, that $f(x)$ is a strictly convex functional. By (6.4), we have

$$f(x) - f(0) = \int_0^1 \langle F(tx), x\rangle\,dt = \langle F(0), x\rangle +$$
$$+ \int_0^1 \langle F(tx) - F(0), x\rangle\,dt = \langle F(0), x\rangle + \int_0^1 \langle F'(t\tau x)tx, x\rangle\,dt.$$

Hence, by (13.5), it follows that

$$f(x) - f(0) \geqslant \langle F(0), x\rangle + 2^{-1}m\|x\|^2 \geqslant$$
$$\geqslant \|x\|\left(\frac{m}{2}\|x\| - \|F(0)\|\right),$$

i. e., $f(x)$ has the m-property and thus, by Corollary 9.1, there exists a unique minimum point x_0, at which $f(x)$ assumes an absolute minimum.

To prove that the Newton approximations converge to x_0, we consider the quadratic functional (13.2) and write

$$\varphi_n(x_{n+1}) = f(x_n) + \langle F(x_n), x_{n+1} - x_n\rangle +$$
$$+ 2^{-1}\langle F'(x_n)(x_{n+1} - x_n), x_{n+1} - x_n\rangle,$$

whence, by (13.1), we have

$$\varphi_n(x_{n+1}) - f(x_n) = -2^{-1}\langle F'(x_n)(x_{n+1}-x_n), x_{n+1}-x_n\rangle.$$

Using (13.5), we now obtain

$$f(x_n) - \varphi_n(x_{n+1}) \geqslant \frac{m}{2}\| x_{n+1}-x_n\|^2$$

or

$$\| x_{n+1}-x_n\|^2 \leqslant \frac{2}{m}[f(x_n) - \varphi_n(x_{n+1})]. \tag{13.9}$$

But, by (13.5),

$$\begin{aligned}\varphi_n(x_{n+1}) = f(x_n) + \langle F(x_n), x_{n+1}-x_n\rangle + \\ + \frac{1}{2}\langle F'(x_n)(x_{n+1}-x_n), x_{n+1}-x_n\rangle \geqslant \\ \geqslant f(x_n) + \langle F(x_n), x_{n+1}-x_n\rangle + \frac{m}{2}\| x_{n+1}-x_n\|^2\end{aligned}$$

or

$$\varphi_n(x_{n+1}) \geqslant f(x_n) + \langle F(x_n), x_{n+1}-x_n\rangle.$$

Into this inequality we substitute the value of $F(x_n)$ from Taylor's formula (see subsection 4.5):

$$F(x_n) = F(x_{n-1}) + F'(x_{n-1})(x_n - x_{n-1}) + r_2(x_{n-1}, x_n),$$

where

$$\begin{aligned}\langle r_2(x_{n-1}, x_n), x_{n+1}-x_n\rangle = \\ = \int_0^1 \langle F''(x_{n-1} + t(x_n - x_{n-1}))(x_n - x_{n-1})^2, x_{n+1}-x_n\rangle (1-t)\,dt,\end{aligned}$$

so that, by (13.6'),

$$|\langle r_2(x_{n-1}, x_n), x_{n+1}-x_n\rangle| \leqslant \frac{1}{2}N\| x_n - x_{n-1}\|^2 \| x_{n+1}-x_n\|.$$

Now, in view of (13.1), we have

$$F(x_{n-1}) + F'(x_{n-1})(x_n - x_{n-1}) = 0,$$

and thus

$$\begin{aligned}\varphi_n(x_{n+1}) \geqslant f(x_n) + \langle F(x_n), x_{n+1}-x_n\rangle = \\ = f(x_n) + \langle r_2(x_{n-1}, x_n), x_{n+1}-x_n\rangle\end{aligned}$$

or

$$f(x_n) - \varphi(x_{n+1}) \leqslant -\langle r_2(x_{n-1}, x_n), x_{n+1} - x_n \rangle \leqslant$$
$$\leqslant \frac{1}{2} N \| x_n - x_{n-1} \|^2 \| x_{n+1} - x_n \|.$$

It thus follows by (13.9) that

$$\| x_{n+1} - x_n \| \leqslant \frac{N}{m} \| x_n - x_{n-1} \|^2,$$

so that, by induction,

$$\| x_{n+1} - x_n \| \leqslant \frac{m}{N} \left[\frac{N}{m} \| x_2 - x_1 \| \right]^{2^{n-1}} = \frac{m}{N} d^{2^{n-1}}.$$

Consequently, for any positive integers n and p,

$$\| x_{n+p} - x_n \| \leqslant \frac{m}{N} \sum_{k=1}^{p} d^{2^{n+k-2}}, \tag{13.10}$$

i. e., the Newton approximations form a fundamental sequence. Since the space E is complete, we have

$$\lim_{n \to \infty} x_n = y_0.$$

It now follows from (13.1) that

$$\lim_{n \to \infty} \Gamma(x_n) F(x_n) = \lim_{n \to \infty} (x_{n+1} - x_n) = 0,$$

whence, by our hypotheses,

$$\Gamma(y_0) F(y_0) = 0. \tag{13.11}$$

Inequality (13.5) then implies that, for fixed x and arbitrary h,

$$\| F'(x) h \| \geqslant m \| h \|, \quad m > 0,$$

and so the inverse $[F'(x)]^{-1} = \Gamma(x)$ exists and is bounded.* The Newton approximations therefore exist for any n (this has already been used in the proof) and it follows from (13.11) that $F(y_0) = 0$.** Since the functional $f(x)$ is strictly convex, its gradient may vanish

* See, for example, Hille and Phillips /1/, Theorem 2.11.6.

** Note that instead of assuming that $\Gamma(x) F(x)$ is continuous, one may assume that $\|F'(x)\| \leqslant C < +\infty$, since then it follows from (13.1) that $F(x_n) = -F'(x_n)(x_{n+1} - x_n)$; consequently, $\|F'(x_n)(x_{n+1} - x_n)\| \leqslant C \|x_{n+1} - x_n\| \to 0$ as $n \to \infty$, since $x_n \to y_0$. Thus $\| F(x_n) \| \to 0$. But the assumption that $\| F'(x) \| \leqslant C$ implies (Lemma 2.3) that $F(x)$ is continuous, so that $F(y_0) = 0$.

at only one point (Lemma 9.1 and Theorem 9.1), and so $y_0 = x_0$. Thus the approximations (13.1) converge to x_0. Letting $p \to \infty$ in (13.10), we now obtain

$$\| x_0 - x_n \| \leqslant \frac{m}{N} \sum_{i=n-1}^{\infty} d^{2^i}.$$

Q. E. D.

Note that if $f(x)$ has a minimum at $x_0 \in E$, we may drop the assumption in Theorem 13.1 that E is reflexive, requiring instead that (13.5) and (13.6') hold for all $x \in D$, where

$$D = \{x \in E : \|x\| \leqslant r;\ r > \|x_0\|\}.$$

Another alternative to the reflexivity assumption in Theorem (13.11) is to assume that E is the dual of a separable normed space (see Theorem 9.3).

Many propositions are known on convergence of the Newton – Kantorovitch approximations to a solution of the equation $F(x) = 0$ (see Kantorovich and Akilov /1/). However, even when $F(x)$ is a potential operator, the roots of the equation $F(x) = 0$ need not be minimum points of the potential $f(x)$ of $F(x)$ and the same certainly applies to the absolute minimum points of $f(x)$. The situation is otherwise if $F(x)$ is the gradient of a convex potential. Here the roots of the equation $F(x) = 0$, and these alone, are the minimum points of $f(x)$ (see Theorem 9.1 and Lemma 12.1). Thus, if $f(x)$ is a convex functional and $F(x) = \operatorname{grad} f(x)$, any theorem on the convergence of Newton's method to a root of the equation $F(x) = 0$ is at the same time a theorem on the convergence of this method to an absolute minimum point of $f(x)$. We may therefore use various well-known propositions, one of which we now present.*

Theorem 13.2. Suppose the gradient $F(x)$ of a convex thrice Gâteaux-differentiable functional $f(x)$ satisfies the conditions

$$\|F(x_1)\| \leqslant \eta, \ \|\Gamma(x)\| \leqslant B, \ \|F'(x)\| \leqslant K,$$

in some ball $D = \{x \in E : \|x - x_1\| \leqslant r\}$ of a Banach space E. If

$$d = B^2 K \eta < 2$$

and

$$R = B\eta \sum_{k=1}^{\infty} \left(\frac{d}{2}\right)^{2^{k-1}-1} < r,$$

* Mysovskikh /1/; see also Kantorovich and Akilov /1/, p.640, Theorem 5.

then $f(x)$ has a minimum point $x_0 \in D$ and the process (13.1) converges to x_0 at the rate

$$\| x_n - x_0 \| \leqslant B\eta \sum_{k=n}^{\infty} \left(\frac{d}{2}\right)^{2^{k-1}-1}.$$

Proof. We first prove that the Newton approximations x_n lie in the ball D. Indeed,

$$\| x_2 - x_1 \| = \| \Gamma(x_1) F(x_1) \| \leqslant B\eta < r,$$

so that $x_2 \in D$. Proceeding by induction, we assume $x_n \in D$ and show that $x_{n+1} \in D$. Using Taylor's formula (subsection 4.5), we write

$$F(x_{k+1}) = F(x_k) + F'(x_k)(x_{k+1} - x_k) + \\ + \int_0^1 F''(x_k + t(x_{k+1} - x_k))(x_{k+1} - x_k)^2 (1 - t)\, dt$$

and, since by (13.1)

$$F(x_k) + F'(x_k)(x_{k+1} - x_k) = 0,$$

it follows from the assumptions of the theorem that

$$\| F(x_{k+1}) \| \leqslant \frac{1}{2} K \| x_{k+1} - x_k \|^2 \quad (k = 1, 2, \ldots, n-1). \tag{13.12}$$

Consequently,

$$\| F(x_2) \| \leqslant \frac{1}{2} K \| x_2 - x_1 \|^2 \leqslant \frac{1}{2} K B^2 \eta^2,$$

$$\| x_3 - x_2 \| \leqslant \| \Gamma(x_2) F(x_2) \| \leqslant \frac{1}{2} K B^3 \eta^2 = B\left(\frac{d}{2}\right)\eta,$$

$$\| F(x_3) \| \leqslant \frac{1}{2} K \| x_3 - x_2 \|^2 \leqslant \frac{1}{2} K B^2 \left(\frac{d}{2}\right)^2 \eta^2,$$

$$\| x_4 - x_3 \| \leqslant \| \Gamma(x_3) F(x_3) \| \leqslant \frac{1}{2} K B^3 \left(\frac{d}{2}\right)^2 \eta^2 = B\eta \left(\frac{d}{2}\right)^{2^2-1}$$

and, continuing these manipulations, we obtain

$$\| x_{k+1} - x_k \| \leqslant B\eta \left(\frac{d}{2}\right)^{2^{k-1}-1} \quad (k = 1, 2, \ldots, n)$$

Hence,

$$\| x_{n+1} - x_1 \| \leqslant \sum_{k=1}^{n} \| x_{k+1} - x_k \| \leqslant B\eta \sum_{k=1}^{n} \left(\frac{d}{2}\right)^{2^{k-1}-1} < r,$$

i.e., $x_{n+1} \in D$, and thus, by induction, $x_m \in D$ for any m. It now follows from inequality (13.12) that

$$\| x_{n+p} - x_n \| \leqslant \sum_{k=n}^{n+p-1} \| x_{k+1} - x_k \| \leqslant B\eta \sum_{k=n}^{n+p-1} \left(\frac{d}{2}\right)^{2^{k-1}-1}, \qquad (13.13)$$

so that the Newton approximations $\{x_n\}$ comprise a fundamental sequence. Since E is a complete space, it follows that the sequence has a limit $x_0 \in D$.

To complete the proof, we observe that, since $F''(x)$ is bounded in the ball D, it follows that $F'(x)$ satisfies a Lipschitz condition (Lemma 2.3) and is bounded in D, so that $F(x)$ is a continuous operator on D, for it also satisfies a Lipschitz condition in D. Applying the operator $F'(x_0)$ to (13.1), we obtain

$$F(x_n) = F'(x_n)(x_n - x_{n+1}),$$

whence

$$\| F(x_n) \| \leqslant \| F'(x_n) \| \| x_n - x_{n+1} \| \to 0$$

as $n \to \infty$. Consequently,

$$\lim F(x_n) = 0$$

and, since $F(x)$ is continuous on D, we have $F(x_0) = 0$. Since $F(x) = \operatorname{grad} f(x)$ and $f(x)$ is a convex functional, we may assert that x_0 is an absolute minimum point of $f(x)$. Letting $p \to \infty$ in (13.13), we conclude

$$\| x_0 - x_n \| \leqslant B\eta \sum_{k=n}^{\infty} \left(\frac{d}{2}\right)^{2^{k-1}-1}.$$

Q. E. D.

Remark. Chapters III and IV lean heavily on the author's paper /23/.

Chapter V

NONLINEAR EQUATIONS OF HAMMERSTEIN TYPE WITH POTENTIAL OPERATORS AND THEOREMS ON SQUARE ROOTS OF LINEAR OPERATORS

To solve the equation $\Phi(x)=0$ by the variational method, one may follow two approaches. The first concerns the case that the range of Φ is contained in a normed space: one selects a minimizing sequence for the functional $f(x)=\|\Phi(x)\|$. The second approach is to construct a functional whose critical points are the roots or pre-images of the roots of $\Phi(x)$, where $\Phi(x)$ is assumed not to be potential. In this context various properties of potential operators and the propositions of Chapter III on the minima of functionals play an essential role. In this chapter we take the second approach, demonstrating in § 17 its application to equations of Hammerstein type $x=BF(x)$ in various function spaces (Banach and locally convex). To be precise, the equation $\Phi=0$, where $\Phi(x)=x-BF(x)$ in a given space E, is replaced by an equivalent equation $\Psi=0$, where Ψ is a potential operator in some Hilbert space H. The passage from the mapping Φ to the mapping Ψ leans on several propositions on square roots of linear operators in Banach or locally convex spaces. For this reason, §§ 15 and 16 of this chapter are devoted to a study of square roots of linear operators.

In § 15 we present theorems on square roots of bounded operators in Banach or locally convex embeddable spaces, and § 16 contains theorems on square roots of unbounded closed operators in (not necessarily embeddable) Banach spaces. In the first section of this chapter (§14), we establish some propositions on equations with potential operators.

§ 14. GENERAL EXISTENCE-UNIQUENESS THEOREMS

In this section we prove propositions on the existence and uniqueness of solutions of the equation

$$F(x)=y, \tag{14.1}$$

where $F(x)$ is a potential operator defined in a Banach space E, and y is a fixed vector in E^*. We assume either that E is reflexive or that it is the dual of a separable normed space X, $E=X^*$. In the first case, the unit ball $\|x\|\leqslant 1$ is weakly compact,* whereas in the

* See Dunford and Schwartz /1/, p. 68, Theorem 28 and p. 425, Theorem 7.

second the unit ball in X is weakly compact.* We denote the potential of the operator $F(x)$ by $f(x)$, $F(x) = \text{grad}\, f(x)$.

14.1. Existence and uniqueness of solution of the homogeneous equation

Theorem 14.1. Let $f(x)$ *be a real finite Gâteaux-differentiable functional, defined in a reflexive Banach space or in the dual of a separable normed space, and let* $F(x) = \text{grad}\, f(x)$. *If* $f(x)$ *is weakly lower semicontinuous and*

$$\overline{\lim_{R\to\infty}}\, f(x) = +\infty \qquad (R = \|x\|),$$

then the equation $F(x) = 0$ *has a solution.*

This theorem follows from Theorem 9.5, the remark following it, and Theorem 9.1.

Theorem 14.2. Let $f(x)$ *be a real finite Gâteaux-differentiable functional defined in a real reflexive Banach space or in the dual of a normed separable space, weakly lower semicontinuous and satisfying the condition*

$$\langle F(x),\, x\rangle > 0, \qquad F(x) = \text{grad}\, f(x),$$

on some sphere $\|x\| = R > 0$. *Then the equation* $F(x) = 0$ *has a solution* x_0 *such that* $\|x_0\| < R$.

This theorem follows from Theorems 9.1 and 9.8. It suffices, of course, that the functional $f(x)$ with the given properties be defined on the ball $\|x\| \leqslant R$. Note that if $E = X^*$, where X is separable, weak continuity of $f(x)$ should be understood as X-weak continuity.

Remark 14.1. If we replace the requirement that $\langle F(x),\, x\rangle > 0$ by the condition that $\langle F(x),\, x - a\rangle > 0$ on the sphere $\|x - a\| = R > 0$, then there exists a solution x_0 of the equation $F(x) = 0$ such that $\|x_0 - a\| < R$ (see Remark 9.2).

Remark 14.2. Let $U(x)$ be a star-shaped neighborhood relative to a of some point a in a normed space E. Let $f(x)$ be a real Gâteaux-differentiable functional which is nondecreasing along any segment $(x - a) \cap U(a)$ beginning at a; then (see Lemmas 9.2 and 9.3)

$$\langle F(x),\, x - a\rangle \geqslant 0, \quad F(x) = \text{grad}\, f(x). \qquad (14.2)$$

* See Hille and Phillips /1/, section 2.10.

Conversely, if (14.2) holds for any $x \in U(a)$, one can prove that the functional $f(x)$ is nondecreasing along $(x-a) \cap U(a)$. Thus, inequality (14.2) is a necessary and sufficient condition for $f(x)$ to be nondecreasing along $(x-a) \cap U(a)$.

If (14.2) holds, of course, $f(x)$ has a minimum at a and thus $F(a)=0$. The converse does not always hold, i. e., if $f(x)$ has a local minimum at the point a, inequality (14.2) need not hold in a preassigned neighborhood $U(a)$.

If, however, $F(x)$ is a monotone operator, then

$$\langle F(x)-F(a),\ x-a\rangle \geqslant 0,$$

and so, if $f(x)$ has a minimum at a, inequality (14.2) holds in any given neighborhood $U(a)$, since $F(a)=0$. Thus, if $F(x)$ is a monotone potential operator, $F(a)=0$ if and only if inequality (14.2) is valid.

In the sequel we shall need

Definition 14.1. An operator $F(x)$ from a normed space E to E^* is said to be weakly integrable if for any $x \in E$ the integral

$$\int_0^1 \langle F(tx),\ x\rangle\, dt$$

exists.

Theorem 14.3. Let $F(x)$ be a monotone potential operator defined in a real reflexive Banach space, and satisfying the following conditions:

1. *$F(x)$ is weakly integrable;*
2. *$\langle F(x), x\rangle \geqslant \|x\|\gamma(\|x\|)$, where $\gamma(t)$ is integrable on $[0, R]$ for any $R>0$;*
3. $$\overline{\lim_{R\to\infty}} \int_0^R \gamma(t)\, dt = c > 0.$$

Then the equation $F(x)=0$ has a solution.

This theorem follows from Theorems 9.1 and 9.7. Alternatively, conditions 2 and 3 of our theorem imply that $\langle F(x),\ x\rangle > 0$ on some sphere $\|x\|=r>0$, and so Theorem 14.3 follows from Theorem 14.2.

Theorem 14.4. Let $F(x)$ be a strictly monotone potential operator, defined in a real reflexive Banach space and satisfying the inequality

$$\langle F(x), x\rangle > 0$$

on some sphere $\|x\|=r>0$. Then the equation $F(x)=0$ has a unique solution x_0 and $\|x_0\|<r$.

P r o o f. According to Theorem 8.4, the fact that $F(x)$ is monotone implies that its potential $f(x)$ ($F(x) = \operatorname{grad} f(x)$) is weakly lower semicontinuous and thus, by Theorem 14.2, the equation has a solution x_0 such that $\|x_0\| < r$. The uniqueness of the solution follows from the following proposition (see Remark 9.1):

L e m m a 14.1. If $F(x)$ is a strictly monotone operator from a normed space E to E^*, the equation

$$F(x) = y \tag{14.1}$$

has at most one solution.

T h e o r e m 14.5. Let $F(x) = \operatorname{grad} f(x)$ be a strictly monotone and weakly integrable operator defined in a reflexive Banach space and satisfying the condition

$$\langle F(x) - F(0),\ x \rangle \geqslant \|x\| \gamma(\|x\|),$$

where $\gamma(z)$ is integrable on $[0, R]$ for any $R > 0$, and the function

$$C(R) = \int_0^R \gamma(z)\, dz$$

satisfies the inequality

$$C(R_0) > R_0 \|F(0)\|$$

for some $R = R_0 > 0$. Then the equation $F(x) = 0$ has a unique solution x_0 and $\|x_0\| < R_0$.

This theorem follows from Theorems 9.1, 9.9 and Lemma 14.1.

14.2. Existence and uniqueness of solution of the general equation

T h e o r e m 14.6. Let $F(x)$ be a monotone potential operator defined in a real reflexive Banach space E, satisfying the condition

$$\lim_{R \to \infty} \frac{\langle F(x), x \rangle}{R} = +\infty \qquad (R = \|x\|), \tag{14.3}$$

and such that $\langle F(tx), x \rangle$ is continuous in t for any $x \in E$.

Then equation (14.1) has a solution for any $y \in E^*$, so that F is a surjective mapping $E \to E^*$. If F is strictly monotone, it is a bijective mapping $E \to E^*$.

Proof. Let $f(x)$ be the potential of $F(x)$ and set $g(x)=\langle y, x\rangle$, where $y \in E^*$. Set

$$\varphi(x)=f(x)-g(x).$$

Then

$$\operatorname{grad} \varphi(x)=F(x)-y.$$

Proceeding as in the proof of Theorem 9.6, we get

$$f(x)-f(0) \geqslant \frac{1}{2}\|x\|\left(\frac{\langle F(t_1 x), t_1 x\rangle}{t_1\|x\|}-\|F(0)\|\right),$$

where $1/2 \leqslant t_1 \leqslant 1$, whence

$$\varphi(x)-\varphi(0) \geqslant \frac{1}{2}\|x\|\left(\frac{\langle F(t_1 x), t_1 x\rangle}{t_1\|x\|}-\|F(0)\|\right)-\langle y, x\rangle \geqslant$$
$$\geqslant \frac{1}{2}\|x\|\left(\frac{\langle F(t_1 x), t_1 x\rangle}{t_1\|x\|}-\|F(0)\|-2\|y\|\right).$$

Now, by (14.3), there exists a sphere

$$S_r=\{x \in E:\|x\|=r>0\},$$

on which $\varphi(x)>\varphi(0)$. By assumption, the ball

$$D_r=\{x \in E:\|x\| \leqslant r\}$$

is weakly closed and weakly compact, and $\varphi(x)$ is weakly lower semicontinuous, as the difference of $f(x)$, which is a weakly lower semicontinuous functional (by Theorem 8.4), and the weakly semicontinuous functional $g(x)=\langle y, x\rangle$. Thus $\varphi(x)$ has the m-property, i. e., it has a local minimum at some interior point $x_0 \in D_r$ and so, by Theorem 9.1,

$$\operatorname{grad} \varphi(x_0)=0$$

or

$$F(x_0)-y=0.$$

We have thus proved that the mapping $F: E \rightarrow E^*$ is surjective. Finally, if F is strictly monotone, Lemma 14.1 entails that the mapping $F: E \rightarrow E^*$ is bijective. Q. E. D.

Theorem 14.7. Let $F(x)$ be a weakly integrable monotone potential operator, defined in a real reflexive Banach space and satisfying the condition

$$\langle F(x),\ x\rangle \geqslant \|x\|\gamma(\|x\|),$$

where $\gamma(z)$ is integrable on $[0, R]$ for any $R>0$ and

$$\overline{\lim_{R\to\infty}}\frac{1}{R}\int_0^R \gamma(z)\,dz = +\infty. \tag{14.4}$$

Then the mapping $F: E\to E^*$ is surjective. If F is strictly monotone, then $F: E\to E^*$ is a bijective mapping.

Proof. Set $F(x) = \operatorname{grad} f(x)$. Then it follows from our assumptions and Theorem 8.4 that the functional $f(x)$ is weakly lower semicontinuous, so that, for any fixed $y\in E^*$, the functional

$$\varphi(x) = f(x) - \langle y,\ x\rangle$$

is weakly lower semicontinuous. Furthermore, by (6.4),

$$f(x) - f(0) = \int_0^1 \langle F(tx),\ tx\rangle\frac{dt}{t},$$

so that, by assumption,

$$f(x) - f(0) \geqslant \int_0^1 \|x\|\gamma(t\|x\|)\,dt = \int_0^R \gamma(z)\,dz, \quad R = \|x\|.$$

Hence

$$\varphi(x) - \varphi(0) \geqslant$$
$$\geqslant \int_0^R \gamma(z)\,dz - \langle y,\ x\rangle \geqslant R\left(\frac{1}{R}\int_0^R \gamma(z)\,dz - \|y\|\right), \tag{14.5}$$

so that, by (14.4), there exists $R = r > 0$, such that for all points of the sphere $S_r = \{x\in E : \|x\| = r\}$ we have

$$\varphi(x) > \varphi(0).$$

As in the proof of Theorem 14.6, this inequality implies the assertions of our theorem.

Remark 14.3. If we replace the inequality $\langle F(x), x\rangle \geqslant \|x\|\gamma(\|x\|)$ in Theorem 14.7 by $\langle F(x)-F(0), x\rangle \geqslant \|x\|\gamma(\|x\|)$, we obtain

$$\varphi(x)-\varphi(0) \geqslant R\left(\frac{1}{R}\int_0^R \gamma(z)\,dz-\|y\|-\|F(0)\|\right),$$

and so all the assertions of Theorem 14.7 remain valid.

Remark 14.4. If we replace condition (14.4) in Theorem 14.7 by

$$\lim_{z\to+\infty} \gamma(z)=+\infty, \tag{14.6}$$

the statement of the theorem remains valid. Indeed, it follows from (14.6) that, whatever the value of $\|y\|$, there exists $r>0$ such that $\gamma(z)>c\geqslant 2\|y\|$ for any $z>r$. Set

$$A=\int_0^r \gamma(z)\,dz.$$

Then, for $R>r$,

$$\frac{1}{R}\int_0^R \gamma(z)\,dz \geqslant \frac{A}{R}+2\|y\|\left(1-\frac{r}{R}\right).$$

Assume now that R is so large that, for $\|y\|>0$,

$$\frac{1}{R}|A|<\frac{1}{2}\|y\|,\quad \frac{r}{R}<\frac{1}{4}.$$

Then

$$\frac{1}{R}\int_0^R \gamma(z)\,dz>\|y\|.$$

Hence, by (14.5), there exists a sphere $\|x\|=R_0>r$ on which

$$\varphi(x)>\varphi(0). \tag{14.7}$$

A similar argument yields (14.7) for $y=0$. Theorem (14.7) now follows from inequality (14.7).

Remark 14.5. Theorems 14.6 and 14.7 furnish sufficient conditions for a potential operator $F: E\to E^*$ to be bijective. In that case, the inverse operator $G: E^*\to E$ certainly exists, and it is also bijective. If we assume in addition that $F(x)$ is continuous and

$$\| F(x) - F(y) \| \geqslant \gamma_0(\| x - y \|), \tag{14.8}$$

where $\gamma_0(t)$ is a continuous increasing function defined for $t \geqslant 0$ and vanishing at zero, then the mapping $F: E \to E^*$ is a homeomorphism. Indeed,* $\gamma_0(t)$ is continuous and increasing, the inverse function γ_0^{-1} exists, so that, by (14.8),

$$\| x - y \| \leqslant \gamma_0^{-1}(\| F(x) - F(y) \|).$$

Let $G: E^* \to E$ be the inverse of F, $F(x) = u$, $F(y) = v$. Then the last inequality implies

$$\| G(u) - G(v) \| \leqslant \gamma_0^{-1}(\| u - v \|),$$

i. e., $G: E^* \to E$ is continuous. Thus $F: E \to E^*$ is a homeomorphism, provided $F(x)$ is a continuous operator.

A sufficient condition for (14.8) to hold is

$$\langle F(x) - F(y), x - y \rangle \geqslant \gamma_0(\| x - y \|) \| x - y \|.$$

14.3. Smooth operators

An operator $F(x)$ is said to be smooth if it has a linear Gâteaux differential $DF(x, h)$. In this subsection we shall assume that the operator $F(x)$ in equation (14.1) is a smooth potential operator.

Theorem 14.8. Let $F(x)$ *be a smooth potential operator, defined in a reflexive Banach space* E, *such that the following conditions hold:*

1. $\langle DF(y + tx, x), x \rangle$ *is an integrable function of* t *on* $[0,1]$ *for any* $x, y \in E$.

2. $\langle DF(x, h), h \rangle \geqslant \gamma(\|x\|) \|h\|^2$ *for any* $x, h \in E$, *where* $\gamma(z)$ *is a positive nonincreasing function for* $z \geqslant 0$, *integrable on* $[0, R]$ *for any* $R > 0$, *such that for some* $R > 0$,

$$\frac{1}{R} \alpha(R) \equiv \frac{1}{R} \int_0^R z\gamma(z)\, dz > \| F(0) \|. \tag{14.9}$$

Then the equation $F(x) = 0$ *has a unique solution.*

This theorem follows from Theorems 9.1 and 9.10.

* See, for example, Vainberg /13/.

Theorem 14.9. Let the conditions of Theorem 14.8 hold, except that inequality (14.9) is replaced by the condition

$$\overline{\lim_{R\to\infty}} \frac{1}{R}\alpha(R) = +\infty. \tag{14.10}$$

Then equation (14.1) has a unique solution for any $y \in E^*$, and the mapping $F: E \to E^*$ is bijective.

Proof. Let $f(x)$ be the potential of $F(x)$, $F(x) = \operatorname{grad} f(x)$. It follows from our assumptions and Theorem 8.5 that the functional $f(x)$ is weakly lower semicontinuous. Hence, for any $y \in E^*$, the functional

$$\varphi(x) = f(x) - \langle y, x\rangle$$

is weakly lower semicontinuous, since $\langle y, x\rangle$ is weakly continuous. Using (9.7), we now get

$$\begin{aligned} &\varphi(x) - \varphi(0) \geqslant \\ &\geqslant R\left(\frac{1}{R}\int_0^R z\gamma(z)\,dz - \|F(0)\| - \|y\|\right), \quad R = \|x\|. \end{aligned}$$

Hence, by (14.10), there exists $R = r > 0$ such that

$$\varphi(x) > \varphi(0)$$

on the sphere $\|x\| = r$. As in the proof of Theorem 14.6, this implies the existence of a vector $x_0\,(\|x_0\| < r)$ for which $\operatorname{grad}\varphi(x_0) = 0$, or

$$F(x_0) - y = 0.$$

Since, by assumption, $\langle DF(x, h), h\rangle > 0$ for $h \neq 0$, the Lagrange formula (2.4) gives

$$\begin{aligned} &\langle F(x_2) - F(x_1),\ x_2 - x_1\rangle = \\ &= \langle DF(x_1 + \tau(x_2 - x_1),\ x_2 - x_1),\ x_2 - x_1\rangle > 0, \end{aligned} \tag{14.11}$$

i. e., $F(x) - y$ is a strictly monotone operator so that by Lemma 14.1, x_0 is the unique solution of (14.1). Q. E. D.

Remark 14.6. If we replace the upper limit in Theorem 14.9 by the limit and assume in addition that $F(x)$ is continuous, then the mapping $F: E \to E^*$ is a homeomorphism. Indeed, as $F(x)$ is bijective, it has an inverse $G: E^* \to E$. We need only prove that G is continuous.

To do this, consider any ball

$$D_\rho = \{x \in E : \|x\| \leqslant \rho,\ \rho > 0\}.$$

By the hypotheses of Theorem 14.9 (see condition 2 of Theorem 14.8), it follows from (14.11) that for any $x_1,\ x_2 \in D_\rho$

$$\langle F(x_2) - F(x_1),\ x_2 - x_1 \rangle \geqslant \gamma(\| x_1 + \tau(x_2 - x_1)\|)\| x_2 - x_1 \|^2$$

and, in view of the inequality $\|(1-\tau)x_1 + \tau x_2\| \leqslant (1-\tau)\rho + \tau\rho = \rho$, we have

$$\gamma(\| x_1 + \tau(x_2 - x_1)\|) \geqslant \gamma(\rho).$$

Setting, $\gamma(\rho)\|x_2 - x_1\| = \gamma_\rho(\|x_2 - x_1\|)$, we have

$$\langle F(x_2) - F(x_1),\ x_2 - x_1 \rangle \geqslant \gamma_\rho(\| x_2 - x_1 \|)\| x_2 - x_1 \|,$$

where $\gamma_\rho(z)$ is an increasing continuous function for $z \in [0,\ 2\rho]$, $\gamma_\rho(0) = 0$ or

$$\langle F(x_2) - F(x_1) \| \geqslant \gamma_\rho(\| x_2 - x_1 \|). \tag{14.12}$$

Now, as in the proof of Theorem 9.10 we see that for $R = \|x\|$

$$\langle F(x),\ x \rangle \geqslant \langle F(0),\ x \rangle + R\int_0^R \gamma(z)\,dz \geqslant$$
$$\geqslant \langle F(0),\ x \rangle + \int_0^R z\gamma(z)\,dz \geqslant R\left(\frac{1}{R}\int_0^R z\gamma(z)\,dz - \| F(0) \|\right),$$

or

$$\| F(x) \| \geqslant \frac{1}{R}\int_0^R z\gamma(z)\,dz - \| F(0) \|,$$

i. e., if $\|x\| \to +\infty$ then also $\|F(x)\| \to +\infty$. Thus, for any ball $D_r^* = \{y \in E^* : \| y \| \leqslant r\}$ there exists a ball D_ρ such that if $x \in D_\rho$ then $F(x) \in D_r^*$. Hence, using (14.12) and the existence of the inverse γ_ρ^{-1}, we have

$$\| G(y_2) - G(y_1) \| \leqslant \gamma_\rho^{-1}(\| y_2 - y_1 \|),$$

where $y_1 = F(x_1)$, $y_2 = F(x_2)$. Thus the inverse G is continuous.

14.4. Case of homogeneous operators

Let $F(x)$ be a strongly potential compact operator, defined in a real Hilbert space H, which is positive-homogeneous of degree $k > 0$. Then (see subsection 9.6, Example 9.3) if its potential

$$f(x) = \frac{1}{k+1}(F(x), x)$$

takes on negative values (for example $F(x) \not\equiv 0$, $F(-x) = F(x)$) and $k \neq 1$, the equation

$$x + F(x) = 0 \tag{14.13}$$

has at least one nontrivial solution.

In fact, we have seen that in this case equation (9.24) has nontrivial solutions corresponding to negative values of the parameter λ. Let x_0 be such a solution, corresponding to $\lambda_0 < 0$, i. e.,

$$F(x_0) = \lambda_0 x_0.$$

Setting $x_0 = tz$, where $t > 0$, we obtain

$$t^k F(z) = t\lambda_0 z,$$

or

$$F(z) = -z$$

for $t^{1-k} = -1/\lambda_0$, i. e., $z + F(z) = 0$ and $z \neq 0$.

Note also that apart from this nontrival solution z equation (14.13) has the trivial solution $z = 0$.

§ 15. SQUARE ROOT OF LINEAR OPERATORS IN VECTOR SPACES

15.1. Preliminary remarks

The problem of extracting the square root of an operator B arises when the variational method or the method of monotone operators (see §§ 17 and 19) is used in investigating a nonlinear equation of Hammerstein type

$$x = BF(x),$$

where F is a nonlinear operator from a space X to a space Y and B a linear operator from Y to X. The following results are due to the author.* Let B be a completely continuous integral operator from $L^q(G)$ to $L^p(G)$, where $p > 2$, $p^{-1} + q^{-1} = 1$ and G is a set of finite measure in euclidean n-space, such that the restriction of B_H of B to $L^2(G)$ is a positive selfadjoint operator, so that**

$$B_H x = \sum_{k=1}^{\infty} \frac{(x, l_k)}{\lambda_k} l_k \quad (x \in L^2,\ l_k \in L^2,\ \lambda_k > 0),$$

where l_k are the eigenfunctions of B_H corresponding to eigenvalues λ_k. Then the following assertions hold. For each k, we have $l_k \in L^p$, and the positive square root

$$B_H^{1/2} x = \sum_{k=1}^{\infty} \frac{(x, l_k)}{\sqrt{\lambda_k}} l_k \quad (x \in L^2)$$

has a continuous extension A from L^q to L^2. This extension may be written

$$Av = \sum_{k=1}^{\infty} \frac{\langle v, l_k \rangle}{\sqrt{\lambda_k}} l_k \quad (v \in L^q),$$

where the series converges in the L^2-norm,

$$\langle v, l_k \rangle = \int_G v(s) l_k(s)\, ds \quad (v \in L^q,\ l_k \in L^p)$$

and the adjoint A^* of A may be written

$$A^* u = \sum_{k=1}^{\infty} \frac{(u, l_k)}{\sqrt{\lambda_k}} l_k \quad (u \in L^2),$$

where the series converges in the L^p-norm, and so

$$A^* A = B. \qquad (15.1)$$

The author has proved the same result for completely continuous operators $B: L^q \to L^p$ such that B_H is a quasidefinite selfadjoint operator in L^2. These results have been extended by Engel'son /1/ to the case of reflexive locally convex spaces. Similar propositions have been

* Vainberg /6,9/.

** See, for example, Akhiezer and Glazman /1/.

established for Banach spaces by Krasnosel'skii and Sobolev /1/. Equality (15.1) has been proved in various other situations by Vainberg and Engel'son /1/, Vainberg and Lavrent'ev /1, 2/, Krasnosel'skii and Krein /1/, Sobolev /1/ and others (see Krasnosel'skii, Zabreiko et al. /1/ for a bibliography).

In this and the following section we prove several theorems concerning (15.1) for bounded operators in linear topological spaces and for unbounded operators $B: Y \to X$ in Banach spaces. We shall confine ourselves to the case that the operator B is defined in a reflexive space with range in the dual space, although reflexivity may be dropped in some of the propositions.

15.2. Some concepts in topology of vector spaces

In order to establish equalities such as (15.1), we shall need some fairly standard concepts.

Let X and Y be linear topological spaces. X is said to be embedded* in Y if there exists a continuous linear operator J (the embedding operator) mapping X in one-to-one fashion onto a linear subspace $Y_1 \subset Y$. We shall deal only with the case that each $x \in X$ is also in Y and J is the identity operator.

If X and Y are normed spaces, the embedding operator J is bounded:

$$\| Jx \|_Y \leqslant a \| x \|_X, \quad a = \text{const.} \tag{15.2}$$

In case J is the identity operator, inequality (15.2) assumes a simpler form:

$$\| x \|_Y \leqslant a \| x \|_X. \tag{15.3}$$

Let E be a real separable locally convex space and E' the space of real continuous linear functionals, endowed with some topology τ. We observe that the γ-topology of the space $E' = \mathscr{L}(E, R)$, where $R = (-\infty, +\infty)$ may be defined in the following way (see subsection 3.4), using the concept of the polar of a set.

Let $\langle y, x \rangle$ (or $\langle x, y \rangle$) denote the value of a linear functional $y \in E'$ at a vector $x \in E$ and let $\mathfrak{M}$ be a set in E. The set of vectors $y \in E'$ such that

$$|\langle y, x \rangle| \leqslant 1$$

* This concept is due to Sobolev /1/. See also Smirnov /1/ and Mikhlin /3/.

for any $x \in \mathfrak{M}$ is called the polar* of $\mathfrak{M}$ in E, denoted by $\mathfrak{M}^0$. The following properties are immediate.

1. The polar of any set is a symmetric convex set.
2. If $\mathfrak{M} \subset \mathfrak{N}$ then $\mathfrak{N}^0 \subset \mathfrak{M}^0$.
3. If $\mathfrak{M} \subset \lambda\mathfrak{N}$ then $\mathfrak{N}^0 \subset \lambda\mathfrak{M}^0$.
4. $(\lambda\mathfrak{M}^0) = \frac{1}{\lambda}\mathfrak{M}^0$ for any scalar $\lambda \neq 0$.
5. $\left(\bigcup_\alpha \mathfrak{M}_\alpha\right)^0 = \bigcap_\alpha \mathfrak{M}_\alpha^0$.
6. $(\overline{\mathfrak{M}})^0 = \mathfrak{M}^0$.
7. $(\Gamma\overline{\mathfrak{M}}^0) = \mathfrak{M}^0$ where $\Gamma\mathfrak{M}$ is the convex circled hull of $\mathfrak{M}$:

$$\Gamma\mathfrak{M} = \left\{ x \in E : x = \sum_{k=1}^{n} \lambda_k x_k,\ x_k \in \mathfrak{M},\ \sum_{k=1}^{n} |\lambda_k| = 1 \right\}.$$

8. If $\mathfrak{M}$ is bounded, its polar $\mathfrak{M}^0$ is an absorbing set.

Now let γ be a system of bounded subsets of E, closed under finite unions, such that the union of all sets in γ is dense in E and $\mathfrak{M} \in \gamma$ implies $\lambda\mathfrak{M} \in \gamma$ for any $\lambda > 0$. Now let $\{V\}$ be a system of convex neighborhoods on the real line $R = (-\infty, +\infty)$, i. e., $V_n = \left[-\frac{1}{n}, +\frac{1}{n}\right]$, $n = 1, 2, 3, \ldots$. Then (see subsection 3.4) the γ-topology of the space $E' = \mathscr{L}(E, R)$ is defined by the system of neighborhoods of zero $W(\mathfrak{M}, V_n)$, where $\mathfrak{M} \in \gamma$.

This topology is locally convex. By the definition of the polar, $W(\mathfrak{M}, V_1) = \mathfrak{M}^0$, so that the polars of all sets in γ comprise a fundamental system of neighborhoods of zero in the γ-topology of E'.

If γ comprises all the bounded sets in E, the corresponding γ-topology on E' is denoted by $\beta(E', E)$ and known as the strong topology of E'. Endowed with the topology $\beta(E', E)$, the space E' is known as the strong dual of E.

We need some additional concepts (Bourbaki /3/).

Let τ_1 and τ_2 be two given topologies, determined respectively by neighborhood systems $W_1(x)$ and $W_2(x)$, $x \in E$. If $W_1(x) \subset W_2(x)$ for any $x \in E$, the topology τ_2 is said to majorize the topology τ_1. If in addition $W_1(x) \neq W_2(x)$, then τ_2 is said to be stronger than τ_1 (or, τ_1 is weaker than τ_2).

Given a topology τ in a space E, determined by neighborhood system $W(x)$ for any $x \in E$, let M be a subset of E. The set M may be converted into a topological space with the aid of the neighborhood system $V(x) = M \cap W(x)$ for each $x \in M$. The topology of M so constructed is said to be induced by the topology of E [it is also

* If the spaces X and Y are paired by some bilinear form, the polar of a set $\mathfrak{M} \subset X$ in Y is defined in the same way. See Bourbaki /1/.

known as the relative topology]. If M also has another topology τ_1, we shall say that the topologies of M and E are compatible if the topology of M majorizes the topology induced in M by that of E. In this case, the identity operator from M to E is clearly continuous, and thus M can be embedded in E. Conversely, if the identity operator from $M \subset E$ to E is continuous, the topologies of M and E are compatible.

15.3. Embeddable spaces*

E will now be a real separable locally convex space.

Definition 15.1. The space E is said to be embeddable if there exists a Hilbert space H such that:

1. E is dense in H, H is dense in E', and the topologies of E and H are compatible (see subsection 15.2).

2. If $(y, x) = \langle z, x\rangle$ for any $x \in E$, where $y \in H$ and $z \in E'$, then $z = y$.

Remark 15.1. Note that if E is contained in H and their topologies are compatible, then every subset of the intersection of E with the ball $(x, x) \leqslant r^2$ in H (r is any positive number) contains some neighborhood of zero in the topology of E. Thus the identity operator from E to H is continuous, and the space E is embeddable in H if the conditions of Definition 15.1 hold. Naturally, instead of assuming that the topologies of E and H are compatible, we may assume that the identity operator from E to H is continuous.

*Lemma 15.1.*** Let E be an embeddable space. Then:*

1. The space H is embeddable in E'.

2. The second strong dual is contained in H.

Proof. Let y be a fixed vector in H and x an arbitrary vector in H. Then the inner product (y, x), as a continuous functional on H, is both a linear and continuous functional on E, since by assumption the topologies of E and H are compatible. Indeed, since the functional (y, x) is continuous at the zero of H, it follows that for any $\varepsilon > 0$ there exists a ball

$$D = \{x \in H : \|x\| \leqslant r\}$$

such that $|(y, x)| < \varepsilon$ for any $x \in D$. But, since the topologies of E and H are compatible, there exists a neighborhood V of zero in E such that $V \subset D \cap E$. Thus $|(y, x)| < \varepsilon$ for any $x \in V$, i.e., (y, x) is continuous at zero in E. Since (y, x) is a continuous linear

* In this and the following subsections we present some results from Vainberg and Engel'son /1/ (see also Engel'son /4/).

** Engel'son /4/.

functional on E, there exists a unique vector $z \in E'$ such that

$$\langle z, x \rangle = (y, x) \quad (\forall x \in E).$$

This equality sets up a linear mapping of H onto some linear subset in E', and since by Definition 15.1 this implies that $z = y$, it follows that this mapping is the identity. The inverse mapping is thus single-valued, because if $z = 0$ then $(y, x) = \langle z, x \rangle = 0$ for all $x \in E$, and since E is dense in H, it follows that $y = 0$.

We claim that the topologies of H and E' are compatible. Let V be a neighborhood of zero in E'. Then V contains the polar M^0 of some bounded set $M \subset E$. Since the topologies of E and H are compatible, the intersection of the unit ball

$$D = \{x \in H \colon \|x\| \leqslant 1\}$$

with E contains a neighborhood $U \subset E$. Hence, by the definition of a bounded set, there exists an $\alpha > 0$ such that $\lambda M \subset U \subset D \cap E$ if $|\lambda| \leqslant \alpha$. Thus, by property 3 of the polar,

$$\lambda (D \cap E)^0 \subset M^0.$$

But since $D \cap E \subset D \subset H$ and $D^0 = D$, it follows that

$$D = D^0 \subset (D \cap E)^0 \subset H,$$

whence

$$\lambda D \subset \lambda (D \cap E)^0 \cap H \subset M^0 \cap H,$$

i. e., the intersection $M^0 \cap H$, hence also $V \cap H$, contains a neighborhood of zero $\alpha D \subset H$. We have shown that the topologies of H and E' are compatible, so that the identity mapping of $H \subset E'$ into E' is continuous and H is embeddable in E'.

In order to prove the second part of the lemma, we fix a vector $x \in E''$ and consider the continuous linear functional $\langle y, x \rangle$ on $E' (y \in E')$. Since H is embedded in E', as has been shown, $\langle y, x \rangle$ is a continuous linear functional on H, so there exists a unique vector $u \in H$ such that $\langle y, x \rangle = (y, u)$ for any $y \in H$. Now, using this equality, we identify the vector $x \in E''$ with $u \in H$; this gives the identity mapping of E'' into H, and it is linear. It has a single-valued inverse, because if $u = 0$ then $\langle y, x \rangle = (y, 0) = 0$ for any $y \in H$, and, since H is dense in E', x is the zero vector of E''. Thus $E'' \subset H$. Q. E. D.

For Banach spaces, we have the following

Definition 15.2. E is an embeddable Banach space if there exists a Hilbert space H such that:

1. E is dense in H, H dense in the dual E^*.
2. The bilinear functional $\langle y, x\rangle$, $y \in E^*$ and $x \in E$, coincides with the inner product in H for $y \in H$.

Lemma 15.1 may be stated as follows for Banach spaces.

Lemma 15.2. Let E be an embeddable Banach space. Then H is embeddable in the dual E^* and the second dual E^{**} is contained in H.

15.4. Examples of embeddable spaces

The simplest example of an embeddable Banach space is $L^p(G)$, where $p>2$ and mes $G<\infty$. For this example, $H = L^2(G)$.

We now give examples* of locally convex separable (not necessarily normed) embeddable spaces.

Example 15.1. Let D be the set of infinitely differentiable functions $\varphi(x)$ with compact support, defined on euclidean n-space R^n. This is a linear set, with a topology introduced as follows:

Let $\{\varepsilon_i\}$ be a decreasing sequence of positive numbers converging to zero, $\{m_i\}$ an increasing sequence of positive integers and $V(\{\varepsilon_i\}, \{m_i\})$ the set of all $\varphi(x) \in D$ such that $|D^p\varphi(x)| \leqslant \varepsilon_i$ for all $p \leqslant m_i$, $\|x\| \geqslant i$ $(i = 0, 1, 2, \dots)$, where $D^p\varphi(x)$ is the partial derivative of order p with respect to $x_1, x_2, \dots, x_n$. As a fundamental system of neighborhoods of zero in D, we take the set of all $V(\{\varepsilon_i\}, \{m_i\})$, for all possible sequences $\{\varepsilon_i\}$ and $\{m_i\}$. D becomes a separable locally convex space with no countable fundamental system of neighborhoods of zero (see Schwartz /1/, p. 68). This space cannot be normed. Each bounded set in D is determined by a compact subset $\omega \subset R^n$ and an increasing sequence $\{M_i\}$, being the set of all functions $\varphi(x) \in D$ with support in ω such that $|D^p\varphi(x)| \leqslant M_i$ for all $p \leqslant i$ and $x \in R^n$.

We claim that D is an embeddable space. In fact, every function $\varphi(x) \in D$ is square summable; thus every function $\psi(x) \in L^2(R^n)$ determines a linear functional

$$l(\varphi) = \int_{R^n} \psi(x)\,\varphi(x)\,dx \quad (\forall \varphi \in D),$$

* Engel'son /4, 5/.

so that ψ can be identified with some distribution $T \in D'$. Since D is dense in D' (see Schwartz /1/, p. 75), it follows that $L^2(R^n)$ is dense in D'. Now, if $\psi \in L^2(R^n)$ and $l(\varphi) = 0$ for all $\varphi \in D$, then $\psi(x) = 0$ almost everywhere (see Gel'fand and Shilov /1/, Chapter II, § 1, subsection 5). Thus D is dense in $H = L^2(R^n)$.

We now prove that the topologies of D and $H = L^2(R^n)$ are compatible. To this end, we consider an arbitrary ball $D_\varepsilon = \{\psi \in L^2(R^n) : \|\psi\| \leqslant \varepsilon\}$ and show that $D_\varepsilon \cap D$ contains a neighborhood of zero $U \subset D$. Indeed, we may take U to be

$$U = U(\{\varepsilon_i\}, \{m_i\}); \; m_i = i, \; \varepsilon_i = \varepsilon 2^{-(i+1)/2} (\text{mes}\, R_{i,\, i+1})^{-1/2},$$

where $i = 0, 1, 2, \ldots$, i. e., the set of all functions $\varphi \in D$ such that

$$|D^p \varphi(x)| \leqslant \varepsilon_i \qquad (\forall p \leqslant i, \; \|x\| \geqslant i).$$

Denote by $R_{i,\, i+1}$ the region of R^n contained between two spheres of radii i and $i+1$ with centers at zero. For every $\varphi \in U$, we have

$$|\varphi(x)| \leqslant \varepsilon \cdot 2^{-(i+1)/2} (\text{mes}\, R_{i,\, i+1})^{-1/2}, \; \|x\| \geqslant i, \; i = 0, 1, 2, \ldots$$

and, since $\varphi \in L^2(R^n)$, it follows that for any $\varphi \in U$

$$\int_{R^n} \varphi^2(x)\, dx = \sum_{i=0}^{\infty} \int_{R_{i,\, i+1}} \varphi^2(x)\, dx \leqslant$$

$$\leqslant \sum_{i=0}^{\infty} \int_{R_{i,\, i+1}} \frac{\varepsilon^2}{2^{i+1}\, \text{mes}\, R_{i,\, i+1}}\, dx = \varepsilon^2 \sum_{i=0}^{\infty} \frac{1}{2^{i+1}} = \varepsilon^2.$$

Hence $U \subset D_\varepsilon \cap D$, so that the topology of D majorizes the relative topology of D in $L^2(R^n)$. We have thus shown that D is an embeddable space.

Example 15.2. Let D_ω be the set of infinitely differentiable functions with compact support in some compact set $\omega \subset R^n$. This set is a reflexive space (F) (see Schwartz /1/), with a fundamental system of neighborhoods of zero comprising the sets $U = U(m, \varepsilon)$ of all functions $\varphi \in D_\omega$ such that

$$|D^p \varphi(x)| \leqslant \varepsilon \qquad (p \leqslant m, \; x \in \omega).$$

The bounded sets in D_ω are defined as in D. By the method of Example 15.1, one can prove that D_ω is an embeddable space, with $H = L^2(\omega)$.

Example 15.3. In euclidean m-space E^m, we consider a sequence of functions $\{M_p(x)\}$ defined for all $x = (x_1, x_2, \ldots, x_m) \in E^m$, taking finite or infinite values, and having the following properties: at all points where they are finite they are continuous, and

$$1 \leqslant M_0(x) \leqslant M_1(x) \leqslant \ldots \leqslant M_p(x) \leqslant \ldots$$

Now let $\{\varphi(x)\}$ be a linear system of infinitely differentiable functions for which the product $M_p(x) D^q \varphi(x)$ (where $|q| \leqslant p$ ($|q| = q_1 + q_2 + \ldots + q_m$, q_i is the order of the partial derivative of $\varphi(x)$ with respect to x_i) is continuous and bounded on E^m. Of course, this means that necessarily $D^q\varphi(x) \equiv 0$ at all points $x \in E^m$ where $M_p(x)$ is infinite-valued. This system* can be made into a topological space $K\{M_p\}$ by introducing a countable family of norms

$$\|\varphi\|_p = \sup_{|q| \leqslant p} M_p(x) |D^q \varphi(x)|, \qquad p = 0, 1, 2, \ldots$$

or a countable family of inner products

$$(\varphi_1, \varphi_2)_p = \sum_{|q| \leqslant p} \int_{\omega} [M_p(x)]^2 D^q \varphi_1(x) \overline{D^q \varphi_2(x)}\, dx, \qquad p = 0, 1, 2, \ldots$$

We may assume here that ω is the set of points $x \in E^m$ at which all the $M_p(x)$ are finite. We also suppose that for any p there exists $p' > p$ such that

$$m_{pp'}(x) = \frac{M_p(x)}{M'_p(x)} \qquad (\forall x \in \omega)$$

converges to zero as $|x| \to \infty$ and is a summable function on ω (at points $x \in (E^m \setminus \omega)$ we stipulate that $m_{pp'}(x) \equiv 0$). It follows from this additional assumption (see Gel'fand and Vilenkin /1/, p. 106) that the space $K\{M_p\}$ is nuclear and that every bounded set in this space is relatively bicompact (Gel'fand and Shilov /1/, p. 113). Since $K\{M_p\}$ is a complete Fréchet space (Gel'fand and Shilov /1/, p. 118), it is a barreled space (Bourbaki /1/, Chapter I, § 1), and thus it is also reflexive. Moreover, $\|\varphi\|_p$ and $(\varphi_1, \varphi_2)_p$ are nondecreasing with respect to p, since the $M_p(x)$ are nondecreasing functions of p.

To prove that $\Phi = K\{M_p\}$ is embeddable, we define a Hilbert space H_p as the completion of Φ with respect to one of the inner products $(\varphi_1, \varphi_2)_p$. The triad Φ, H, Φ' is called a *rigged Hilbert space*** (Gel'fand and Vilenkin /1/, p. 138): Φ is embeddable in H, H is a closed subset of the strong dual Φ'.

* The space $K\{M_p\}$ has been studied in Gel'fand and Shilov /1/, Gel'fand and Vilenkin /1/.

** [The Russian term used here has also been translated as "equipped"; our "rigged" follows the English translation of Gel'fand and Vilenkin /1/ (New York, Academic Press, 1964).]

15.5. The square root of a bounded positive operator

Here and in the next subsection E is a real separable reflexive locally convex space, E' is its strong dual and B is a bounded self-adjoint linear operator from E' to E. Recall that a linear operator from E' to E is bounded* if it maps a neighborhood of zero U in E' onto a bounded set $M=B(U)$ in E. A bounded operator B is clearly continuous, since, if V is an arbitrary neighborhood of zero and M is bounded, there exists $\lambda_0>0$ such that $\lambda M\subset V$ for $|\lambda|\leqslant\lambda_0$. Hence $\lambda B(U)\subset V$, or $B(\lambda U)\subset V$, so that B is indeed continuous. Now, since $B:E'\to E$ is a bounded operator, its adjoint B': $E'\to E$ exists and is bounded.** Indeed if $B(U)\subset M$, then $B'(M^0)\subset U^0$, where U^0 and M^0 are the polars of U and M, respectively, and if U is a neighborhood of zero in E' and M a bounded set in E, then M^0 is a neighborhood of zero in E' and U^0 a bounded set in $E''=E$. A linear operator $B:E'\to E$ is *selfadjoint* if

$$\langle Bx,\ y\rangle=\langle x,\ By\rangle\qquad(\forall x,\ y\in E'). \tag{15.4}$$

If E is an embeddable space and B a continuous operator, then, by the extension theorem for identities (see Bourbaki /3/, Chapter I, § 3, Proposition 8), equality (15.4) holds if

$$(B_Hx,\ y)=(x,\ B_Hy),\qquad(\forall x,\ y\in H), \tag{15.5}$$

where B_H is the restriction of B to $H\subset E'$, and, because of the density of H in E', the inequality

$$(B_Hx,\ x)\geqslant 0\qquad(\forall x\in H) \tag{15.6}$$

implies that

$$\langle Bx,\ x\rangle\geqslant 0\qquad(\forall x\in E'), \tag{15.7}$$

so that B is a positive operator from E' to E. We shall assume, moreover, that if E is embeddable, the bilinear functional $\langle y, x\rangle$, $x\in E$ and $y\in E'$, coincides with the inner product in H if $y\in H$. We shall later have need of the following well-known proposition (Bourbaki /3/, Chapter II, § 3, Theorem 1).

Lemma 15.3. Let A be a continuous linear operator defined on a dense linear subset M of a separable locally convex space E_x, with range in a complete separable locally convex linear space E_y. Then there exists a linear operator T from E_x to E_y such that

* Raikov /1/.

** We shall also need this theorem later.

$$Tx = Ax \quad (\forall x \in M).$$

The operator T is known as an e x t e n s i o n of A.

T h e o r e m 15.1. *Let E be an embeddable reflexive space and B a bounded positive selfadjoint operator from the strong dual E' to E. Then the positive square root $B_H^{1/2}$ of the restriction B_H of B to H can be extended to a bounded linear operator T from E' to H, such that*

$$B = T'T, \tag{15.8}$$

where the adjoint of T is $T' = B_H^{1/2}$.

P r o o f. By assumption, B maps a neighborhood of zero $V \subset E'$ onto a bounded set $M = B(V) \subset E$, and since the topologies of H and E' are compatible, there exists in H a neighborhood of zero $D \subset H \cap V$ such that $B(D) \subset M$; moreover, M is bounded in H, since the topologies of E and H are also compatible. The boundedness of M in H implies that

$$\|x\| \leqslant C = \text{const} \quad (\forall x \in M).$$

Considering the sphere $\|x\| = r > 0$ in D, we get

$$\|B_H x\| \leqslant \frac{C_1}{r}\|x\| \quad (\|x\| = r,\ C_1 = \text{const}),$$

i. e., the restriction of B to H is a bounded positive selfadjoint operator in H.

Let $B_H^{1/2}$ be the positive square root of B_H in H; this operator is also bounded, positive and selfadjoint in H. Since H is a dense linear subset of E', it follows from Lemma 15.3 that $B_H^{1/2}$, as a continuous linear operator from H to H, has a continuous extension T from E' to H; and, as H is normed, T is a bounded operator from E' to H. Hence its adjoint T' is a bounded operator from H to $E'' = E$. We claim that $T' = B_H^{1/2}$. Indeed, for every fixed $x \in H$ and any $y \in H$, the bilinear form $(B_H^{1/2}x, y)$ is a continuous linear functional on H, which is a dense subset of E', and thus by Lemma 15.3 it has a continuous extension $\langle z_x, y\rangle$ defined on the entire space E', $y \in E'$, where $z_x \in E'' = E$, and the vectors $y \in H$ satisfy

$$(B_H^{1/2}x,\ y) = \langle z_x,\ y\rangle = (z_x,\ y).$$

Consequently, $B_H^{1/2}x = z_x \in E$, so that $B^{1/2}$ is a linear operator from H to E. For any fixed $x \in H$, the linear functionals $\langle B_H^{1/2}x,\ y\rangle$ and (x, Ty) are both continuous on E' and they coincide on a dense subset H of E'; hence, by the extension theorem for identities,

$$\langle B_H^{1/2}x,\ y\rangle = (x,\ Ty) \quad (\forall x \in H,\ y \in E').$$

Hence, by the definition of the adjoint, $T' = B_H^{1/2}$.

Now consider the operator

$$A = T'T.$$

Since T is continuous from E' to H and T' is continuous from H to E, it follows from the theorem on composite functions (Bourbaki /3/, Chapter I, § 4) that the linear operator A is continuous from E' to E. But for any vector $x \in H$

$$Ax = T'Tx = B_H^{1/2}(Tx) = B_H^{1/2}(B_H^{1/2})x = B_H x = Bx,$$

where A and B are continuous from E' to E. Since A and B coincide on the dense linear subset H of E', it follows from the extension theorem for identities that $Ay = By$ for any $y \in E'$, as required.

Remark 15.2. Theorem 15.1 clearly remains valid if E is an embeddable reflexive Banach space and B a linear bounded positive selfadjoint operator from the dual E^* to E.

15.6. The square root of a bounded indefinite operator

In this subsection, B will be a bounded operator from E' to E, whose restriction B_H to H is a selfadjoint operator in H. As in subsection 6.6, we set

$$B_H^+ = \frac{1}{2}(|B_H| + B_H), \quad B_H^- = \frac{1}{2}(|B_H| - B_H),$$

$$A = (B_H^+)^{1/2} + (B_H^-)^{1/2}, \quad C = (B_H^+)^{1/2} - (B_H^-)^{1/2},$$

where $|B_H|$ is the absolute value of B_H and $(\ )^{1/2}$ the positive square root of the corresponding positive selfadjoint operator in H. C is known as the principal square root of B_H.

Definition 15.3. The operator B is said to be regular if B_H^+ and B_H^- have bounded extensions B^+ and B^- from E' to E.

Theorem 15.2. Let E be an embeddable reflexive space and B a regular bounded operator from E' to E. Then B is representable in the form

$$B = V'W = W'V, \tag{15.9}$$

where V and W are bounded extensions of A and C, respectively, from E' to H, $V'=A$ and $W'=C$.

Proof. The operators B^+ and B^- satisfy the hypotheses of Theorem 15.1, since they are bounded from E' to E, and their restrictions B_H^+ and B_H^- to H are positive and selfadjoint, so that B^+ and B^- are positive selfadjoint operators from E' to E. Hence, A and C have bounded extensions V and W, respectively, from E' to H, so that the adjoints V' and W' are bounded from H to E, and moreover $V'=A$ and $W'=C$.

Consider the operator

$$T=V'W=AW=\left[(B_H^+)^{1/2}+(B_H^-)^{1/2}\right]W.$$

It is continuous from E' to E, and for $y\in H$

$$Ty=\left[(B_H^+)^{1/2}+(B_H^-)^{1/2}\right]\left[(B_H^+)^{1/2}-(B_H^-)^{1/2}\right]y=$$
$$=(B_H^+-B_H^-)\,y=B_Hy$$

since

$$(B_H^+)^{1/2}(B_H^-)^{1/2}=0.$$

Consequently,

$$Ty=By\quad(\forall y\in H).$$

Since H is dense in E', the extension theorem for identities gives $T=B$.

One proves similarly that $W'V=B$. Q. E. D.

Remark 15.3. Theorem 15.2 remains valid if E is an embeddable reflexive Banach space and B a regular bounded linear operator from the dual E^* to E.

§ 16. SQUARE ROOT OF LINEAR OPERATORS IN BANACH SPACES*

In this section we continue our investigation of the square root of a linear operator $B:E^*\to E$, where E is a reflexive Banach space. As opposed to the previous section, here we shall not assume that E is embeddable, nor that B is bounded. The results of this section

* Vainberg and Lavrent'ev /2/.

will be applied in § 17 to the investigation of equations of Hammerstein type with unbounded operators.

16.1. Auxiliary propositions

Let E be a reflexive Banach space and $\{x_n\}$ a sequence of elements of E. As before, we write $x_n \to x_0$ if the sequence $\{x_n\}$ converges strongly to x_0, and $x_n \rightharpoonup x_0$ if it converges weakly to x_0. We cite a simple proposition for later use.

Lemma 16.1. *Let $\{x_n\}$ be a sequence in E such that:*

1. $\|x_n\| \leqslant C = \text{const}$ $(n = 1, 2, 3, \ldots)$.

2. There exists $x_0 \in E$ such that any weakly fundamental subsequence $\{x_{n_k}\}$ in E converges weakly to x_0.

Then $x_n \rightharpoonup x_0$.

Proof. It suffices to prove that $\{x_n\}$ is weakly fundamental in E, since E is weakly complete. Suppose the contrary. Then there exist $h \in E^*$ and $\varepsilon > 0$ such that we can select a sequence of pairs $\{x'_n, x''_n\}$, of terms $\{x_n\}$ such that

$$|\langle h, x'_n - x''_n \rangle| \geqslant \varepsilon. \tag{16.1}$$

Since $\{x_n\}$ is bounded, we may extract from $\{x'_n, x''_n\}$ a subsequence $\{x'_{n_k}, x''_{n_k}\}$ such that

$$x'_{n_k} \rightharpoonup x'_0, \quad x''_{n_k} \rightharpoonup x''_0.$$

By (16.1), $x'_0 \neq x''_0$. This contradiction proves the lemma.

Let E_x, E_y be reflexive Banach spaces and A a linear operator from E_x to E_y with dense domain $D(A)$. We know that if A is bounded and $x_n \rightharpoonup x_0$, then $Ax_n \rightharpoonup Ax_0$. This is not true for unbounded operators. However, one can prove

Lemma 16.2. *If $x_n \rightharpoonup x_0$, $\|Ax_n\| \leqslant C$ (where the constant C depends on the sequence) and $A^{**} = A$, then $x_0 \in D(A)$ and $Ax_n \rightharpoonup Ax_0$.*

Proof. Let $y_n = Ax_n$. The sequence $\{y_n\}$ is bounded, and it therefore contains weakly convergent subsequences. Consider two such subsequences $\{y'_n\}$ and $\{y''_n\}$, $y'_n \rightharpoonup y'_0$, $y''_n \rightharpoonup y''_0$.

For any $h \in D(A^*)$ we write

$$\langle y'_0, h \rangle = \lim_{n \to \infty} \langle y'_n, h \rangle = \lim_{n \to \infty} \langle x'_n, A^* h \rangle = \langle x_0, A^* h \rangle,$$

$$\langle y''_0, h \rangle = \lim_{n \to \infty} \langle y''_n, h \rangle = \lim_{n \to \infty} \langle x'_n, A^* h \rangle = \langle x_0, A^* h \rangle.$$

Now $D(A^*)$ is dense in E_y^*, because A is a closed operator; it follows that $y_0' = y_0''$. Hence, by (16.1), $Ax_n \to y_0$.

Again, let h be any vector in $D(A^*)$. Then

$$\langle y_0, h\rangle = \lim_{n\to\infty} \langle Ax_n, h\rangle = \lim_{n\to\infty} \langle x_n, A^*h\rangle = \langle x_0, A^*h\rangle, \tag{16.2}$$

so that $x_0 \in D(A^{**}) = D(A)$ and $Ax_0 = A^{**}x_0 = y_0$. Q. E. D.

The next proposition is valid for closed operators with closed domain.

Lemma 16.3. If $x_n \longrightarrow x_0$ and $\|A^*x_n\| \leqslant C$ (the constant C may depend on $\{x_n\} \subset D(A^*)$), then $x_0 \in D(A^*)$ and $Ax_n \longrightarrow Ax_0$.

The proof is the same as that of Lemma 16.2, except that the vector h in (16.2) is taken in $D(A)$.

16.2. The square root of an operator with positive extension

Let E be a Banach space satisfying the following condition (α_1): There exists a Hilbert space H with a linear subset $H_0 \subset H \cap E^*$ which is dense both in H and in E^*, and the value $\langle y, x\rangle$ of a linear functional $y \in E^*$ at a vector $x \in E$ coincides with the inner product (y, x) in H if $x, y \in H$.

Examples of such spaces were given in § 15 and have been cited elsewhere.* We shall say that a linear operator B from E^* to E possesses property (β_1) if B is a closed operator with dense domain $D(B)$ and its restriction B_0 to a dense subset $\tilde{H}_0 \subset D(B) \cap H_0$ of H_0 (see condition (α_1)) is a symmetric operator which has a positive self-adjoint extension B_H to H.

Set $A = B_H^{1/2}$ and let

$$D_0 = \{h \in D(A) : Ah \in E\}. \tag{16.3}$$

Definition 16.1. A vector $x \in E^*$ is called an A-limit if there exists a sequence $\{x_n\} \subset D(A) \cap E^*$ such that $x_n \to x$ in E^* and Ax_n is a fundamental sequence in H.

Every vector $x \in \tilde{H}_0$ is of course an A-limit, so that the set of all A-limit vectors is dense in E^*.

Theorem 16.1. Let E be a space satisfying condition (α_1), B an operator having property (β_1), and D_0 a dense subset of H. Then the positive square root $A = B_H^{1/2}$ can be extended from $\tilde{H}_0$ to a closed operator T from E^* to H such that

$$T^*Tx = Bx \quad (\forall x \in \tilde{H}_0).$$

* See Vainberg and Shragin /1, 2/ and also Lavrent'ev /3/.

Proof. Let G be the set of all A-limit vectors. Then if $x \in G$, there exists a sequence $\{x_n\} \subset D(A) \cap E^*$ such that $x_n \to x$ in E^* and $Ax_n \to y$ in H. We claim that y is uniquely determined by $x \in G$. It will suffice to show that, for any sequence

$$\{h_n\} \subset D(A) \cap E^* \quad (\|h_n\|_{E^*} \to 0)$$

such that $\{Ah_n\}$ is a fundamental sequence in H, we have $\|Ah_n\|_H \to 0$ as $n \to \infty$. Let h be an arbitrary element of D_0 (see 16.3)) and

$$y_0 = \lim_{n\to\infty} Ah_n.$$

Then

$$(y_0, h) = \lim_{n\to\infty} (Ah_n, h) = \lim_{n\to\infty} \langle h_n, Ah\rangle = 0,$$

since $Ah \in E$ and $\|h_n\|_{E^*} \to 0$. Hence, since D_0 is dense in H, we have $y_0 = 0$, [proving our assertion]. For each vector $x \in G$, we now have a unique vector $y \in H$, thereby defining a linear operator $Tx = y$ (linearity is obvious from the construction) with domain $D(T) = G$ dense in E^*.

We now show that the operator T from E^* to H is closed. In fact, let $\{x_n\} \subset D(T)$, $\|x_n - x_0\|_{E^*} \to 0$ and suppose that $\|Tx_n - y_0\|_H \to 0$ for some $y_0 \in H$. Since $D(T) = G$, so that every vector x_n is an A-limit point, it follows that there exists $g_n \in D(A) \cap E^*$ such that

$$\|x_n - g_n\|_{E^*} < 1/2^n, \quad \|Tx_n - Ag_n\| < 1/2^n.$$

Since

$$g_n \in D(A) \cap E^*, \quad \|g_n - x_0\|_{E^*} \to 0, \quad \|Ag_n - y_0\|_H \to 0,$$

we have that x_0 is an A-limit point, so that $x_0 \in D(T)$ by the definition of $D(T)$ and $Tx_0 = y_0$.

It remains to show that for any $x \in \tilde{H}_0$

$$T^*Tx = Bx.$$

Indeed, let $x \in \tilde{H}_0$. Then for any $h \in D(T)$ there exists a sequence $\{h_n\} \subset D(A) \cap E^*$ such that

$$(Th, Tx) = (Th, Ax) = \lim_{n\to\infty} (Th_n, Ax) = \lim_{n\to\infty} (Ah_n, Ax) =$$
$$= \lim_{n\to\infty} (h_n, Bx) = \lim_{n\to\infty} \langle h_n, Bx\rangle = \langle h, Bx\rangle.$$

Hence, by the definition of the adjoint,

$$Tx \in D(T^*), \quad T^*Tx = Bx.$$

Q. E. D.

It is essential for the above theorem that D_0 be dense in H. We now present sufficient conditions for this to be so.

Lemma 16.4. Let E *be a space satisfying condition* (α_1), B *an operator having property* (β_1); *let* $B[\tilde{H}_0]$ *be dense in* H *and*

$$A[\tilde{H}_0] \subset \tilde{H}_0.$$

Then D_0 *is dense in* H.

Proof. Suppose that D_0 is not dense in H. Then there exists $y_0 \in H (y_0 \neq 0)$ such that $y_0 \perp D_0$. We claim that $A[\tilde{H}_0] \subset D_0$. In fact, if $x \in \tilde{H}_0$ then $Ax \in D(A)$, $A(Ax) = B_H x = B_0 x \in E$, so that $Ax \in D_0$. The assumption that $y_0 \perp D_0$ now implies that, if A_0 is the restriction of A to $\tilde{H}_0$, then

$$(A_0 x,\ y_0) = 0 \quad (\forall x \in \tilde{H}_0). \tag{16.4}$$

From this equality we get $y_0 \in D(A_0^*)$ and $A_0^* y_0 = 0$. We now take any vector $h \in \tilde{H}_0$ and show that $(y, Bh) = 0$. Since A_0 is a symmetric operator, it follows (see Dunford and Schwartz /2/, p. 1245) that

$$A_0 \subset \bar{A}_0 = A_0^{**}.$$

Hence,

$$0 = (AA_0^* y_0,\ h) = (A_0^* y_0,\ Ah) = (y_0,\ A_0^{**} Ah) =$$
$$= (y_0,\ A_0 Ah) = (y_0,\ B_0 h) = (y_0,\ Bh).$$

Since $\overline{B[\tilde{H}_0]} = H$, the equality $(y_0, Bh) = 0$ implies that $y_0 = 0$. This contradiction proves the lemma.

Remark 16.1. We have in fact proved an even stronger assertion: $A[\tilde{H}_0] \subset D_0$ and moreover $\overline{A[\tilde{H}_0]} = H$.

Remark 16.2. Let H_1 be the set of all zeros of the operator A_0, $S = H \ominus \tilde{H}_1$, P_S the projection from H onto S and $\tilde{H}_0^s = P_S \tilde{H}_0$. Then it can be shown that $\tilde{H}^s$ is dense in S. Thus the restriction B_S of B_H to S satisfies condition (β_1) with $\tilde{H}_0$ replaced by $\tilde{H}_0^s$. Moreover, if $B_S[\tilde{H}_0^S]$ is dense in S and $A_s[\tilde{H}_0^s] \subset \tilde{H}_0^s$, where A_S is the positive square root of B_S, then the statement of Lemma 16.4 remains true for the set D_0 defined by (16.3). To verify this, it suffices to note that $H_1 \subset D_0$.

Let $G \subset D(B)$. For $x \in G$, we set

$$|||x||| = \|x\|_{E^*} + \|Bx\|_{E}. \tag{16.5}$$

Definition 16.2. An operator B is said to be G-perfect if there exists a dense linear subset G of H whose closure in the metric (16.5) contains $D(B)$.

Theorem 16.2. If the assumptions of Theorem 16.1 hold and B is a G-perfect operator, then

$$T^*Tx = Bx \qquad (\forall x \in D(B)).$$

Proof. Consider any vectors $x \in D(B)$ and $h \in D(T)$. Then there exist sequences $\{x_n\}$ and $\{h_n\}$ such that

$$x_n \in \tilde{H}_0, \quad \|x_n - x\|_{E^*} \to 0, \quad \|Bx_n - Bx\|_E \to 0,$$
$$h_n \in D(A), \quad \|h_n - h\|_{E^*} \to 0, \quad \|Ah_n - Th\|_H \to 0.$$

Hence it follows that $x \in D(T)$. Indeed, $x_n \in D(A)$, $\|x_n - x\|_{E^*} \to 0$ and

$$(A(x_n - x_m), A(x_n - x_m)) = (x_n - x_m, B_H(x_n - x_m)) =$$
$$= \langle x_n - x_m, B(x_n - x_m)\rangle \leqslant \|x_n - x_m\|_{E^*} \|B(x_n - x_m)\|_E \to 0$$

as $m, n \to 0$. Thus $\{Ax_n\}$ is a fundamental sequence in H. Now,

$$(Th, Tx) = \lim_{k\to\infty} [\lim_{n\to\infty} (Ah_k, Ax_n)] =$$
$$= \lim_{k\to\infty} [\lim_{n\to\infty} (h_k, B_H x_n)] = \lim_{k\to\infty} [\lim_{n\to\infty} \langle h_k, Bx_n\rangle] =$$
$$= \lim_{k\to\infty} \langle h_k, Bx\rangle = \langle h, Bx\rangle.$$

Since h is any vector of $D(T)$, the last equality yields that $Tx \in D(T^*)$ and

$$T^*Tx = Bx \qquad (\forall x \in D(B))$$

Theorem 16.3. Let E be a space satisfying condition (α_1) and B a bounded operator from E^* to E having property (β_1). Then the positive square root $A = B_H^{1/2}$ has a bounded extension T from E^* to H and $T^*T = B$.

Remark 16.3. The assumptions of Theorem 16.3 do not imply that B_H is bounded in H.*

* Examples of such operators will be considered in subsection 16.5.

Proof. For any vector $x \in H_0$, we have

$$(Ax, Ax) = (A^2x, x) = (B_H x, x).$$

Since $B_H = B$ on H_0, it follows that

$$(Ax, Ax) = \langle Bx, x \rangle \leqslant \| Bx \|_E \| x \|_{E^*}$$

or

$$\| Ax \|_H^2 \leqslant M^2 \| x \|_{E^*}^2,$$

where

$$M^2 = \| B \|_{E^* \to E}.$$

Thus,

$$\| Ax \|_H \leqslant M \| x \|_{E^*} \quad (\forall x \in E^*). \tag{16.6}$$

Since H_0 is dense in E^*, inequality (16.6) implies that A has a continuous extension T from E^* to H such that

$$\| Tx \|_H \leqslant M \| x \|_{E^*} \quad (\forall x \in E^*).$$

It follows that the adjoint T^* of T is continuous from H to E and

$$\| T^*x \|_E \leqslant M \| x \|_H \quad (x \in H).$$

Let x be any vector in E^* and $\{x_n\}$ a sequence in H_0 such that $\| x - x_n \|_{E^*} \to 0$. Then

$$T^*Tx = \lim_{n \to \infty} T^*Tx_n = \lim_{n \to \infty} B_H x_n = \lim_{n \to \infty} Bx_n = Bx,$$

i. e.,

$$T^*T = B.$$

Let E be a Banach space E satisfying condition (α_2): There exists a Hilbert space H such that the set $H_0 \subset H \cap E$ is dense both in H and in E, and the value $\langle y, x \rangle$ of a linear functional $y \in E^*$ at a vector $x \in E$ coincides with the inner product (y, x) in H if $x, y \in H$.

Theorem 16.4. Let condition (α_2) *be satisfied and let* B *be a bounded operator from* E *to* E^* *whose restriction to* H_0 *is a* *symmetric operator admitting a positive selfadjoint extension* B_H. *Then the positive square root* $A = B_H^{1/2}$ *has a continuous extension* T *from* E *to* H *such that*

$$T^*T = B.$$

The proof is similar to that of Theorem 16.3.

Theorem 16.9 is an analog of Theorem 16.3. If E satisfies condition (α_2), the analogs of Theorems 16.1 and 16.2 also hold, and they may be proved in similar fashion.

16.3. The square root of an operator with indefinite extension

We shall say that a linear operator B from E^* to E possesses property (β_2) if B is a closed operator (with dense domain $D(B)$) whose restriction to a dense subset of H, $\mathring{H}_0 \subset D(B) \cap H_0$ (where H_0 is dense both in E^* and in H), is a symmetric operator admitting a selfadjoint extension B_H.

As in § 15, we set

$$B_H^+ = \frac{1}{2}(|B_H| + B_H), \quad B_H^- = \frac{1}{2}(|B_H| - B_H),$$
$$A = (B_H^+)^{1/2} + (B_H^-)^{1/2}, \qquad C = (B_H^+)^{1/2} - (B_H^-)^{1/2},$$

and consider the sets

$$D_A = \{h \in D(A) : Ah \in E\}, \quad D_C = \{h \in D(C) : Ch \in E\}.$$

These sets are analogous to D_0 (see (16.3)).

Theorem 16.5. Let E *be a space satisfying condition* (α_1), B *an operator having property* (β_2), *and let the sets* D_A *and* D_C *be dense in* H. *Then the operators* A *and* C *can be extended from* $\mathring{H}_0$ *to closed operators* V *and* W, *respectively, such that*

$$V^*Wx = W^*Vx = Bx \qquad (\forall x \in \mathring{H}_0). \tag{16.7}$$

Proof. The construction of the operators V and W is as in the case of the operator T in the proof of Theorem 16.1, as is the proof that they are closed.

We need only prove (16.7). To this end, consider any vectors $x \in \mathring{H}_0$ and $h \in D(V) \cap E^*$. Then by the definition of V (see Definition 16.1), there exists a sequence $\{h_n\} \subset D(A) \cap E^*$ such that

$$(Vh, Wx) = (Vh, Cx) = \lim_{n\to\infty} (Vh_n, Cx) = \lim_{n\to\infty} (Ah_n, Cx) =$$
$$= \lim_{n\to\infty} (h_n, ACx) = \lim_{n\to\infty} (n_n, B_H x) = \lim_{n\to\infty} (h_n, Bx) =$$
$$= \lim_{n\to\infty} \langle h_n, Bx\rangle = \langle h, Bx\rangle.$$

Hence, by definition of the adjoint,

$$Wx \in D(V^*), \quad V^*Wx = Bx \quad (\forall x \in \tilde{H}_0).$$

Similarly, one can prove that

$$Vx \in D(W^*), \quad W^*Vx = Bx \quad (\forall x \in \tilde{H}_0).$$

Q. E. D.

Lemma 16.5. Let E *be a space satisfying condition* (α_1) *and* B *an operator having property* (β_2); *let* $B[\tilde{H}_0]$ *be dense in* H, $(B_H^+)^{1/2}[\tilde{H}_0] \subset \tilde{H}_0$ *and* $(B_H^-)^{1/2}[\tilde{H}_0] \subset \tilde{H}_0$. *Then the sets* D_A *and* D_C *are dense in* H.

It suffices to prove the assertion for D_A, since the proof for D_C is similar. Suppose that $\bar{D}_A \neq H$. Then there exists a nonzero vector $y_0 \in H$ orthogonal to D_A. If we show that $y_0 = 0$, this will prove the lemma.

By assumption, $C[\tilde{H}_0] \subset \tilde{H}_0$. It turns out that $C[\tilde{H}_0] \subset D_A$ since, if $x \in \tilde{H}_0$, then $B_H x = ACx$. Hence, if A_0 and C_0 are the restrictions of A and C, respectively, to $\tilde{H}_0$, then for any $x \in \tilde{H}_0$ we have $C_0 x \in D_A$, and, as y_0 is orthogonal to D_A,

$$(C_0 x, y_0) = 0.$$

It follows that $y_0 \in D(C_0^*)$ and $C_0^* y_0 = 0$.

We now consider any vector $h \in \tilde{H}_0$ and show that $(y_0, Bh) = 0$. Indeed, as C_0 is a symmetric operator, we have*

$$C_0 \subset \dot{\bar{C}}_0 = C^{**}.$$

Hence, since $C_0^* y_0 = 0$,

$$0 = (A_0 C_0^* y_0, h) = (C_0^* y_0, A_0 h) = (y_0, C_0^{**} A_0 h) =$$
$$= (y_0, C_0 A_0 h) = (y_0, B_0 h) = (y_0, B_H h) = (y_0, Bh).$$

Here we have used the assumption that $A_0[\tilde{H}_0] \subset \tilde{H}_0$. Since $(y_0, Bh) = 0$ for any $h \in \tilde{H}_0$ and $\overline{B[\tilde{H}_0]} = H$ by assumption, it follows that $y_0 = 0$. Q. E. D.

Remarks 16.1 and 16.2 are also valid in this situation.

Let $U \subset D(B) \cap H$ be a set such that $B[U] \subset H$. For $x \in U$, we set

$$|x| = \|x\|_{E^*} + \|Bx\|_E + \||B_H|^{1/2}\|_H. \tag{16.8}$$

Definition 16.3. An operator B is said to be indefinite U-perfect if there exists a dense linear subset U of H, whose closure in the metric (16.8) contains $D(B)$.

Theorem 16.6. Under the assumptions of Theorem 16.5, if B is indefinite $\tilde{H}_0$-perfect, then for any vector $x \in D(B)$,

$$V^*Wx = W^*Vx = Bx.$$

Proof. We first show that

$$D(B) \subset D(V) \cap D(W). \tag{16.9}$$

Let $x \in D(B)$. Then there exists a sequence $\{x_n\} \subset \tilde{H}_0$ such that

$$\|x_n - x\|_{E^*} \to 0, \quad \|Bx_n - Bx\|_E \to 0,$$
$$\||B_H|^{1/2} x_n - |B_H|^{1/2} x_m\|_H \to 0$$

as $n \to \infty$ and $m \to \infty$. If we set $h_{mn} = x_n - x_m$, then

$$\begin{aligned}(Ah_{mn}, Ah_{mn}) &= \\ &= \left(\left((B_H^+)^{1/2} + (B_H^-)^{1/2}\right) h_{mn}, \left((B_H^+)^{1/2} + (B_H^-)^{1/2}\right) h_{mn}\right) = \\ &= (B_H^- h_{mn}, h_{mn}) + (B_H^- h_{mn}, h_{mn}) = (|B_H| h_{mn}, h_{mn}) = \\ &= (|B_H|^{1/2} h_{mn}, |B_H|^{1/2} h_{mn}) \to 0\end{aligned}$$

as $n \to \infty$, $m \to \infty$. Consequently, $\{Ah_{mn}\}$ is a fundamental sequence in H, and so x is an A-limit point, i. e., $x \in D(V)$. Similarly, one shows that $x \in D(W)$. Thus (16.9) holds.

For any $h \in D(V)$ and $x \in D(B)$, there exist sequences $\{x_n\}$ and $\{h_n\}$ such that

$$x_n \in \tilde{H}_0, \quad \|x_n - x\|_{E^*} \to 0, \quad \|Bx_n - Bx\|_E \to 0,$$

$h_n \in D(A)$, $\|h_n - h\|_{E^*} \to 0$, $\|Ah_n - Vh\|_H \to 0$ as $n \to \infty$.

Using all the above relations, we have

$$\begin{aligned}(Vh, Wx) &= \lim_{k \to \infty} [\lim_{n \to \infty} (Ah_k, Cx_n)] = \\ &= \lim_{k \to \infty} [\lim_{n \to \infty} (h_k, B_H x_n)] = \lim_{k \to \infty} [\lim_{n \to \infty} \langle h_k, Bx_n \rangle] = \\ &= \lim_{k \to \infty} \langle h_k, Bx \rangle = \langle h, Bx \rangle.\end{aligned}$$

Hence it follows that $Wx \in D(V^*)$ and

$$V^*Wx = Bx \quad (\forall x \in D(B)).$$

Similarly, one proves that $W^*Vx = Bx$ $(\forall x \in D(B))$.

16.4. The square root of a bounded operator with indefinite extension

As in subsection 16.3, we assume here that the space E satisfies condition (α_1) and that B has property (β_2).

Definition 16.4. An operator B with property (β_2) is said to be *semibounded* if, for any $x \in H_0$,

$$(B_H^+ x, x) \leqslant K \| x \|_{E^*}^2, \quad (B_H^- x, x) \leqslant K \| x \|_{E^*}^2.$$

Theorem 16.7. Let E be a space satisfying condition (α_1), and B a bounded operator from E^* to E which is also semibounded. Then the operators A and C have bounded extensions V and W, respectively, from E^* to H, such that

$$W^*V = V^*W = B. \tag{16.10}$$

Proof. Let x be any vector in E^* and $\{x_n\}$ a sequence in H_0 converging in E^* to x. Then

$$\| Ah_{mn} \|_H^2 = (Ah_{mn}, Ah_{mn}) = \\ = (B_H^+ h_{mn}, h_{mn}) + (B_H^- h_{mn}, h_{mn}) \leqslant 2K \| h_{mn} \|_{E^*}^2, \tag{16.11}$$

where

$$h_{mn} = x_m - x_n.$$

Thus $\{Ax_n\}$ is a fundamental sequence in H. Consider the operator V from E^* to H defined by

$$Vx = \lim_{n \to \infty} Ax_n.$$

From (16.11) it follows that

$$\| Vx \|_H \leqslant \sqrt{2K} \| x \|_{E^*},$$

i. e., V is a bounded operator from E^* to H.

Similarly, one can prove that the operator C can be extended from H_0 to a bounded operator W from E^* to H, such that

$$\| Wx \|_H \leqslant \sqrt{2K} \| x \|_{E^*}.$$

To prove (16.10), we take any vector $x \in E^*$ and consider a sequence $\{x_n\} \subset H_0$ converging in E^* to x. Then, for any $h \in H_0$,

$$\begin{aligned} \langle W^* V x, h \rangle &= (Vx, Wh) = \lim_{n\to\infty} (Vx_n, Wh) = \\ &= \lim_{n\to\infty} (Ax_n, Ch) = \lim_{n\to\infty} (B_H x_n, h) = \lim_{n\to\infty} (B_0 x_n, h) = \\ &= \lim_{n\to\infty} \langle Bx_n, h \rangle = \langle Bx, h \rangle. \end{aligned}$$

Since H_0 is dense in E^*,

$$W^* V = B.$$

Similarly one shows that $V^* W = B$. Q. E. D.

Definition 16.5. An operator B with property (β_2) is said to split if there exist closed operators B^+ and B^- from $D(B)$ to E for which

$$B^+ x + B^- x = Bx \quad (\forall x \in D(B)),$$

and their restrictions to H_0 (see condition (α_1)) coincide with B_H^+ and B_H^-, respectively, i. e.,

$$(B^+)_H = B_H^+, \quad (-B^-)_H = B_H^- \quad (\forall x \in H_0).$$

The operators B^+ and B^- are then called the splitting operators.

Theorem 16.8. Suppose an operator B splits and its splitting operators B^+ and B^- are bounded from E^* to E. Then the operators $A = (B_H^+)^{1/2} + (B_H^-)^{1/2}$ and $C = (B_H^+)^{1/2} - (B_H^-)^{1/2}$ have bounded extensions V and W, respectively, from E^* to E, such that

$$V^* W = W^* V = B.$$

Proof. For any $x \in H_0$, we may write

$$\begin{aligned} (Ax, Ax) &= (B_H^+ x, x) + (B_H^- x, x) = \\ &= ((B^+)_H x, x) + ((-B^-)_H x, x) = \\ &= \langle B^+ x, x \rangle + \langle -B^- x, x \rangle \leqslant (\| B^+ x \|_E + \| B^- x \|_E) \| x \|_{E^*} \end{aligned}$$

or

$$\| Ax \|_H^2 \leqslant M^2 \| x \|_{E^*}^2,$$

where

$$M^2 = \| B^+ \|_{E^* \to E} + \| B^- \|_{E^* \to E},$$

i. e.,

$$\| Ax \|_E \leqslant M \| x \|_{E^*}. \tag{16.12}$$

Since H_0 is dense in E^*, it follows from (16.12) that the operator A has a continuous extension V from E^* to H, such that

$$\| Vx \|_H \leqslant M \| x \|_{E^*} \quad (\forall x \in E^*).$$

Hence the adjoint V^* is continuous from H to E and

$$\| V^*x \|_E \leqslant M \| x \|_H \quad (\forall x \in H).$$

One can similarly show that the operator C has a continuous extension W from E^* to H such that

$$\| Wx \|_H \leqslant M \| x \|_{E^*}, \quad \| W^*y \|_E \leqslant M \| y \|_H \quad (\forall x \in E^*, \ y \in H).$$

Now let x be any vector in E and $\{x_n\}$ a sequence in H_0 such that

$$\lim_{n \to \infty} \| x - x_n \|_{E^*} = 0.$$

Then, as in the proof of Theorem 16.7, we find that

$$V^*W = W^*V = B.$$

Q. E. D.

Theorems analogous to Theorems 16.5, 16.6, 16.7 and 16.8 are valid if we interchange E and E^*.

16.5. Example of a linear integral operator*

We give an example of a linear integral operator, bounded from L^q to $L^p (p > 2, \ p^{-1} + q^{-1} = 1$, but not bounded in L^2.

* Lavrent'ev /3/.

Let $R^+ = [0, +\infty)$ and let $\varphi(t)$ be a nonnegative function such that

$$\int_0^{+\infty} \varphi^p(t)\,dt < \infty, \quad \int_0^{+\infty} \varphi^2(t)\,dt = +\infty.$$

Let $f(t) \in L^2(R^+)$ be a nonnegative function such that

$$\int_0^{\infty} \varphi(t) f(t)\,dt = +\infty.$$

Then there exists an increasing sequence of numbers ν_i, $\nu_i \uparrow +\infty$, such that

$$\bigcup_{i=0}^{\infty} \sigma_i = R^+, \quad \sigma_i = [\nu_i, \nu_{i+1})$$

and

$$\int_{\sigma_i} \varphi(t) f(t)\,dt = 1. \tag{16.13}$$

Set

$$K(s, t) = \sum_{i=0}^{\infty} \varphi_i(s)\varphi_i(t),$$

where

$$\varphi_i(t) = \begin{cases} \varphi(t), & t \in \sigma_i \\ 0, & t \overline{\in} \sigma_i. \end{cases}$$

Since $\varphi(t) \in L^p(R^+)$, it follows that $\varphi_i(t) \in L^2(R^+)$.

The real function $K(s, t)$ is measurable on $R^+ \times R^+$, and for any fixed $s \in R^+$

$$K(s, t) = \varphi_k(s)\varphi_k(t), \text{ where } s \in \sigma_k,$$

so that, for almost all s,

$$k^2(s) = \int_0^{+\infty} K^2(s, t)\,dt = \varphi_k^2(s) \int_{\sigma_k} \varphi_k^2(t) < \infty.$$

Thus $K(s, t)$ is a Carleman kernel.*

* See, for example, Akhiezer /1/.

Let A be the integral operator

$$Au = \int_0^{+\infty} K(s, t) u(t) dt,$$

with the following domain $D(A)$:

$$D(A) = \{u(t):\ Au \in L^2(R^+)\}.$$

This is a linear subset of $L^2(R^+)$ and it contains the dense subset of $L^2(R^+)$ consisting of the functions $v(t)$ for which

$$\int_0^{+\infty} k(s) |v(s)| ds < \infty.$$

The operator A is symmetric, since for any $u,\ v \in D(A)$

$$(Au, v) = \int_0^{+\infty} v(s) ds \int_0^{+\infty} K(s, t) u(t) dt =$$

$$= \sum_{i=0}^{\infty} \int_{\sigma_i} v(s) ds \int_0^{+\infty} \varphi_i(s) \varphi_i(t) u(t) dt =$$

$$= \sum_{i=0}^{\infty} \int_{\sigma_i} v(s) ds \int_{\sigma_i} \varphi_i(s) \varphi_i(t) u(t) dt =$$

$$= \sum_{i=0}^{\infty} \int_{\sigma_i} u(t) dt \int_{\sigma_i} \varphi_i(s) \varphi_i(t) v(s) ds =$$

$$= \int_0^{+\infty} u(t) dt \int_0^{+\infty} K(s, t) v(t) dt = (u, Av).$$

It can be shown that A is a selfadjoint operator. Indeed, since $D(A)$ is dense in $L^2(R^+)$, the adjoint exists, so that for any $h \in D(A)$ and $g \in D(A^*)$,

$$(Ah, g) = (h, A^* g).$$

But

$$(Ah, g) = \int_0^{+\infty} g(s) ds \int_0^{+\infty} K(s, t) h(t) dt =$$

$$= \sum_{i=0}^{\infty} \int_{\sigma_i} g(s) ds \int_{\sigma_i} \varphi_i(s) \varphi_i(t) h(t) dt =$$

$$= \int_0^{+\infty} h(t) dt \int_0^{+\infty} K(s, t) g(s) ds,$$

whence

$$\int_0^{+\infty} h(t)\,dt\left(\int_0^{+\infty} K(s, t)\, g(s)\, ds - A^* g\right) = 0,$$

so that by the density of $D(A)$ in $L^2(R^+)$,

$$A^* g = Ag.$$

Note that A is a positive operator, since for any $\psi \in D(A)$

$$(A\psi, \psi) = \sum_{i=0}^{\infty}\left(\int_{\sigma_i} \varphi_i(t)\,\psi(t)\,dt\right)^2 \geqslant 0.$$

We now show that A is unbounded in $L^2(R^+)$. Consider the sequence

$$f_n(s) = \begin{cases} f(s), & s \in [0, \nu_n) \\ 0, & s \overline{\in} [0, \nu_n) \end{cases}, \qquad n = 1, 2, 3, \ldots$$

By (16.13), for any n the function $f_n(s)$ is nonzero on a set of positive measure. We set $e_n = f_n/\| f_n \|$ and write

$$Ae_n = \frac{1}{\|f_n\|}\int_0^{+\infty} K(s, t)\, f_n(t)\, dt = \frac{1}{\|f_n\|}\int_0^{\nu_n} K(s, t)\, f(t)\, dt =$$

$$= \frac{1}{\|f_n\|}\sum_{i=0}^{n-1}\int_{\sigma_i} K(s, t)\, f(t)\, dt = \frac{1}{\|f_n\|}\sum_{i=0}^{n-1}\int_{\sigma_i} \varphi_i(s)\,\varphi_i(t)\, f(t)\, dt =$$

$$= \frac{1}{\|f_n\|}\sum_{i=0}^{n-1} \varphi_i(s).$$

Consequently,

$$Ae_n = \begin{cases} \dfrac{1}{\|f_n\|}\varphi(s), & s \in [0, \nu_n), \\ 0, & s \overline{\in} [0, \nu_n), \end{cases}$$

and

$$\| Ae_n \|^2 = \frac{1}{\|f_n\|^2}\int_0^{\nu_n} \varphi^2(s)\, ds \to +\infty$$

as $n \to \infty$, so that A is unbounded in $L^2(R^n)$.

Finally, we show that A is a bounded operator from L^q to L^p. By the definition of $K(s, t)$,

$$\int_0^{+\infty}\int_0^{+\infty} K^p(s, t)\,ds\,dt = \sum_{i=0}^{\infty}\int_{\sigma_i}\int_{\sigma_i} \varphi_i^p(s)\,\varphi_i^p(t)\,ds\,dt = \sum_{i=0}^{\infty}\left(\int_{\sigma_i}\varphi_i^p(s)\,ds\right)^2 < +\infty,$$

since

$$\sum_{i=0}^{\infty}\int_{\sigma_i}\varphi_i^p(s)\,ds = \int_0^{+\infty}\varphi^p(s)\,ds < +\infty.$$

Thus, for any $v \in L^q$,

$$\| Av \|_p = \left\| \int_0^{+\infty} K(s, t)\,v(t)\,dt \right\|_p \leqslant \left\| \left(\int_0^{+\infty} K^p(s, t)\,dt\right)^{1/p} \| v \|_q \right\|_p = \left(\int_0^{+\infty}\int_0^{+\infty} K^p(s, t)\,ds\,dt\right)^{1/p} \| v \|_q,$$

or

$$\| Av \|_p \leqslant M \| v \|_q \quad (\| u \|_p = \| u \|_{L^p}, \quad \| v \|_q = \| v \|_{L^q}).$$

§ 17. NONLINEAR EQUATIONS OF HAMMERSTEIN TYPE

In this section, we shall use the theorems in § 14 and the theorems on square roots (§§ 15, 16) to establish various propositions concerning equations of Hammerstein type

$$x = BF(x), \tag{17.1}$$

where F is a potential operator and B a linear operator. This method was applied in VM to the investigation of these equations in real Hilbert space and in real L-spaces.* Here we shall study equation (17.1) in certain real reflexive spaces (Banach and locally convex spaces).

* See VM for a history of the problem.

17.1. Equations with a positive operator

We let E be a reflexive embeddable Banach space, F a potential operator defined in E, and B a positive bounded selfadjoint operator from E to E^* (see subsection 15.5).

Theorem 17.1. Assume that:

1. E is a real reflexive embeddable Banach space $(E \subset H \subset E^*)$.

2. B is a positive bounded selfadjoint operator from E^* to E and $A = B_H^{1/2}$, where B_H is the restriction of the linear operator B to H.

3. F is a potential operator defined on E, whose potential $f(x)$ is weakly upper semicontinuous.

4. On some sphere $\|u\| = r > 0$ in the Hilbert space H,

$$\langle F(Au), Au\rangle < (u, u). \tag{17.2}$$

Then equation (17.1) has a solution lying in E.

Proof. Consider the functional

$$\varphi(u) = \frac{1}{2}(u, u) - f(Au), \quad u \in H, \tag{17.3}$$

defined on H. Since the operator $A = T^*$ (see Theorem 15.1 and Remark 15.2) is bounded from H to E and $(-1)f(x)$ is, by assumption, weakly lower semicontinuous in E, the functional $\varphi(u)$ is lower weakly semicontinuous as a sum of lower weakly semicontinuous functionals. Moreover (see subsection 6.6), since $A^* = T$,

$$\Phi(u) = \operatorname{grad} \varphi(u) = u - TF(Au),$$

whence

$$(\Phi(u), u) = (u, u) - (TF(Au), u) = (u, u) - \langle F(Au), Au\rangle > 0.$$

Consequently, by Theorem 14.2, there exists a solution u_0 $(\|u_0\| < r)$ of the equation $\Phi(u) = 0$, or

$$u_0 = TF(Au_0).$$

Applying the operator A and recalling that, by Theorem 15.1, $B = AT$, we get

$$x_0 = BF(x_0), \quad x_0 = Au_0.$$

Q. E. D.

Remark 17.1. Sufficient conditions for (17.2) to hold may be formulated as follows. It is clear that if $F(x)$ is a negative operator, i. e..

$$\langle F(x), x\rangle \leqslant 0 \quad (\forall x \in E),$$

then (17.2) holds on any sphere $\|u\| = r > 0$. Another sufficient condition for (17.2) is that

$$\langle F(x), x\rangle \leqslant a(x, x), \quad (\forall x \in E),$$

where $a\|B_H\| < 1$. In fact, we have

$$\langle F(Ax), Ax\rangle \leqslant a(Au, Au) = \\ = a(B_H u, u) \leqslant a\|B_H\|(u, u) < (u, u)$$

on any sphere $\|u\| = r > 0$.

Theorem 17.2. Let conditions 1, 2, and 4 of Theorem 17.1 hold and suppose that:

3′. The potential operator $F(x)$, defined on E, satisfies the inequality

$$\langle F(x) - F(y), x - y\rangle \leqslant b(x, y)\|x - y\|_H^2,$$

where $b(x, y)\|B_H\| < 1$.

Then equation (17.1) has a unique solution lying in E.

Proof. As in the proof of Theorem 17.1, we consider the functional (17.3). The gradient $\Phi(x)$ of this functional $\varphi(u)$ satisfies the inequalities

$$(\Phi(u) - \Phi(v), u - v) = \\ = (u - v, u - v) - \langle F(Au) - F(Av), Au - Av\rangle \geqslant \\ \geqslant (u - v, u - v) - b(Au, Av)(A(u - v), A(u - v)) \geqslant \\ \geqslant (1 - b(Au, Av)\|B_H\|)(u - v, u - v) \geqslant 0,$$

with equality possible only if $u = v$. Thus $\Phi(u)$ is a strictly monotone operator. Therefore, by Theorem 8.4, $\varphi(u)$ is weakly lower semicontinuous, and thus the existence of a solution u_0 of the equation

$$u = TF(Au) \tag{17.4}$$

follows from Theorem 17.1. Since $\Phi(u)$ is strictly monotone, this solution is unique by Lemma 14.1. Finally, since equations (17.1) and (17.4) are equivalent (see subsection 6.6), equation (17.1) also has a unique solution. Q. E. D.

Other, similar theorems may be proved by combining the theorems of § 14 with theorems on the square root of a positive operator $B: E^* \to E$.

17.2. Equations with an indefinite operator

As in the previous subsection, E will again be a real reflexive embeddable Banach space. In studying equation (17.1) when B is an indefinite regular operator (Definition 15.3), we may use an idea already used in other cases.* This yields several propositions, of which we shall cite only one, which amply illustrates the method.

Theorem 17.3. *Suppose that:*

1. *E is a real reflexive embeddable Banach space* ($E \subset H \subset E^*$) *and H is a Hilbert space.*

2. *B is a regular bounded selfadjoint operator from E^* to E.*

3. *F is a hemicontinuous potential operator, defined on E, such that for any $u, v \in H$*

$$\langle F(Au) - F(Av),\ Au - Av\rangle \geqslant \\ \geqslant (1+\alpha)\| P_1(u-v)\|_H^2, \qquad (17.5)$$

where $\alpha > 0$, $A = (B_H^+)^{1/2} + (B_H^-)^{1/2}$ and P_1 is the projection from H onto the subspace $H_1 \subset H$ determined by the positive part of the spectrum of B_H.

Then equation (17.1) has a unique solution lying in E.

Proof. Let B_H be the restriction of B to H. By assumption

$$\inf_{\|u\|=1} (B_H u, u) = a > -\infty, \quad \sup_{\|u\|=1} (B_H u, u) = b < +\infty,$$

where $a < 0$, $b > 0$. Let E_t be a resolution of the identity for the operator B_H; then $E(\Delta) = E_b - E_0 = P_1$ is an orthogonal projection from H onto an invariant subspace $H_1 \subset H$ which reduces B_H. Let P_2 be the orthogonal projection from H onto $H \ominus H_1$. Set

$$((u))^2 = \| P_1 u\|^2 - \| P_2 u\|^2$$

and consider the functional

$$\varphi(u) = 2f(Au) - ((u))^2,$$

* See VM, §§ 10, 24.

where f is the potential of the operator F. Then

$$\Phi(u) = \operatorname{grad} \varphi(u) = 2VF(Au) - 2(P_1 u - P_2 u),$$

where $V = A^*$ is a bounded operator from E^* to H (see subsection 15.6). We claim that Φ is strictly monotone. Indeed,

$$\begin{aligned}\frac{1}{2}(\Phi(u) - \Phi(v),\ u - v) &= (V[F(Au) - F(Av)],\ u - v) - \\ &- [(P_1(u - v),\ u - v) - (P_2(u - v),\ u - v)] = \\ &= \langle F(Au) - F(Av),\ Au - Av\rangle - \\ &\qquad - [\| P_1(u - v)\|^2 - \| P_2(u - v)\|^2].\end{aligned}$$

Hence, by (17.5),

$$\begin{aligned}\frac{1}{2}(\Phi(u) - \Phi(v),\ u - v) &\geqslant (1 + \alpha)\| u - v\|^2 - \| P_1(u - v)\|^2 + \\ &+ \| P_2(u - v)\|^2 \geqslant \alpha \| u - v\|^2,\end{aligned}$$

and thus, by Theorem 8.4, the functional $\varphi(u)$ is weakly lower semicontinuous. Using (6.4), we may now write

$$\begin{aligned}\varphi(u) - 2f(0) &= \int_0^1 (\Phi(tu),\ u)\, dt = (\Phi(0),\ u) + \\ &+ \int_0^1 (\Phi(tu) - \Phi(0),\ tu) \frac{dt}{t}.\end{aligned}$$

In view of the previous results, therefore,

$$\begin{aligned}\varphi(u) &\geqslant 2f(0) + (\Phi(0),\ u) + 2\alpha \| u\|^2 \int_0^1 t\, dt \geqslant \\ &\geqslant \alpha \| u\|^2 - \| \Phi(0)\| \| u\| + 2f(0) = \\ &= \| u\| (\alpha \| u\| - \| \Phi(0)\|) + 2f(0).\end{aligned}$$

Consequently, on some sphere $\|u\| = r > 0$,

$$\varphi(u) > \varphi(0) = 2f(0),$$

i. e., the functional $\varphi(u)$ has the m-property, so that there exists a vector $u_0 \in H$ $(\|u_0\|) < r)$, such that $\Phi(u_0) = 0$ or

$$VF(Au_0) - (P_1 - P_2) u_0 = 0.$$

From the fact that Φ is strictly monotone, we have that this vector is unique (see Lemma 14.1). Applying the operator $P_1 - P_2$ to the last inequality and noting that $(P_1 - P_2)A = C$ (see subsection 15.6), so that $(P_1 - P_2)V = W$, we obtain

$$WF(Au_0) - u_0 = 0.$$

Finally, applying the operator $V^* = A$ to this inequality and setting $Au_0 = x_0$, we find, by Theorem 15.2, that

$$BF(x_0) - x_0 = 0.$$

Our theorems on the equivalence of equations (see subsection 6.6) now imply that x_0 is the unique solution of equation (17.1). Q. E. D.

Remark 17.2. Since the spectrum of the selfadjoint operator B_H lies outside some interval $(0, m)$, $m > 0$, inequality (17.5) will hold if

$$\langle F(x) - F(y),\ x - y\rangle \geqslant \frac{1+\alpha}{m}\| x - y\|_H^2 \qquad (\forall x,\ y \in E). \qquad (17.6)$$

where $\alpha > 0$.

In fact, since the projections P_1 and P_2 reduce the operator A, it follows that

$$\| Au\|_H^2 = \| AP_1u\|_H^2 + \| AP_2u\|_H^2 \geqslant \| AP_1u\|_H^2 =$$
$$= \int_{\sqrt{m}}^{\sqrt{b}} t^2 d(E_tP_1u,\ P_1u) \geqslant m\| P_1u\|^2.$$

Hence

$$\| A(u - v)\|_H^2 \geqslant m\| P_1(u - v)\|^2, \qquad (17.7)$$

so that, by (17.6),

$$\langle F(Au) - F(Av),\ Au - Av\rangle \geqslant$$
$$\geqslant \frac{1+\alpha}{m}\| Au - Av\|_H^2 \geqslant (1+\alpha)\| P_1(u - v)\|_H^2.$$

Note that if B_H is a quasinegative operator (see VM), i. e., H_1 has finite positive dimension, then B is a regular operator.

17.3. Equations of Hammerstein type in locally convex spaces

We now study equation (17.1) on the assumption that the potential operator $F(x)$ is defined in a real separable locally convex space E and that B is a positive selfadjoint operator from E' to E. We give two propositions, illustrating a method of proof of existence-uniqueness theorems for equation (17.1) in linear topological spaces.

Theorem 17.4. Suppose that:

1. The space E and operator B satisfy the assumptions of Theorem 15.1.

2. The potential $f(x)$ of the potential operator $F(x)$, defined on E, is weakly upper semicontinuous and satisfies the inequality

$$2f(x) \leqslant a(x, x) + b(x, x)^{\alpha/2} + c, \qquad x \in E, \tag{17.8}$$

where $a\|B_H\|_H < 1$, b and c are arbitrary positive numbers, $0 < \alpha < 2$.

Then there exists a solution of (17.1), lying in E.

Proof.* Consider the functional

$$\varphi(u) = (u, u) - 2f(Au), \quad u \in H, \quad A = B_H^{1/2}.$$

As we have shown, A is a bounded operator from H to E (Theorem 15.1), and thus, for any sequence $u_n \rightharpoonup u_0$ ($u_n, u_0 \in H$) and any vector $h \in E'$, we have (see the proof of Theorem 15.1)

$$\langle Au_n, h\rangle = (u_n, Th) \to (u_0, Th) = \langle Au_0, h\rangle,$$

i. e., $Au_n \rightharpoonup Au_0$ as $n \to \infty$. Now, since the functional $(-1)f(x)$ is weakly lower semicontinuous by assumption, i. e., for any transfinite sequence $x_\alpha (\alpha \in \mathfrak{A})$ weakly convergent to u_0

$$f(x_0) \leqslant \varliminf_{\alpha \in \mathfrak{A}} f(x_\alpha),$$

we have that for any ordinary sequence

$$f(Au_0) \leqslant \varliminf_{n \to \infty} f(Au_n).$$

Consequently, the functional $\varphi(u)$ is weakly lower semicontinuous in H, being a sum of two such functionals. We claim that $\varphi(u)$ has the m-property.

* See Vainberg /9/, Theorem 10.1, Engel'son /4/, Theorem 1.

Indeed, by (17.8),

$$\varphi(u) \geqslant (u,\ u) - a\,(Au,\ Au) - b\,(Au,\ Au)^{\alpha/2} - c.$$

But

$$(Au,\ Au) = (B_H u,\ u) \leqslant \| B_H \|_H (u,\ u),$$

whence

$$\varphi(u) \geqslant \| u \|^{\alpha} \left[(1 - a \| B_H \|_H) \| u \|^{2-\alpha} - b \| B_H \|_H^{\alpha/2} \right] - c.$$

This inequality implies that $\varphi(u) > \varphi(0)$ on a sphere $\|u\| = r$ of sufficiently large radius r, and thus, by Theorem 9.4, there exists a vector $u_0 \in H\,(\|u_0\| < r)$ such that grad $\varphi(u_0) = 0$, or

$$2u_0 - 2TF(Au_0) = 0 \quad (T = A')$$

Applying $T' = A$ to the left-hand side and noting that $AT = B$ (see Theorem 15.1), we have

$$x_0 = BF(x_0),$$

where $x_0 = Au_0 \in E$. Q. E. D.

Theorem 17.5. Suppose that:

1. The space E and operator B satisfy the assumptions of Theorem 15.1.

2. The potential operator $F(x)$, defined on E, is hemicontinuous and satisfies the inequality

$$\langle F(x) - F(y),\ x - y \rangle \leqslant b \| x - y \|_H^2 \qquad (\forall x,\ y \in E),$$

where

$$b \| B_H \| < 1, \qquad b = \text{const}.$$

Then equation (17.1) has a unique solution lying in E.

Proof. As in the proof of Theorem 17.4, we consider the functional $\varphi(u)$, whose gradient is

$$\Phi(u) = 2\,[u - TF(Au)], \quad T' = A = B_H^{1/2}.$$

Then

$$\begin{aligned}(\Phi(u)-\Phi(v),\ u-v)&=\\ &=2(u-v,\ u-v)-2(T[F(Au)-F(Av)],\ u-v)=\\ &=2(u-v,\ u-v)-2\langle F(Au)-F(Av),\ Au-Av\rangle\geqslant\\ &\geqslant 2(u-v,\ u-v)-2b(A(u-v),\ A(u-v))\geqslant\\ &\geqslant 2(1-b\|B_H\|_H)(u-v,\ u-v)=\gamma(u-v,\ u-v),\end{aligned}$$

i. e., Φ is a strictly monotone operator, and so, by Theorem 8.4, the functional $\varphi(u)$ is weakly lower semicontinuous in H. Furthermore, by (6.4),

$$\begin{aligned}\varphi(u)-\varphi(0)&=\int_0^1(\Phi(tu),\ u)\,dt=\\ &=(\Phi(0),\ u)+\int_0^1(\Phi(tu)-\Phi(0),\ tu)\frac{dt}{t}\geqslant\\ &\geqslant(\Phi(0),\ u)+\frac{\gamma}{2}(u,\ u)\geqslant\|u\|\left(\frac{\gamma}{2}\|u\|-\|\Phi(0)\|\right).\end{aligned}$$

Thus, on some sphere $\|u\|=r>0$ in the space H, we have

$$\varphi(u)>\varphi(0),$$

i. e., $\varphi(u)$ has the m-property, whence, by Theorem 9.4, grad $\varphi(u_0)=0$ $(\|u_0\|<r)$ or

$$2u_0-2TF(Au_0)=0.$$

Since $\Phi(u)$ is strictly monotone, the vector u_0 is unique. Applying the operator $A/2$ to the last equality and taking into account that $B=AT$ by Theorem 15.1, we get

$$x_0=BF(x_0),\quad x_0=Au_0\in E,$$

i. e., x_0 is a solution of equation (17.1); uniqueness follows from our propositions on the equivalence of equations (see subsection 6.6).

17.4. Equation with an indefinite operator in locally convex spaces

*Theorem 17.6.** *Suppose that:*

1. *E is a real embeddable locally convex reflexive space.*
2. *B is a bounded selfadjoint operator from E' to E, whose restriction B_H to $H(E\subset H\subset E')$, is a quasinegative operator.*

* Vainberg /16/, Theorem 2; compare Vainberg /9/, Theorem 10.3 and Engel'son /4/, Theorem 3.

3. $F(x)$ is a monotone potential operator defined on E with potential $f(x)$ satisfying the inequality

$$f(x) \geqslant \frac{1}{m}(x, x) + b(x, x)^{\alpha/2} + c \qquad (\forall x \in E),$$

where m is the infimum of the positive spectrum of B_H, $0 \leqslant \alpha < 2$, b and c are arbitrary negative numbers.

Then equation (17.1) has a solution lying in E.

Proof. Since B_H is quasinegative, i. e. (see VM), the positive part of its spectrum consists of a finite number of eigenvalues of finite multiplicity, it follows that B is a regular operator (see Definition 15.3), so that Theorem 15.2 applies. Consider the functional

$$\varphi(u) = 2f(Au) + \| P_2 u \|^2 - \| P_1 u \|^2 \qquad (\forall u \in H),$$

where

$$A = (B_H^+)^{1/2} + (B_H^-)^{1/2},$$

P_1 is the projection from H onto the subspace $H_1 \subset H$ determined by the positive part of the spectrum of B_H, and P_2 is the projection from H onto $H \ominus H_1$. Then $\varphi(u)$ is weakly lower semicontinuous on H. In fact, $f(x)$ is weakly lower semicontinuous on E and, since by Theorem 15.2 A is a bounded operator from H to E, it follows (as in the proof of Theorem 17.4) that the functional $f(Au)$ is weakly lower semicontinuous on H. Next, the functional $\|Px\|^2$ is weakly lower semicontinuous, since its gradient is a monotone operator, and the functional $\|P_1 x\|^2$ is weakly continuous as its gradient is completely continuous. We claim that $\varphi(u)$ has the m-property in H.

Indeed, by assumption,

$$\varphi(u) \geqslant$$
$$\geqslant \frac{2}{m}(Au, Au) + 2b(Au, Au)^{\alpha/2} + 2c + \| P_2 u \|^2 - \| P_1 u \|^2.$$

But, as in the proof of inequality (17.7), we find that

$$(Au, Au) \geqslant m \| P_1 u \|_H^2.$$

Hence, noting that $b < 0$ and

$$(Au, Au) = (B_H u, u) \leqslant \| B_H \|_H \| u \|^2,$$

we have

$$\begin{aligned}\varphi(u) &\geqslant \\ &\geqslant 2\| P_1 u\|^2 + \| P_2 u\|^2 - \| P_1 u\|^2 - 2|b| \| B_H\|_H^{\alpha/2} \| u\|^\alpha + 2c \geqslant \\ &\geqslant \| u\|^2 - 2\| B_H\|^{\alpha/2} \| u\|^\alpha + 2c = \\ &= \| u\|^\alpha [\| u\|^{2-\alpha} - 2\| A\|^{\alpha/2}] + 2c.\end{aligned}$$

This inequality shows that, on some sphere $\|u\| = r > 0$ in H,

$$\varphi(u) > \varphi(0),$$

i. e., $\varphi(x)$ has the m-property. Consequently, there exists a vector $u_0 (\|u_0\| < r)$ in H such that $\operatorname{grad}\varphi(u_0) = 0$, or

$$VF(Au_0) - (P_1 - P_2) u_0 = 0.$$

If we apply the operator $P_1 - P_2$ to this equality, noting that $(P_1 - P_2)A = C$ and $(P_1 - P_2)V = W$ (see subsection 15.6), we get

$$WF(Au_0) - u_0 = 0.$$

Applying the operator $V^* = A$ to this equality, we see, via Theorem 15.2, that

$$BF(x_0) - x_0 = 0, \quad x_0 = Au_0 \in E.$$

Q. E. D.

17.5. Equations of Hammerstein type in a special case*

Here we study equation (17.1) on the assumption that the reflexive Banach space E satisfies either condition (α_1) or condition (α_2) and that the operator B has property (β_1) (see subsection 16.2).

Theorem 17.7. Suppose that:

1. *E is a reflexive space satisfying condition (α_1) and B a bounded operator from E^* to E having property (β_1)*
2. *$F(x)$ is a potential operator defined on E, whose potential $f(x)$ is weakly upper semicontinuous and satisfies the inequality*

$$f(x) \leqslant -\omega(\|x\|), \tag{17.9}$$

* Vainberg and Lavrent'ev /1/.

where $\omega(t) < 0$ *for* $t \geqslant 0$ *and, for some positive* R *and* $\delta = \delta(R)$,

$$\inf_{[0,\, a]} R^{-2}\omega(t) \geqslant \frac{1}{2} + \delta, \quad R^2\delta > -f(0), \quad a = R\| B \|^{1/2}.$$

Then equation (17.1) has a solution lying in E.

Proof. Let $A = B_H^{1/2}$ and let T be the bounded extension of A from E^* to E which exists by virtue of Theorem 16.3. Consider the functional

$$\varphi(u) = \frac{1}{2}(u,\, u) - f(T^*u) \qquad (\forall u \in H).$$

Since T^* is bounded, the functional $\varphi(u)$ is weakly lower semicontinuous in the Hilbert space H. Next, by (17.9),

$$\varphi(u) \geqslant \frac{1}{2}(u,\, u) + \omega(\| T^*u \|),$$

so that, by the assumption of the theorem, we see that on the sphere $\| u \| = R$

$$\varphi(u) \geqslant \left[\frac{1}{2} + R^{-2}\omega(\| T^*u \|)\right] R^2 \geqslant$$
$$\geqslant R^2\left[\frac{1}{2} + \inf_{[0,\, a]} \frac{\omega(t)}{R^2}\right] > R^2\delta > \varphi(0),$$

i. e., $\varphi(u)$ has the m-property. Hence there exists a vector $u_0 \in H$ ($\| u_0 \| < R$) such that grad $\varphi(u_0) = 0$, or

$$u_0 - T^{**}F(T^*u_0) = 0.$$

Since T is bounded, it follows (Hille and Phillips /1/, Theorem 2.11.10) that $T^{**} = T$ and so

$$u_0 - TF(T^*u_0) = 0$$

If we now apply the operator T^*, Theorem 16.3 yields

$$x_0 - BF(x_0) = 0, \quad x_0 = T^*u_0 \in E.$$

Q. E. D.

Theorem 17.8. *Suppose that:*

1. E *is a reflexive space satisfying condition* (α_1) *and* B *a bounded operator from* E^* *to* E *having property* (β_1).

2. $F(x)$ *is a potential operator defined on* E, *whose potential is weakly upper semicontinuous, and*

$$\langle F(x), x\rangle \leqslant a(x, x) + b(x, x)^{\gamma} + c \quad (\forall x \in E), \tag{17.10}$$

where

$$a > 0, \quad b > 0, \quad c > 0, \quad a\|B\| < 1.$$

Then equation (17.1) has a solution lying in E.

Proof. As in the proof of the previous theorem, we consider the functional

$$\varphi(u) = \frac{1}{2}(u, u) - f(T^{*}u) \qquad (\forall u \in H),$$

which is weakly lower semicontinuous due to the assumptions of the theorem. We then have

$$\Phi(u) = \operatorname{grad} \varphi(u) = u - T^{**}F(T^{*}u) = u - TF(T^{*}u),$$

since T is a bounded operator. Now,

$$(\Phi(u), u) = (u, u) - (TF(T^{*}u), u) = (u, u) - \langle F(T^{*}u), T^{*}u\rangle \geqslant$$
$$\geqslant (u, u) - a(T^{*}u, T^{*}u) - b(T^{*}u, T^{*}u)^{\gamma} - c.$$

As we saw in the proof of Theorem 16.3, however,

$$(T^{*}u, T^{*}u) = \|T^{*}u\|^{2} \leqslant M\|u\|_{H}^{2},$$

where $M^{2} = \|B\|_{E^{*}\to E}$. Thus

$$(\Phi(u), u) \geqslant$$
$$\geqslant (u, u) - aM^{2}(u, u) - bM^{2\gamma}(u, u)^{\gamma} - c =$$
$$= (u, u)^{\gamma}\left[(1 - aM^{2})(u, u)^{1-\gamma} - b\right] - c.$$

Hence it follows that on some sphere $\|u\| = r > 0$

$$(\Phi(u), u) > 0,$$

and, by Theorem 9.8, there exists a vector $u_0 \in H$ ($\|u_0\| < r$) such that $\Phi(u_0) = 0$, i. e.,

$$u_0 - TF(T^{*}u_0) = 0.$$

Applying the operator T^{*}, we get

$$x_0 = BF(x_0), \quad x_0 = T^{*}u_0 \in E,$$

because by Theorem 16.3 $T^{*}T = B$. Q. E. D.

Remark 17.3. This theorem remains valid if we replace (17.10) by the inequality

$$\langle F(x), x\rangle \leqslant a(x)(x, x) \qquad (\forall x \in E),$$

where $a(x)$ is a functional such that $a(x)\|B\| < 1$.

Theorem 17.9. Suppose that:

1. E is a reflexive space satisfying condition (α_2).

2. B is a bounded operator from E to E^*, whose restriction to H_0 is a symmetric operator admitting a positive selfadjoint extension B_H.

3. $F(x)$ is a potential operator, defined on E^*, whose potential $f(x)$ satisfies the assumptions of Theorem 17.7 for $x \in E^*$.

Then equation (17.1) has a solution lying in E^*.

The proof is analogous to that of Theorem 17.7, using Theorem 16.4.

17.6. Generalized solutions of Hammerstein equations

In what follows we assume that $\tilde{H}_0$ (see subsection 16.2, property (β_1)) is a maximal set.

Let E_0 be the closure of $\tilde{H}_0$ in the metric

$$|x| = \|x\|_{E^*} + \|B^*x\|_E \qquad (\forall x \in \tilde{H}_0).$$

Definition 17.1. A vector x_0 in a reflexive Banach space E is called a generalized solution of equation (17.1) if

$$\langle h, x_0\rangle = \langle B^*h, F(x_0)\rangle, \tag{17.11}$$

where h is an arbitrary vector in E_0 and the set $B^*[E_0]$ is dense in E.

The following proposition provides a sufficient condition for the generalized solution to be a classical solution. We shall use Definition 16.2.

Lemma 17.1. If B^* is an $\tilde{H}_0$-perfect operator, a generalized solution of (17.1) is a classical solution.

Proof. Let x_0 be a generalized solution of equation (17.1) and x any vector in $D(B^*)$. Then there exists a sequence in $\tilde{H}_0$ such that

$$\|x_n - x\|_{E^*} \to 0, \quad \|B^*x_n - B^*x\| \to 0$$

as $n \to \infty$. This shows that $D(B^*) = E_0$. Hence, via (17.11), we see that $F(x_0) \in D(B)$ and so

$$\langle h, x_0\rangle = \langle h, BF(x_0)\rangle.$$

Since h is an arbitrary vector in a dense subset of E^*, it follows that $BF(x_0) = x_0$. Q. E. D.

In what follows we use the set D_0 defined by (16.3).

Theorem 17.10. Let E be a reflexive space satisfying condition (α_1) and B an operator having property (β_1). Furthermore, let the sets D_0 and $B(\tilde{H}_0)$ be dense in H and E, respectively. If the potential $f(x)$ of $F(x)$ is weakly upper semicontinuous and satisfies the inequality

$$f(x) \leqslant -\omega(\|x\|_E) \qquad (x \in E), \tag{17.12}$$

where $\omega(t)$ is bounded below for $t \geqslant 0$ and

$$\lim \omega(t) = +\infty,$$

then (17.1) has a generalized solution in E.

Proof. Consider the functional

$$\varphi(u) = \frac{1}{2}(u, u) - f(T^*u) \qquad (\forall u \in D(T^*)),$$

where T is a closed operator from E^* to H (see Theorem 16.1). By (17.12),

$$\varphi(u) \geqslant \frac{1}{2}(u, u) + \omega(\|T^*u\|_E),$$

and $\varphi(u)$ is bounded below. Set

$$m = \inf_{u \in D(T^*)} \varphi(u)$$

and let $\{u_n\} \subset D(T^*)$ be a sequence such that

$$\lim_{n\to\infty} \varphi(u_n) = m. \tag{17.13}$$

We claim that the sequences $\{u_n\}$ and $\{T^*u_n\}$ are bounded in H and E, respectively. In fact, if this is not the case, there exists a subsequence $\{u_{n_k}\}$ such that

$$\varphi(u_{n_k}) \geqslant \left[\frac{1}{2}(u_{n_k}, u_{n_k}) + \omega\left(\|T^*u_{n_k}\|_E\right) \to \infty,\right.$$

but this contradicts (17.13). Since the two sequences are bounded, there exists a subsequence $\{u_{m_k}\}$ such that $u_{m_k} \to u_0$ as $k \to \infty$ and

$$\|T^* u_{m_k}\| \leqslant C,$$

where C is a constant depending on the sequence. By Lemma 16.3, then, $T^* u_{m_k} \rightharpoonup T^* u_0$ in E, and since $\varphi(u)$ is weakly lower semicontinuous,

$$m \leqslant \varphi(u_0) \leqslant \varliminf_{k\to\infty} \left[\frac{1}{2}(u_{m_k}, u_{m_k}) - f(T^* u_{m_k})\right] = \\ = \varliminf_{k\to\infty} \varphi(u_{m_k}) = m.$$

Thus $\varphi(u_0) = m$. Now, since $\varphi(u)$ is a functional defined on a dense subset of E and it has a minimum at a point $u_0 \in D(T^*)$, it follows that $D\varphi(u_0, h) = 0$ for any $h \in D(T^*)$. But

$$D\varphi(u_0, h) = (u_0, h) - Df(T^* u_0, h) = (u_0 h) - \langle F(T^* u_0), T^* h\rangle,$$

whence

$$(u_0, h) = \langle F(T^* u_0), T^* h\rangle. \tag{17.14}$$

Let g be an arbitrary vector in $\tilde{H}_0$, and set $h = Tg$. By Theorem 16.1,

$$Tg \in D(T^*), \quad T^* Tg = Bg.$$

By (17.14), it now follows that

$$\langle T^* u_0, g\rangle = (u_0, Tg) = \langle F(T^* u_0), T^* Tg\rangle = \\ = \langle F(T^* u_0), Bg\rangle = \langle F(T^* u_0), B^* g\rangle,$$

since the operators B and B^* coincide on H. If we set $T^* u_0 = x_0 \in E$, then $\langle x_0, g\rangle = \langle F(x_0), B^* g\rangle$; this equality holds for any $g \in \tilde{H}_0$ so that it is also valid on E_0. However, by assumption, $B^*[E_0]$ is dense in E, since

$$B^*[E_0] \supset B^*[\tilde{H}_0] = B[\tilde{H}_0],$$

and so the above equality proves the theorem.

We present one more proposition on the existence of a generalized solution in the indefinite case. We shall use the notation of subsection 16.3: B_H, B_H^+, B_H^-, A, C, D_A, D_C, V, W, property (β_2) of the operator B, Definition 14.1, and in addition the following

Definition 17.2. An operator B is said to possess property (β_3) if it possesses property (β_2) and, for any $x \in D(V^*)$ and $\varepsilon > 0$, there exist $\delta > 0$ and $z_0 \in \tilde{H}_0$ such that

$$\| x - z_0 \|_H < \delta \quad \text{and} \quad \| V^* x - V^* z_0 \|_E < \varepsilon.$$

*Theorem 17.11.** *Suppose that:*

1. E is a reflexive space satisfying condition (α_1) *and B on operator with property* (β_3).

2. The sets D_A, D_c and $B[\mathring{H}_0]$ are dense in H and E, respectively, and $D(B) \cap D(B^*) \supset \mathring{H}_0$.

3. The spectrum of B_H lies outside some interval $(0, m)$, $m > 0$.

4. The potential operator $F(x)$ defined on E is weakly integrable and

$$\langle F(x) - F(y), x - y \rangle \geqslant$$
$$\geqslant \frac{1+\alpha}{m} \| x - y \|_H^2 + \beta \| x - y \|_E^2 \qquad (x, y \in D),$$

where $\alpha > 0$, $\beta > 0$, $D = A[\mathring{H}_0] \subset E$ is dense in H and the potential $f(x)$ is uniformly continuous on every bounded subset of $D(V^*)$.

Then equation (17.1) has a generalized solution in E, which is a classical solution if B^* is an $\mathring{H}_0$-perfect operator.

17.7. Equations with quasipotential operators

Definition 17.3. A weakly integrable mapping $F: E \to E^*$ is said to be quasipotential, if the functional

$$f(x) = f_0 + \int_0^1 \langle F(tx), x \rangle \, dt \tag{17.15}$$

has the property that, for any $x, y \in E$,

$$f(x) - f(y) = \int_0^1 \langle F(y + t(x - y)), x - y \rangle \, dt. \tag{17.16}$$

The functional (17.15) is called the potential of the mapping F.

Without loss of generality, we may assume that $f_0 = 0$. A quasipotential operator on a convex set $\omega \subset E$ is defined in the same way.

The Nemytskii operator is a monotone quasipotential operator, if its generating function $g(u, x)$ satisfies (5.1) and the following conditions:** it is nondecreasing and right continuous in u for almost every x, and measurable in x for any fixed u.

* Vainberg and Lavrent'ev /3/.

** In this case (see Shragin, I /6/) $g(u(x), x)$ is measurable for any measurable function $u(x)$.

Note that if the mapping F is hemicontinuous, it follows from (17.16) that the integral is independent of the path and so, by Theorem 6.2, a hemicontinuous quasipotential operator is potential.

Moreover, if the mapping F is locally bounded, it follows from (17.16) that f is continuous. Finally, if F is monotone then, by Theorem 8.7, its potential f is convex and weakly lower semicontinuous.

Equations with quasipotential operators have been studied by V. N. Pavlenko. We present one proposition,

Theorem 17.12. Suppose that:

1. F is a monotone quasipotential operator, defined in a real Banach space E, and its potential f is continuous.

2. $B: E^* \to E$ is a linear operator such that

$$B = AA^*, \tag{17.17}$$

where A is a completely continuous operator from a real Hilbert space H to E.

3. There exists a sphere $\|x\| = r > 0$ in H on which the functional

$$\varphi(u) = (u, u) - 2f(Au), \quad u \in H \tag{17.18}$$

is positive.

Then the equation $x = BF(x)$ is solvable in E.

Proof. Since (u, u) is a weakly lower semicontinuous functional in H and $f(Au)$ is a weakly continuous functional it follows that φ is weakly lower semicontinuous in H. This functional has the m-property, since by assumption $\varphi(u) > \varphi(0)$ for $\|u\| = r$, so that $\varphi(u)$ has a minimum point u_0 $(\|u_0\| < r)$,

$$\varphi(u_0 + \tau h) - \varphi(u_0) \geqslant 0$$

for sufficiently small $\tau > 0$ and any unit vector $h \in H$. This inequality yields

$$2\tau(u_0 - A^*F(Au_0), h) + \tau^2 - 2I \geqslant 0,$$

where

$$I = \int_0^1 \langle F(Au_0 + \tau t Ah) - F(Au_0), \tau Ah \rangle \, dt.$$

Consequently,

$$(u_0 - A^*F(Au_0), h) \geqslant 0$$

and in view of the arbitrary choice of the unit vector h,

$$u_0 = A^* F(Au_0).$$

Applying A to this equality and using the fact that $AA^* = B$, we conclude that $x_0 = BF(x_0)$, where $x_0 = Au_0$. Q. E. D.

The problem of expressing an operator in the form (17.17) was considered in § 16 (see also VM). For condition 3 of the theorem to hold, it is sufficient, for example, that $\langle F(x), x\rangle \leqslant c\|x\|^{\gamma+1}$ where $c > 0$, $0 < \gamma < 1$. Of course, if A is a positive operator, condition 3 holds, modulo certain restrictions on f (see VM, Theorem 10.1). The assertion of Theorem 17.12 remains valid if we require that the mapping $F(x)$, rather than being monotone, should satisfy the condition: for any $x, h \in E$ and any $\varepsilon > 0$, there exists $\delta(x, h, \varepsilon) > 0$ such that $\langle F(x + \tau h) - F(x), h\rangle > -\varepsilon$ for $0 < \tau < \delta$.

We make one final remark. In investigating equation (17.1) we have resorted to reducing the equation to equivalent equations, as in subsection 6.6. The motivation for this approach lies in the fact that, even if F is potential, BF need not be potential, and the variational method can be applied directly only to equations with potential operators. As far as uniqueness theorems are concerned, the essential tools here are Lemma 6.1 (equivalence of equation (6.12) to an equation (6.13) with a superposition operator $TF(A)$), and Lemma 6.2 (equivalence of equation (6.12) to an equation (6.14) with a superposition operator $WF(A)$. The idea of replacing BF by a superposition operator is also used in studying eigenvectors and bifurcation points of BF via variational methods (see VM and other publications of the author listed in the bibliography of that book).

Chapter VI

THE METHOD OF MONOTONE OPERATORS

In previous chapters we established existence theorems for nonlinear equations of the form $F(x) = y$ and $x = BF(x)$, under the assumption that F is a potential operator. The proofs were essentially based on the construction of a nonlinear functional and the determination of its minimum or critical points. Hence the name "variational method." We now examine these equations without assuming that F is a potential operator. We suppose that F is a monotone operator, and employ various monotonicity properties of F in order to prove uniqueness and existence theorems. This method of investigation is now known as the "method of monotone operators."

Various problems of nonlinear analysis have led to the use of monotonicity properties of nonlinear operators. For example, even before the advent of the actual theory, Vishik /1 – 5/ used monotonicity in proving the solvability of boundary-value problems for quasilinear differential equations of strongly elliptic type (strongly elliptic operators are monotone). The present author (see, for example, Vainberg /8, 10/) used monotonicity, discarding the assumption that the Hammerstein operator is completely continuous or the corresponding functional weakly continuous. Progress in the theory of monotone operators, formulated only recently, is due to the efforts of many authors, chief among these being Minty, Browder, Kato, Leray and Lions.

In the first section of this chapter, we prove theorems, due mainly to Minty and Browder, concerning the equation $F(x) = y$, where F is a monotone operator. Although the work of these authors represents a major contribution to the theory, they were not the first in the field. Their research was anticipated, in particular, by several Soviet mathematicians, who established some less general results on nonlinear equations with monotone operators. Worthy of mention in this connection are Vainberg /8, 9/, (1955 and 1956), Vainberg and Shragin /3/, Engel'son /2, 4/, Vainberg and Kachurovskii /1/, Kirpotina /1/, Kachurovskii /1, 2/, and also Vishik /1–4/ and Lavrent'ev /1/. The Theorems of §18 are used in §§ 19, 20 to investigate equations of Hammerstein type $x = BF(x)$ and equations $Bx = F(x)$, where B is linear and F monotone. It is common knowledge that Hammerstein's original studies of the equation $x = BF(x)$ were continued by Nemytskii, Golomb and others. All these investigations were summarized in VM and in Krasnosel'skii /1/. In §§ 19, 20, equations of Hammerstein type are examined via the method of monotone operators in Banach and locally convex spaces, and several propositions on these equations are established. We shall not consider other methods for investigation of the equation $x = BF(x)$, referring the reader to Krasnosel'skii /1/ and Krasnosel'skii, Zabreiko et al. /1/.

In § 19 we assume that the operator B is bounded, while in §20 we assume only that it is closed.

§ 18. GENERAL THEOREMS ON MONOTONE OPERATORS

In this section we shall prove basic existence theorems for solutions of the equations $F(x) = 0$ and $F(x) = y$, where F is a monotone operator from E to E^*. E will be a Banach space of dimension greater than unity.

18.1. Auxiliary propositions

Lemma 18.1. Let E be a real Banach space and T a hemicontinuous operator from E to E^*. Suppose that for some pair of vectors $u_0 \in E$, $v_0 \in E^*$,*

$$\langle Tu - v_0, u - u_0 \rangle \geqslant 0, \tag{18.1}$$

where u is any vector in E; then $v_0 = Tu_0$.

Proof. Suppose that $Tu_0 \neq v_0$. Then, by the definition of the norm, there exists a nonzero vector $z \in E$ such that

$$\langle Tu_0 - v_0, z \rangle > \frac{1}{2} \| z \| \| Tu_0 - v_0 \| > 0. \tag{18.2}$$

Since T is hemicontinuous, it follows that for sufficiently small t

$$| \langle T(u_0 - tz) - Tu_0, z \rangle | \leqslant \frac{1}{3} \| z \| \| Tu_0 - v_0 \|. \tag{18.3}$$

Now, by (18.1), we may write

$$\langle T(u_0 - tz) - v_0, (u_0 - tz) - u_0 \rangle \geqslant 0,$$

giving

$$\langle T(u_0 - tz) - Tu_0, -tz \rangle + \langle Tu_0 - v_0, -tz \rangle \geqslant 0,$$

or

$$\langle T(u_0 - tz) - Tu_0, -z \rangle \geqslant \langle Tu_0 - v_0, z \rangle.$$

Hence, by 18.2, it follows that

$$| \langle T(u_0 - tz) - Tu_0, z \rangle | > \frac{1}{2} \| z \| \| Tu_0 - v_0 \|.$$

* This lemma was proved by Minty /1/ for Hilbert spaces, and by Browder /5/ and Minty /2/ (independently) for Banach spaces.

This inequality contradicts (18.3). Q. E. D.

Remark 18.1.* The assumptions of the lemma may be somewhat weakened. Let the set $B \subset E$ contain the vector u_0 which it surrounds densely, i. e., for any $z \in G$, where G is a dense linear subset of E, there exists $t_0 > 0$ such that $(u_0 + tz) \subset B$ for $0 \leqslant t < t_0$. Then, if T is an operator defined on B and hemicontinuous at u_0, the assertion of the lemma is valid.

The lemma is also true for locally convex spaces.**

Lemma 18.2.† *Let D be an open bounded set in euclidean n-space $E^{(n)}$, containing zero, and G a continuous mapping of the closure $\bar{D}$ into $E^{(n)}$. If*

$$(G(x), x) > 0 \qquad (\forall x \in D'),$$

on the boundary D' of D (where $(\ ,\ )$ denotes the inner product), then the equation $G(x) = 0$ has at least one solution $x_0 \in D$.

Proof. Let

$$G(x) = (G_1(x), G_2(x), \ldots, G_n(x))$$

and $x = (x_1, x_2, \ldots, x_n)$. Consider the mapping

$$G_t(x) = (G_{t1}(x), G_{t2}(x), \ldots, G_{tn}(x)),$$

where

$$G_{ti}(x) = tx_i + (1 - t) G_i(x) \qquad (0 \leqslant t \leqslant 1,\ x \in \bar{D}).$$

Since

$$(G_t(x), x) = t\sum_{i=1}^{n} x_i^2 + (1 - t)(G(x), x) > 0$$

on D', it follows that $\| G_t(x) \| > 0$ for all $x \in D'$ and $t \in [0, 1]$, so that the degree of the mapping $G_t(x)$ relative to zero, i. e., $d(G_t(x), D, 0)$, is the same for all $t \in [0, 1]$. But $d(G_1(x), D, 0) = 1$, so that

$$d(G(x), D, 0) \equiv d(G_0(x), D, 0) = 1.$$

* Minty /2/.

** J. Gossez /1/ and Nguyen-Phong-Chau /1/.

† Vishik /1/ and /5/, Lemma 3.

Hence, by a well known theorem of topology,* the equation $G(x) = 0$ has a solution $x_0 \in D$. Q. E. D.

In what follows we consider the ball

$$D_r = \{x \in E: \|x\| \leqslant r\}$$

in a reflexive Banach space E as a topological space (D_r, τ), with the topology τ induced by the weak topology of E. Since every ball in a reflexive Banach space is weakly compact, the topological space (D_r, τ) is bicompact.

Definition 18.1. A family of sets in a topological space is said to have the finite intersection property (f. i. p.) if any finite subfamily has nonempty intersection.

The following proposition is well known.**

Lemma 18.3. A topological space is bicompact if and only if every family with the f.i.p. has nonempty intersection.

Example 18.1. Let E be a Banach space and

$$D_r = \{x \in E: \|x\| \leqslant r, \ r > 0\}.$$

Consider the set

$$E_x(y) = \{y \in D_r: \langle G(x), x - y \rangle \geqslant 0\}, \tag{18.4}$$

where x is a fixed vector in E and G a mapping from E to E^* such that $\langle G(z), z \rangle \geqslant 0$ if $\|z\| > r$. The set $E_x(y)$ is not empty, for if $x \in D_r$ then $y = x \in E_x(y)$, and if $x \notin D_r$ then $0 \in E_x(y)$. This set is convex since, if $y_1, y_2 \in E_x(y)$ then, for any $t \in (0, 1)$,

$$\begin{aligned}\langle G(x), x - ty_1 - (1-t)y_2 \rangle &= \\ &= \langle G(x), t(x - y_1) + (1-t)(x - y_2) \rangle = \\ &= t\langle G(x), x - y_1 \rangle + (1-t)\langle G(x), x - y_2 \rangle \geqslant 0.\end{aligned}$$

This set is closed. Indeed, for any sequence $\{y_n\} \subset E_x(y)$ such that $y_n \to y_0$ as $n \to \infty$,

$$\langle G(x), x - y_0 \rangle = \lim_{n \to \infty} \langle G(x), x - y_n \rangle \geqslant 0,$$

so that $y_0 \in E_x(y)$. Since $E_x(y)$ is convex and closed, it is weakly closed.†

* Leray and Schauder /1/ or Krasnosel'skii /1/.

** Dunford and Schwartz /1/, 1. 5. 6.

† See Dunford and Schwartz /1/, Theorem 5.3.13. This theorem is sometimes called Mazur's theorem, although Mazur /2/ proved only that a convex closed set is sequentially weakly closed.

Lemma 18.4. Let $G(x)$ be a hemicontinuous monotone mapping from a Banach space E to E^*, satisfying the following condition: there exists $M > 0$ such that, for $\|x\| \geqslant M$,

$$\langle G(x), x\rangle > 0. \tag{18.5}$$

Then the family of all sets (18.4) (x ranging over E) is a family of weakly closed sets with the f. i. p.

Proof. In Example 18.1 we proved that each set $E_x(y)$ is weakly closed. It remains to show that any finite subfamily $E_{x_1}(y)$, $E_{x_2}(y), \ldots, E_{x_n}(y)$ $(x_i \neq x_k;\ i,\ k \leqslant n)$ has nonempty intersection (see Definition 18.1). Consider the linear system of vectors

$$z = a_1x_1 + a_2x_2 + \ldots + a_nx_n \tag{18.6}$$

for all real $a_1, a_2, \ldots, a_n$, where $a = (a_1, a_2, \ldots, a_n)$ is treated as a vector in euclidean n-space $E^{(n)}$. Clearly, if a $a \in E^{(m)}$, then $z \in E^{(m)}$, where $E^{(m)}$ is a subspace of E of dimension $m \leqslant n$. It follows from (18.6) that there exists a number $R > 0$ such that if $|a| = R$ then $\|z\| \geqslant r$. Consider the mapping

$$Q(a) = (Q_1(a),\ Q_2(a),\ \ldots,\ Q_n(a));\ Q_i(a) = \langle G(z),\ x_i\rangle$$

of the space $E^{(n)}$ into itself. By Theorem 1.2, this mapping is continuous. Moreover,

$$(Q(a),\ a) = \langle G(z),\ z\rangle,$$

where the parentheses denote the inner product in $E^{(n)}$, and thus, on the boundary of the ball $|a| \leqslant R$

$$(Q(a),\ a) > 0.$$

By Lemma 18.2, this inequality implies the existence of a vector $a_0 = (a_1^0, a_2^0, \ldots, a_n^0)$ for which $Q(a_0) = 0$. Thus,

$$\left.\begin{aligned} &\langle G(z_0),\ z_0\rangle = (Q(a_0),\ a_0) = 0, \\ &\langle G(z_0),\ x_i\rangle = Q_i(a_0) = 0 \qquad (i = 1,\ 2,\ \ldots,\ n), \end{aligned}\right\} \tag{18.7}$$

where $z_0 = a_1^0x_1 + a_2^0x_2 + \ldots + a_n^0x_n$. It is important to note that, for any distinct vectors $x_1, x_2, \ldots, x_n (n > 1)$, the number $R > 0$ can be chosen so that $\|z_0\| < r$, where r is a fixed positive number, $r < 2M$. Since $G(x)$ is by hypothesis monotone,

$$\langle G(x_i) - G(z_0),\ x_i - z_0\rangle \geqslant 0 \qquad (i = 1,\ 2,\ \ldots,\ n).$$

Hence, by (18.7),

$$\begin{aligned}\langle G(x_i),\ x_i - z_0\rangle = \\ = \langle G(x_i),\ x_i - z_0\rangle - \langle G(z_0),\ x_i\rangle + \langle G(z_0),\ z_0\rangle = \\ = \langle G(x_i) - G(z_0),\ x_i - z_0\rangle \geqslant 0,\end{aligned}$$

i. e., the sets $E_{x_i}(y)$ have a common point $y = z_0 \in D_r$. Q. E. D.

This lemma is essentially due to Minty /2/.

18.2. Fundamental theorems on monotone operators

Theorem 18.1. *Let $F(x)$ be a hemicontinuous monotone operator mapping a reflexive Banach space E into E^*, which satisfies the following condition: there exists a positive number M such that for any $x \in E$, $\|x\| \geqslant M$,*

$$\langle F(x),\ x\rangle > 0.$$

Then there exists a solution to the equation $F(x) = 0$.

Proof. By Lemma 18.4, there exists a positive number r such that the family

$$E_x(y) = \{y \in D_r:\ \langle F(x),\ x - y\rangle \geqslant 0\},$$

where $D_r = \{y \in E : \|y\| \leqslant r\}$ and x is any vector in E, is a family of weakly closed sets in D_r, having the f. i. p. Viewing D_r as a topological space (see subsection 18.1) (D_r, τ), we conclude via the reflexivity of E that (D_r, τ) is a bicompact space. Hence, by Lemma 18.3, the family $E_x(y)$ has nonempty intersection, i. e., there exists a vector $y = x_0 \in D_r$ such that

$$\langle F(x),\ x - x_0\rangle \geqslant 0 \quad (\forall x \in E).$$

Lemma 18.1 now gives that $F(x_0) = 0$. Q. E. D.

Theorem 18.2. *Let $F(x)$ be a hemicontinuous, monotone, and coercive operator mapping a reflexive Banach space E into E^*. Then the mapping $F: E \to E^*$ is surjective, i. e., the equation $F(x) = v$ has a solution for any $v \in E^*$.*

Proof. Since the operator $F(x)$ is coercive, there exists a real function $\gamma(t)$ of the nonnegative variable t such that $\gamma(t) \to +\infty$ as $t \to +\infty$, and

$$\langle F(x),\ x\rangle \geqslant \|x\|\gamma(\|x\|).$$

Consider the mapping

$$G_v(x) = F(x) - v,$$

where v is an arbitrary but fixed vector in E^*. This mapping is hemicontinuous and monotone. Now,

$$\langle G_v(x), x\rangle = \langle F(x), x\rangle - \langle v, x\rangle \geqslant \|x\|(\gamma(\|x\|) - \|v\|).$$

There exists, therefore, a positive number M_v such that $\langle G_v(x), x\rangle > 0$ whenever $\|x\| \geqslant M_v$. The operator $G_v(x)$ thus satisfies the hypotheses of Theorem 18.1, so that there exists a vector x_0 such that $F(x_0) = v$. Q. E. D.

Remark 18.2. Under the assumptions of Theorem 18.2, if $F(x)$ is also assumed to be strictly monotone, it follows from Lemma 14.1 that the mapping $F: E \to E^*$ is bijective, i.e., the inverse mapping F^{-1} exists and it is also bijective (see Remark 14.5). If, moreover, $F(x)$ is continuous and

$$\|F(x) - F(y)\| \geqslant \gamma_0(\|x - y\|), \tag{18.8}$$

where $\gamma_0(t)$ is a positive increasing function defined for $t \geqslant 0$ such that $\gamma(0) = 0$, then by Remark 14.5 the mapping $F: E \to E^*$ is a homeomorphism of E onto E^*.

Observe that (18.8) will hold if

$$\langle F(x) - F(y), x - y\rangle \geqslant \gamma_0(\|x - y\|)\|x - y\|.$$

To conclude this subsection, we note that Theorems 18.1 and 18.2 were proved independently by Browder /5/ and Minty /2/. Under some additional restrictions (that F is a potential mapping or a strongly accretive mapping satisfying (11.8)), similar theorems were established previously by the present author (see Vainberg /8–13/, Vainberg and Kachurovskii /1/).

18.3. Special case

Suppose that T is a monotone hemicontinuous operator in a real reflexive Banach space E such that

$$\langle T(x), x\rangle \geqslant \|x\|\gamma(\|x\|),$$

where $\gamma(t) > 0$ for $t > 0$, $\gamma(t) \to +\infty$ as $t \to +\infty$, $T(0) = 0$, and T is continuous at zero. If the norm in E is Gâteaux-differentiable, the operator $U_\alpha(x)$ (see Example 5.7) satisfies these conditions

Theorem 18.3. Suppose that:

1. T is an operator defined in E, with the above properties.

2. F is a bounded, monotone hemicontinuous operator from E to E^*, and*

$$\langle F(x), x\rangle \geqslant 0 \tag{18.9}$$

if $\|x\| > M > 0$.

Then there exists a solution to the equation $F(x) = 0$.

Proof: Consider the sequence of operators

$$G_n(z) = T\left(\frac{z}{n}\right) + F(z) \qquad (n = 1, 2, 3, \ldots; \ z \in E).$$

For any fixed n, the operator $G_n(z)$ satisfies the hypotheses of Theorem 18.2, and so there exists a vector z_n such that $G_n(z_n) = 0$, i. e.,

$$T\left(\frac{z_n}{n}\right) + F(z_n) = 0. \tag{18.10}$$

Hence

$$\left\langle T\left(\frac{z_n}{n}\right), z_n\right\rangle + \langle F(z_n), z_n\rangle = 0,$$

and since $\left\langle T\left(\frac{z_n}{n}\right), z_n\right\rangle > 0$ for $z_n \neq 0$, it follows that

$$\langle F(z_n), z_n\rangle < 0.$$

Hence, by (18.9),

$$\|z_n\| < M \quad (M > 0), \qquad n = 1, 2, 3, \ldots$$

Since the sequence $\{z_n\}$ is bounded, the hypotheses of the theorem entail that $\{F(z_n)\}$ is also bounded, and thus, as E is reflexive, there exists a subsequence $\{z_{n_k}\}$ such that

$$z_{n_k} \rightharpoonup x_0, \ F(z_{n_k}) \rightharpoonup y_0 \in E^* \ \text{as} \ k \to \infty.$$

Now $z_n/n \to 0$ as $n \to \infty$, and thus, performing the limit passage in (18.10) over the subsequence z_{n_k} and taking into account that

$$T(z_{n_k}/n_k) \to 0 \quad \text{as} \quad k \to \infty,$$

* I. e., it maps any bounded set into a bounded one.

we conclude that $y_0=0$. Next, since for any $x \in E$

$$\langle F(x)-F(z_{n_k}),\ x-z_{n_k}\rangle \geqslant 0,$$

it follows that

$$\langle F(x),\ x-z_{n_k}\rangle-\langle F(z_{n_k}),\ x\rangle \geqslant \langle -F(z_{n_k}),\ z_{n_k}\rangle.$$

But by (18.10), $T(z_{n_k}/n_k)=-F(z_{n_k})$, so that

$$\langle -F(z_{n_k}),\ z_{n_k}\rangle=\langle T(z_{n_k}/n_k),\ z_{n_k}\rangle \geqslant 0.$$

Consequently,

$$\langle F(x),\ x-z_{n_k}\rangle-\langle F(z_{n_k}),\ x\rangle \geqslant 0.$$

Hence, letting $k\to\infty$ and using the fact that $z_{n_k}\rightharpoonup x_0,\ F(z_{n_k})\rightharpoonup y_0=0$, we get

$$\langle F(x),\ x-x_0\rangle \geqslant 0.$$

Lemma 18.1 thus yields that $F(x_0)=0$. Q. E. D.

Remark 18.3. Theorem 18.3 remains valid without the assumption that $\gamma(t)\to+\infty$ as $t\to+\infty$. In this case the equality $G_n(z_n)=0$ follows from Theorem 18.1.

18.4. The case of a complex space

Let E be a complex normed space, E^* its dual, and let $\langle y, x\rangle$ denote the value of a linear functional $y\in E^*$ at a vector $x\in E$. Set

$$\langle\langle y,\ x\rangle\rangle=\operatorname{Re}\langle y,\ x\rangle. \tag{18.11}$$

Definition 18.2.* An operator T from E to E^* is said to be monotone if

$$\langle\langle Tx-Ty,\ x-y\rangle\rangle \geqslant 0 \quad (\forall x,\ y,\ \in E),$$

strictly monotone if equality is possible only when $x=y$, and coercive if

$$\lim_{R\to\infty}\frac{\langle\langle Tx,\ x\rangle\rangle}{R}=+\infty \quad (R=\|x\|).$$

* See §1.

It turns out that the fundamental propositions of the previous subsections remain valid for complex spaces. As an example, we prove Lemma 18.1.

Lemma 18.1'. Let T *be a hemicontinuous operator from a complex Banach space* E *to* E^*. *Let* X *be a dense linear subset of* E, *and* $u_0 \in E$, $v_0 \in E^*$ *fixed vectors such that, for any* $u \in X$,

$$\langle\langle Tu - v_0, u - u_0 \rangle\rangle \geqslant 0. \tag{18.12}$$

Then $v_0 = Tu_0$.

Proof. Consider an arbitrary vector $u \in E$. Since X is dense in E, there exists a sequence $\{u_k\} \subset X$ such that $u_k \to u$ as $k \to \infty$. Since T is hemicontinuous, it is also demicontinuous (Theorem 1.2), so that $Tu_k \rightharpoonup Tu$ as $k \to \infty$. But, by (18.12), for each $u_k \in X$,

$$\langle\langle Tu_k - v_0, u_k - u_0 \rangle\rangle \geqslant 0.$$

Letting $k \to \infty$ and recalling that then

$$(Tu_k - v_0) \rightharpoonup Tu - v_0, \quad (u_k - u_0) \longrightarrow u - u_0,$$

we see that (18.12) holds for any $u \in E$. Suppose now that $Tu_0 \neq v_0$. Then there exists $z \in E$ such that

$$\langle\langle v_0 - Tu_0, z \rangle\rangle > 0. \tag{18.13}$$

Setting $u_t = u_0 + tz$ in (18.12) and dividing by t, we see that $\langle\langle Tu_t - v_0, z \rangle\rangle \geqslant 0$. If we now add the left-hand side of (18.13) to both sides of this inequality, we obtain

$$\langle\langle Tu_t - Tu_0, z \rangle\rangle \geqslant \langle\langle v_0 - Tu_0, z \rangle\rangle > 0.$$

Letting $t \to 0$, we arrive at a contradiction. Q. E. D.

Lemma 18.2 also remains valid if we assume that $\mathrm{Re}(G(x), x) > 0$.

Lemma 18.4 and its proof remain intact if we replace $\langle , \rangle$ by $\langle\langle , \rangle\rangle$. Finally, Lemma 18.3 also holds in the complex case. Thus, the fundamental Theorems 18.1 and 18.2 on monotone operators remain valid for complex spaces.

18.5. Semimonotone operators

Definition 18.3. A mapping T from a normed space E to E^* is said to be semimonotone if there exists a strongly continuous mapping $C: E \to E^*$ with domain $D(C) \supset D(T)$ such that the mapping $G = T + C$ is monotone.

Here we shall prove a general proposition due to Browder /5/. Following the lines of Browder's proof, we first establish a few auxiliary propositions.

Lemma 18.5. Let G be a continuous mapping of a finite-dimensional Banach space X into X^* such that

$$\mathrm{Re}\langle G(x), x\rangle \geqslant \gamma(\|x\|)\|x\|, \tag{18.14}$$

where $\gamma(t) \to +\infty$ as $t \to +\infty$. Then G is surjective.

Proof. Suppose first that $X = E_n$, where E_n is euclidean n-space, so that $E_n^* = E_n$,

$$\langle G(x), x\rangle = (G(x), x) \quad \text{and} \quad (x, x) = |x|^2.$$

We must show that for any $y_0 \in E_n^* = E_n$, there is a vector $x_0 \in E$ such that $G(x_0) = y_0$. To this end, we set

$$F(x) = G(x) - y_0.$$

Then, by (18.14), we have

$$\mathrm{Re}(F(x), x) = \mathrm{Re}(G(x), x) - \mathrm{Re}(y_0, x) \geqslant$$
$$\geqslant \gamma(|x|)|x| - \mathrm{Re}(y_0, x) \geqslant (\gamma(|x|) - |y_0|)|x|.$$

By the properties of $\gamma(t)$, there thus exists a sphere, $S = \{x \in E_n : |x| = R > 0\}$, on which

$$\mathrm{Re}(F(x), x) > 0.$$

By Lemma 18.2 (see also subsection 18.4), this inequality implies the existence of a vector $x_0 \in E_n$ such that $F(x_0) = 0$, i. e., $G(x_0) = y_0$. The lemma is thus proved if $X = E_n$. In the general case, since X is of finite dimension, there exists a homeomorphism I of E_n onto X such that the norms $|x|$ and $\|x\|$ are equivalent. Now, the transformation $G' = I^*GI$ maps E_n into itself, and

$$(G'u, u) = (I^*GIu, u) = \langle GIu, Iu\rangle.$$

Thus, by (18.14),

$$\mathrm{Re}(G'u, u) \geqslant \gamma(\|Iu\|)\|Iu\|,$$

and so, as we have shown, the mapping $G' : E_n \to E_n$ is surjective. Hence G is also surjective. Q. E. D.

Let E be a Banach space and X a closed subspace.

Definition 18.4. An operator P is a projection of E onto X if P is a bounded linear operator from E onto X, i. e., its range $R(P)$ coincides with X, and $P^2 = P$.

Definition 18.5. A sequence of projections $\{P_i\}$ is said to be commutative increasing if $i < k$ implies $P_iP_k = P_kP_i = P_i$.

Lemma 18.6. Let $\{E_i\}$ be a sequence of finite-dimensional subspaces of a space E such that $E_i \subset E_{i+1}$ for any i, and let P_1 be a given projection from E onto E_1. Then there exists a commutative increasing family of projections $\{P_i\}$, generated by P_1, such that $P_iE = E_i$ and, if $R(P_i^*) = E_i'$, then $\{P_i^*\}$ is a commutative increasing family of projections from E^* onto E_i' and $E_i' \subset E_{i+1}'$ for each i.

Proof. Suppose that, given the projection P_1, we have already constructed projections $P_2, P_3, \ldots, P_r$ with

$$P_iP_k = P_kP_i = P_i, \quad i < k \leqslant r.$$

Without loss of generality, we may take the dimension of the factor space E_{i+1}/E_i to be equal to 1 for any i. The projection P_{r+1} must be so constructed that $P_iP_{r+1} = P_i$ for any $i < r+1$; this will be the case if $P_rP_{r+1} = P_r$, for $P_iP_{r+1} = (P_iP_r)P_{r+1} = P_i(P_rP_{r+1}) = P_iP_r = P_i$ for $i < r$. Since P_r maps E onto E_r and the dimension of the factor space E_{r+1}/E_r is 1, there exists a nonzero vector $u_0 \in (E_{r+1}/E_r)$. By a corollary* of the Hahn-Banach theorem, there exists a vector $v \in E^*$ such that

$$\langle v, u_0\rangle = 1, \quad \langle v, u\rangle = 0 \quad (\forall u \in E_r).$$

Set

$$P_{r+1}u = \langle v, u\rangle u_0 + P_ru \quad (\forall u \in E).$$

Then

$$P_{r+1}u_0 = u_0 + P_ru_0 = u_0,$$
$$P_{r+1}u = 0 + P_ru = u \quad (\forall u \in E_r),$$
$$P_{r+1}u = (\alpha u_0 + P_ru) \in E_{r+1} \quad (\forall u \in E),$$

i. e., P_{r+1} is the projection from E onto E_{r+1}, and $P_rP_{r+1} = P_r$ since

$$P_rP_{r+1}u = P_r(\langle v, u\rangle u_0 + P_ru) = 0 + P_r^2u = P_ru.$$

* Dunford and Schwartz /1/, Corollary II. 3. 13.

We have thus proved the first assertion of the lemma. To prove the second, we note that for any i the adjoint P_i^* maps E^* into itself, since $P_i : E \to E$. Furthermore, since $P_i^2 = P_i$ and $P_iP_k = P_kP_i = P_i$ for $i < k$, it follows that for $i < k$

$$(P_i^*)^2 = (P_i^2)^* = P_i^*, \qquad P_i^*P_k^* = P_k^*P_i^* = P_i^*.$$

Thus $\{P_i^*\}$ is a commutative increasing family of projections from E^* onto $\{E_i'\}$. Now, for any $u \in E_i'$ we have that, for any $k > i$,

$$P_k^*u = P_k^*P_i^*u = P_i^*u = u,$$

and so $E_i' \subset E_k'$.

Now consider a sequence of continuous linear mappings $K_i : E_i' \to E_i^*$ such that

$$\langle K_i w, u\rangle = \langle w, u\rangle \quad (\forall u \in E_i, \ \forall w \in E_i').$$

Observe that K_i is a one-to-one mapping, since $K_i w_0 = 0$ implies that for any $u \in E$

$$0 = \langle K_i w_0, P_i u\rangle = \langle w_0, P_i u\rangle = \langle P_i^* w_0, u\rangle,$$

i. e., $P_i^* w_0 = 0$, and, as $P_i^* w = w$ for any $w \in E_i'$, we have $w_0 = 0$. Since K_i vanishes only at zero, it is one-to-one. We now claim that $K_i[E_i'] = E_i^*$; this will imply that E_i' is a closed subspace, since K_i is a closed operator.

Indeed, consider any vector $w_1 \in E_i^*$. Then, by the above-mentioned corollary to the Hahn-Banach theorem, there exists a vector $w_2 \in E^*$ such that $\|w_2\| = \|w_1\|$ and $\langle w_2, u\rangle = \langle w_1, u\rangle$ for any $u \in E_i$. But

$$\langle K_i P_i^* w_2, u\rangle = \langle P_i^* w_2, u\rangle = \langle w_2, P_i u\rangle = \langle w_2, u\rangle = \langle w_1, u\rangle,$$

whence $K_i(P_i w_2) = w_1$, i. e., $K_i[E_i'] = E_i^*$.

We note, moreover, that $\|K_i w\| \leqslant \|w\|$, since $K_i w$ is the restriction of w to a subspace. Q. E. D.

Lemma 18.7. Suppose that:

1. G_0 is a monotone hemicontinuous operator from E to E^*, C is a strongly continuous operator from E to E^* and $G = G_0 + C$.

2. $\{E_i\}$ is an increasing family of finite-dimensional subspaces of E, whose union is dense in E, and $\{P_i\}$ is a commutative increasing family of projections such that $P_i[E] = E_i$.

3. $u_k \in E_k$, $u_k \rightharpoonup u_0 \in E$ as $k \to \infty$ and $P_k^* G(u_k) \to w \in E^*$.

Then $w = G(u_0)$.

Proof. Let $u \in E_i$, so that $P_i u = u$. In view of the monotonicity of G_0, we have

$$\mathrm{Re}\,\langle G_0(u_k) - G_0(P_i u),\ u_k - P_i u\rangle \geqslant 0,$$

and since $u_k \rightharpoonup u_0$ as $k \to \infty$, it follows that

$$\mathrm{Re}\,\langle G(P_i u),\ u_k - P_i u\rangle \to \mathrm{Re}\,\langle G(P_i u),\ u_0 - P_i u\rangle.$$

Since $u_k \in E_k$, $P_k u_k = u_k$ and $P_k P_i = P_i$ for $k > i$, we may write

$$\begin{aligned}\mathrm{Re}\,\langle G_0(u_k),\ u_k - P_i u\rangle &= \\ = \mathrm{Re}\,\langle G(u_k),\ P_k u_k - P_k P_i u\rangle &- \mathrm{Re}\,\langle C(u_k),\ u_k - P_i u\rangle = \\ = \mathrm{Re}\,\langle P_k^* G(u_k),\ u_k - P_i u\rangle &- \mathrm{Re}\,\langle C(u_k),\ u_k - P_i u\rangle.\end{aligned}$$

But $u_k \rightharpoonup u_0$ as $k \to \infty$, so that, by the strong continuity of C,

$$\mathrm{Re}\,\langle C(u_k),\ u_k - P_i u\rangle \to \mathrm{Re}\,\langle C(u_0),\ u_0 - P_i u\rangle.$$

Finally, since $P_k^* G(u_k) \to w$ as $k \to \infty$,

$$\mathrm{Re}\,\langle P_k^* G(u_k),\ u_k - P_i u\rangle \to \mathrm{Re}\,\langle w,\ u_0 - P_i u\rangle.$$

Combining all our results, we obtain

$$\mathrm{Re}\,\langle [w - C(u_0)] - G_0(u),\ u_0 - u\rangle \geqslant 0 \tag{18.15}$$

for any $u \in \bigcup_j E_i$. Since $\overline{\bigcup_i E_i} = E$, it follows from (18.15) by Lemma 18.1'. (see subsection 18.4) that $w - C(u_0) = G_0(u_0)$, i. e., $w = G(u_0)$. Q. E. D.

It is clear from the proof that the assertion of the lemma still holds if the strong continuity of C is replaced by the assumption that C is weakly continuous and $\langle C(u), u\rangle = 0$ for $\forall u \in E$.

Theorem 18.4. Let T be a semimonotone hemicontinuous coercive operator from a reflexive Banach space E to E^*. Then the equation $Tx = y$ has a solution for any $y \in E^*$, i. e., the mapping $T: E \to E^*$ is surjective.

We present a proof for the case of a separable infinite-dimensional complex space E. Let E_1' be a one-dimensional subspace of E^* containing the vector y and let P_1' be the projection of E^* onto E_1'. Set $P_1 = (P_1')^*$ and let E_1 denote the range of P_1. Starting from E_1, we can construct an increasing sequence $\{E_i\}$ of finite-dimensional subspaces of E such that $\bigcup E_i$ is dense in E. By Lemma 18.6, we can construct a commutative increasing sequence of projections $\{P_i\}$ generated by the projection P_1 such that E_i is the range of P_i. For any fixed k, define a mapping T_k: $E_k \to E'_k$ (E'_k is defined as in Lemma 18.6) by

$$T_k u = P^*_k Tu \quad (\forall u \in E_k).$$

Since the operator T is coercive by hypothesis, there exists a real function $\gamma(t)$, defined for nonnegative t, such that $\gamma(t) \to \infty$ only as $t \to \infty$, and

$$\operatorname{Re}\langle T(x), x\rangle \geqslant \|x\| \gamma(\|x\|) \quad (\forall x \in E).$$

Hence, invoking our previous results, we have that for any $u \in E_k$

$$\operatorname{Re}\langle T_k(u), u\rangle = \operatorname{Re}\langle P^*_k Tu, u\rangle = \operatorname{Re}\langle Tu, P_k u\rangle = \\ = \operatorname{Re}\langle Tu, u\rangle \geqslant \|u\| \gamma(\|u\|).$$

Since T is hemicontinuous and the space E'_k is finite-dimensional, we conclude that the mapping T_k (see Theorem 1.2) is continuous. Thus, by Lemma 18.5, T_k is surjective, and there therefore exists a vector $u_k \in E_k$ such that

$$T_k u_k = P^*_k T u_k = y \quad (y \in E'_1; \quad k = 1, 2, 3, \ldots).$$

Since $E'_1 \subset E'_k$, it follows that for any k, $y \in E'_k$ and so

$$\operatorname{Re}\langle y, u_k\rangle = \operatorname{Re}\langle P^*_k T u_k, u_k\rangle = \operatorname{Re}\langle T u_k, u_k\rangle \geqslant \|u_k\| \gamma(\|u_k\|).$$

But

$$\operatorname{Re}\langle y, u_k\rangle \leqslant |\langle y, u_k\rangle| \leqslant \|y\| \|u_k\|,$$

whence $\gamma(\|u_k\|) \leqslant \|y\|$, and thus $\|u_k\| \leqslant M$, where the constant M depends only on the fixed vector y. The sequence $\{u_k\}$ is thus bounded. By the reflexivity of E, there exists a subsequence $\{u_{n_k}\}$ such that $u_{n_k} \rightharpoonup u_0$ as $k \to \infty$. Then, in view of the previous results, $P^*_{n_k} T u_{n_k} = y$, and thus the sequence $P^*_{n_k} T u_{n_k}$ is strongly convergent. Consequently, by Lemma 18.7, we have $T(u_0) = y$. Q. E. D.

The assertion of the theorem holds if T is a hemicontinuous and monotone operator such that $T = G + C$, where G is monotone and C weakly continuous, with $\langle C(u), u\rangle = 0$ for $\forall u \in E$.

Remark 18.4. If the space E is not required to be separable, the proof of Theorem 18.4 is more difficult. In this case, one must consider the set $\mathscr{E} = \{E_\alpha\}$ of all finite-dimensional subspaces $E_\alpha \subset E$ containing E_1. The set $\mathscr{E}$ is partially ordered by inclusion, and each of its linearly ordered subsets has an upper bound;

thus, by Zorn's lemma, there exists a maximal element. Repeating the entire proof, we obtain a transfinite sequence of vectors $\{u_\alpha\}$ such that $\|u_\alpha\| \leqslant M$. From this sequence we extract a transfinite subsequence of vectors $u_\gamma \longrightarrow u_0$, with $P^*_\beta T u_\beta = y$. These relations now yield the assertion of the theorem, as before.

18.6. The case of strongly monotone operators

A mapping $F: E \to E^*$ is said to be strongly monotone if

$$\langle F(x+h) - F(x),\ h\rangle \geqslant \|h\| \gamma(\|h\|), \tag{18.16}$$

where $\gamma(t)$ is an increasing function such that $\gamma(0) = 0$ and $\gamma(t) \to \infty$ as $t \to \infty$. In particular, if $\gamma(t) = ct$, where $c > 0$, then any operator F satisfying (18.16) with this function γ is strongly monotone.

Theorem 18.5. Let F be a hemicontinuous strongly monotone operator mapping a real reflexive Banach space E into E^*. Then the mapping $F: E \to E^*$ is bijective. If, in addition, F is continuous, the mapping $F: E \to E^*$ is a homeomorphism of E onto E^*.

Proof. It follows from inequality (18.16) that F is strictly monotone. The same inequality also implies that

$$\langle F(x),\ x\rangle = \langle F(x) - F(0),\ x\rangle + \langle F(0),\ x\rangle \geqslant$$
$$\geqslant \|x\| \gamma(\|x\|) - |\langle F(0),\ x\rangle| \geqslant \|x\| (\gamma(\|x\|) - \|F(0)\|),$$

and since $(\gamma(\|x\|) - \|F(0)\|) \to +\infty$ as $\|x\| \to \infty$, F is coercive. Since F is a hemicontinuous, monotone and coercive operator, the mapping $F: E \to E^*$ is bijective by Theorem 18.2 and Remark 18.2, so that it has a bijective inverse F^{-1}. Finally, by Remark 18.2, inequality (18.16) implies that F is a homeomorphism of E onto E^*. Q. E. D.

18.7. Pseudomonotone operators

Definition 18.6. An operator F from a real or complex Banach space E to E^* is said to be pseudomonotone if, for any vectors x, y in its domain $D(F)$,

$$|\langle F(x) - F(y),\ x - y\rangle| \geqslant \|x - y\| \gamma(\|x - y\|),$$

where $\gamma(t)$ is an increasing function such that $\gamma(0) = 0$, $\gamma(t) \to +\infty$ as $t \to +\infty$.

Any pseudomonotone operator satisfies the condition

$$\lim_{R \to \infty} \frac{|\langle F(x),\ x\rangle|}{R} = +\infty \qquad (R = \|x\|).$$

Indeed, let $x_0 \in D(F)$. Then

$$|\langle F(x), x-x_0\rangle| = |\langle F(x)-F(x_0), x-x_0\rangle + \langle F(x_0), x-x_0\rangle| \geqslant \|x-x_0\|\gamma(\|x-x_0\|) - |\langle F(x_0), x-x_0\rangle| \geqslant \geqslant \|x\|\left[\frac{\|x-x_0\|}{\|x\|}(\gamma(\|x-x_0\|) - \|F(x_0)\|)\right].$$

The first result for equations with pseudomonotone operators in Hilbert space was contained in a manuscript by Zarantonello /1/. We present it here.

Theorem 18.6. Let $F: H \to H$ be a bounded continuous operator on a Hilbert space H, satisfying the inequality

$$|(F(x)-F(y), x-y)| \geqslant c\|x-y\|^2 \quad (c>0)$$

for all $x, y \in H$.

Then F is a homeomorphism of H onto itself.

This proposition was strengthened and extended to more general spaces by Browder /10, 12, 13, 30/. Of the relevant results we cite the following theorem.*

Theorem 18.7. Let F be a hemicontinuous operator mapping a real reflexive Banach space E into E^* and satisfying the following conditions:

1. $|\langle F(x), x\rangle| \geqslant \|x\|\gamma(\|x\|)$, where γ is a real function such that $\gamma(t) \to +\infty$ as $t \to +\infty$.

2. For every $m>0$, there exists a nondecreasing function $k_m(t)$ of a real variable, vanishing at zero, such that

$$|\langle F(x)-F(y), x-y\rangle| \geqslant k_m(\|x-y\|)\|x-y\|$$

for all x and y in the ball $\|x\| \leqslant m$.

Then F is a one-to-one mapping of E onto E^* which has a continuous inverse.

To conclude this section, we present a simple proof** of the following proposition on monotone operators.

Theorem 18.8. Let F be a bounded hemicontinuous monotone coercive operator from a reflexive separable real space E to E^*. Then the mapping $F: E \to E^*$ is surjective.

Proof. Let $\{\varphi_i\}$ be a complete system in E, i. e., a system of vectors whose linear span is dense in E. We denote by E^m the subspace spanned by $\varphi_1, \varphi_2, \ldots, \varphi_m$, and now seek an approximate solution $u_m \in E^m$ of the equation

$$F(u) = f, \quad f \in E^* \tag{18.17}$$

* Browder /10/, Theorem 1.

** Leray and Lions /1/, Dubinskii /2/, Lions /2/.

as a solution of the system

$$\langle F(u_m), \varphi_i\rangle = \langle f, \varphi_i\rangle, \qquad i = 1, 2, \ldots, m, \tag{18.18}$$

i. e., we use the projection method. This system is solvable (see Lemma 18.2 and the proof of Lemma 18.4), since the fact that F is coercive implies that the expression

$$I_m = \langle F(u_m), u_m\rangle - \langle f, u_m\rangle,$$

obtained from (18.18), satisfies the inequality

$$I_m \geqslant \langle F(u_m), u_m\rangle - c\| u_m \| > 0, \ c = \| f \|,$$

if $\|u_m\| = \rho$, where ρ is a sufficiently large number.

Let v_m be a solution of system (18.18), so that

$$\langle F(v_m), v_m\rangle = \langle f, v_m\rangle \leqslant \| f \| \| v_m \|, \tag{18.19}$$

and thus, if $v_m \neq 0$,

$$\langle F(v_m), v_m\rangle \| v_m \|^{-1} \leqslant \| f \|.$$

Hence, since F is coercive, we have $\|v_m\| \leqslant C_1$, where the constant C_1 does not depend on m. Since F is bounded, it follows from this uniform estimate that $\|F(v_m)\| \leqslant C_2$. We thus have

$$\| v_m \| \leqslant C, \quad \| F(v_m) \| \leqslant C \quad (C = \max(C_1, C_2)).$$

By the reflexivity of E, these estimates imply the existence of a subsequence $\{v_{n_k}\}$ such that

$$v_{n_k} \rightharpoonup v \in E, \qquad F(v_{n_k}) \rightharpoonup w \in E^*. \tag{18.20}$$

Setting $m = n_k$ in (18.18) and letting $k \to \infty$, we obtain

$$\langle w, \varphi_i\rangle = \langle f, \varphi_i\rangle, \ \forall i,$$

and so

$$w = f. \tag{18.21}$$

It then follows from (18.19) that

$$\langle F(v_{n_k}), v_{n_k}\rangle = \langle f, v_{n_k}\rangle \to \langle f, v\rangle = \langle w, v\rangle,$$

i. e., as $k \to \infty$,

$$\langle F(v_{n_k}), v_{n_k} \rangle \to \langle w, v \rangle. \tag{18.22}$$

To complete the proof, we use the monotonicity of F and write

$$\langle F(u) - F(v_{n_k}), u - v_{n_k} \rangle \geqslant 0.$$

Letting $k \to \infty$ and using (18.20) and (18. 22), we see that

$$\langle F(u) - w, u - v \rangle \geqslant 0.$$

By Lemma 18.1, therefore, $w = F(v)$, and (18.21) now gives $F(v) = f$. Q. E. D.

Remark 18.5. The assertion of Theorem 18.8 and its proof remain valid if, instead of assuming that F is monotone, we suppose that it may be represented as $F = \Phi + C$, where Φ is a monotone operator and C is either 1) strongly continuous or 2) weakly continuous and such that $\langle C(u), u \rangle = 0$, $\forall u \in E$. To complete the proof, one starts with the inequality

$$\langle \Phi(u) - \Phi(v_{n_k}), u - v_{n_k} \rangle \geqslant 0 \quad (\forall u \in E)$$

and employs reasoning similar to that leading to inequality (18.15).

§ 19. NONLINEAR EQUATIONS OF HAMMERSTEIN TYPE IN BANACH AND LOCALLY CONVEX SPACES

19.1. Equation in special spaces*

Definition 19.1. A real Banach space E is said to have property (π) if: 1) E is reflexive; 2) there exists a monotone hemicontinuous operator $A: E^* \to E$ such that $A0 = 0$, A is continuous at zero, and $\langle Au, u \rangle \geqslant c\|u\|^{\alpha}$, where $c > 0$, $\alpha > 1$.

Examples of such spaces are reflexive Banach spaces E such that the norm in E^* is Fréchet-differentiable. In these spaces, a suitable operator A may be defined by

$$Ax = \|x\|^{\alpha} \operatorname{grad} \|x\|, \ \alpha > 0, \ x \neq 0, \ x \in E^*, \qquad A0 = 0,$$

* See A.M. Vainberg /1/.

since (see Example 5.7)

$$\langle Ax, x\rangle = \|x\|^{\alpha+1};$$
$$\langle Ax - Ay, x - y\rangle \geqslant (\|x\| - \|y\|)(\|x\|^{\alpha} - \|y\|^{\alpha}) \geqslant 0.$$

Theorem 19.1. Let E be a Banach space having property (π), *F a monotone hemicontinuous bounded operator from E to E^* such that*

$$\langle F(u), u\rangle \geqslant 0 \quad \text{if} \quad \|u\| > \lambda > 0 \quad (\lambda = \text{const}),$$

and B a positive bounded linear operator from E^* to E. Then the equation

$$u + BF(u) = 0$$

has a solution in E. This solution is unique if B is strictly positive or F strictly monotone.

Proof. Consider the sequence of operators $(n = 1, 2, 3, \ldots)$

$$\Phi_n z = A\frac{z}{n} + B^* z + BFB^* z \quad (z \in E^*),$$

from E^* to E. For each operator Φ_n, we have

$$\langle \Phi_n z_1 - \Phi_n z_2, z_1 - z_2\rangle = \left\langle A\frac{z_1}{n} - A\frac{z_2}{n}, z_1 - z_2\right\rangle +$$
$$+ \langle B(z_1 - z_2), z_1 - z_2\rangle + \langle FB^* z_1 - FB^* z_2, B^* z_1 - B^* z_2\rangle \geqslant 0,$$

i. e., Φ_n is monotone for every n. Now,

$$\langle \Phi_n z, z\rangle = n\left\langle A\frac{z}{n}, \frac{z}{n}\right\rangle + \langle Bz, z\rangle + \langle FB^* z, B^* z\rangle$$

and, since A is coercive, B is positive, and the last term is nonnegative for large values of $\|B^* z\|$, it follows that the operator Φ_n is coercive. By Theorem 18.2, then, the equation $\Phi_n z = 0$ has a solution, z_n say. Consequently,

$$A\frac{z_n}{n} + B^* z_n + BFB^* z_n = 0. \tag{19.1}$$

Hence

$$\left\langle A\frac{z_n}{n}, z_n\right\rangle + \langle Bz_n, z_n\rangle + \langle FB^* z_n, B^* z_n\rangle = 0. \tag{19.2}$$

Since the first term in (19.2) is positive for $z_n \neq 0$ and the second nonnegative, it follows that the third term is negative, whence, according to the hypotheses of the theorem, $\|B^* z_n\| \leqslant \lambda$ for any n. Since F is bounded, this inequality implies that $\|FB^* z_n\| \leqslant C = \text{const}$, so that $|\langle FB^* z_n, B^* z_n\rangle| \leqslant \lambda C$. Since this sequence of numbers is bounded and $\langle Bz_n, z_n\rangle$ is nonnegative, we conclude from (19.2) that

$\left\langle A \frac{z_n}{n}, z_n \right\rangle \leqslant d =$ const for any n, or

$$\frac{c}{n^{\alpha-1}} \| z_n \|^{\alpha} \leqslant d.$$

Hence, as $n \to \infty$,

$$\frac{\| z_n \|^{\alpha}}{n^{\alpha}} \leqslant \frac{d}{nc} \to 0, \quad \text{i. e.,} \quad \frac{\| z_n \|}{n} \to 0.$$

By the continuity of the operator A at zero, we have

$$\lim_{n \to \infty} A \frac{z_n}{n} = 0. \tag{19.3}$$

Since the sequences $\{B^* z_n\}$ and $\{FB^* z_n\}$ are bounded, respectively, in the spaces E and E^*, it follows from our reflexivity assumption that suitably chosen subsequences will converge weakly to x_0 and y_0, respectively, i. e.,

$$B^* z_{n_k} \rightharpoonup x_0 \in E, \qquad FB^* z_{n_k} \rightharpoonup y_0 \in E^*. \tag{19.4}$$

Going to the limit in (19.2) through the subsequence n_k, we obtain by (19.3) and (19.4) that

$$x_0 + By_0 = 0, \tag{19.5}$$

since the bounded operator B transforms weakly convergent sequences into weakly convergent sequences.

To complete the proof, it remains to show that $y_0 = Fx_0$ in (19.5). As F is monotone, we have for any $x \in E$

$$\langle Fx - FB^* z_{n_k}, x - B^* z_{n_k} \rangle \geqslant 0$$

or

$$\langle Fx, x - B^* z_{n_k} \rangle - \langle FB^* z_{n_k}, x \rangle \geqslant \langle FB^* z_{n_k}, -B^* z_{n_k} \rangle.$$

Hence, by (19.1),

$$\langle Fx, x - B^*z_n\rangle - \langle FB^*z_{n_k}, x\rangle \geqslant$$
$$\geqslant \left\langle FB^*z_{n_k}, A\frac{z_{n_k}}{n_k}\right\rangle + \langle FB^*z_{n_k}, BFB^*z_{n_k}\rangle.$$

Letting $k\to\infty$ in this inequality and recalling that $B^*z_{n_k}\rightharpoonup x_0$, $FB^*z_n\rightharpoonup y_0$ and, moreover,

$$\left|\left\langle FB^*z_{n_k}, A\frac{z_{n_k}}{n_k}\right\rangle\right| \leqslant C\left\| A\frac{z_{n_k}}{n_k}\right\| \to 0,$$
$$\varliminf_{k\to\infty} \langle FB^*z_{n_k}, BFB^*z_{n_k}\rangle \geqslant \langle y_0, By_0\rangle,$$

since (see Example 8.6) the functional $\langle z, Bz\rangle$ is weakly lower semicontinuous, we get

$$\langle Fx, x - x_0\rangle - \langle y_0, x\rangle \geqslant - \langle y_0, x_0\rangle,$$

or

$$\langle Fx - y_0, x - x_0\rangle \geqslant 0.$$

It now follows from Lemma 18.1 that $y_0 = Fx_0$ and (19.5) becomes

$$x_0 + BFx_0 = 0.$$

The first part of our theorem is thus proved.

Suppose now that B is strictly positive or that F is strictly monotone. Then, if x_1 and x_2 denote two solutions of the equation $u + BFu = 0$, subtraction gives

$$x_2 - x_1 + BFx_2 + BFx_1 = 0,$$

whence

$$\langle Fx_2 - Fx_1, x_2 - x_1\rangle + \langle BFx_2 - BFx_1, Fx_2 - Fx_1\rangle = 0.$$

If $x_1 \neq x_2$, then one of the terms in the last equality is positive and the other is nonnegative, so we have a contradiction. Thus, the equation $u + BFu = 0$ cannot have more than one solution. Q. E. D.

19.2. Some properties of Hammerstein operators

Given a nonlinear mapping $F: E_x \to E_y$ and a linear mapping $B: E_y \to E_x$, where E_x and E_y are vector spaces, we shall call the mapping $\Gamma = BF : E_x \to E_x$ an operator of Hammerstein type. If F is a Nemytskii operator and B a linear integral operator, then Γ is called a Hammerstein operator. Either of the equations

$$x = BF(x) \tag{19.6}$$

or $x + BF(x) = 0$ is called an equation of Hammerstein type.

We now present some properties of the operator Γ which will be useful both in investigating the uniqueness of the solution to equation (19.6) and in reducing this equation to an equation with monotone operators.

Let E be a real embeddable space (see Definition 15.2) and let H be the corresponding Hilbert space. Then (see Lemma 6.1) if $F : E \to E'$ and B is a positive bounded selfadjoint operator from the strong dual E' to E, equation (19.6) is equivalent to

$$y = TF(Ay), \qquad y \in H, \tag{19.7}$$

where $A = B_H^{1/2}$, B_H is the restriction of B to H, $T : E' \to H$ is a bounded extension of the operator A and $T' = A$ (see Theorem 15.1). Corresponding solutions of (19.6) and (19.7) are related to one another by

$$x = Ay, \ \ y = TF(x).$$

Let X be the solution set of equation (19.6) and $x_1, x_2 \in X$. Then by the above relations, if $x_1 \neq x_2$, then

$$\begin{aligned} 0 < (y_1 - y_2, y_1 - y_2) &= (y_1 - y_2, TF(Ay_1) - TF(Ay_2)) = \\ &= \langle T'y_1 - T'y_2, F(Ay_1) - F(Ay_2)\rangle = \\ &= \langle x_1 - x_2, F(x_1) - F(x_2)\rangle. \end{aligned}$$

Consequently, F is strictly monotone on the solution set of equation (19.6), so that the latter has at most one solution if F is a dissipative operator. This uniqueness theorem can be generalized.

We cite one result, relevant for Hilbert spaces. Let H be a real Hilbert space and let B be a selfadjoint operator from H to H with dense domain. Consider equation (19.6), where F is a nonlinear operator from H to H.

Theorem 19.2. Suppose the spectrum of the operator B has no points on some interval $(0, m)$, $m > 0$. Let X be the solution set of (19.6). Then for any $u_1, u_2 \in X$

$$(F(u_1) - F_2(u_2), u_1 - u_2) \leqslant \frac{1}{m} \| u_1 - u_2 \|^2. \tag{19.8}$$

Proof. Let E_t be a resolution of the identity for the operator B and $P_1 = I - E_m$ $(Ix = x)$. Set $H_1 = P_1 H$, $H_2 = H \ominus H_1$ and let P_2 denote the projection on H_2, so that for any $x \in H$

$$x = P_1 x + P_2 x \quad (P_1 x \perp P_2 x).$$

Set

$$B_+ = \frac{1}{2}(|B| + B), \quad B_- = \frac{1}{2}(|B| - B),$$
$$A = B_+^{1/2} + B_-^{1/2}, \quad C = B_+^{1/2} - B_-^{1/2}.$$

Then $(B_+^{1/2} x, B_-^{1/2} x) = 0$, so that

$$(Cx, Cx) = (C^2 x, x) = ((B_+ + B_-)x, x) =$$
$$= ((B_+ + B_-)x, P_1 x + P_2 x) = (B_+ x, P_1 x) + (B_- x, P_2 x),$$

since

$$(B_+ x, P_2 x) = 0, \quad (B_- x, P_1 x) = 0.$$

Now, $(B_- x, P_2 x) \geqslant 0$, and so

$$(Cx, Cx) \geqslant (B_+ x, P_1 x) =$$
$$= \int_m^{+\infty} t\, d(E_t x, P_1 x) \geqslant m \int_m^{+\infty} d(E_t x, P_1 x) =$$
$$= m(P_1 x, P_1 x) = m \| P_1 x \|^2.$$

Thus

$$\frac{1}{m} \| Cx \|^2 \geqslant \| P_1 x \|^2 \quad (\forall x \in D(C)). \tag{19.9}$$

Let u_1 and u_2 be two solutions of equation (19.6). Then

$$u_i = BF(u_i) = CAF(u_i) \quad (i = 1, 2).$$

Setting $v_i = AF(u_i)$, we have

$$u_i = Cv_i, \quad v_i = AF(Cv_i) \quad (i = 1, 2).$$

Therefore, since $(P_1 - P_2) A = C$, we have

$$((P_1 - P_2)(v_2 - v_1), v_2 - v_1) = (C[F(Cv_2) - F(Cv_1)], v_2 - v_1) =$$
$$= (F(Cv_2) - F(Cv_1), Cv_2 - Cv_1) = (F(u_2) - F(u_1), u_2 - u_1),$$

or

$$(F(u_2)-F(u_1),\ u_2-u_1)=$$
$$=(P_1(v_2-v_1),\ v_2-v_1)-(P_2(v_2-v_1),\ v_2-v_1)=$$
$$=\| P_1(v_2-v_1)\|^2-\| P_2(v_2-v_1)\|^2\leqslant\| P_1(v_2-v_1)\|^2.$$

Inequality (19.8) now follows from (19.9). Q. E. D.

This theorem yields

Theorem 19.3. Let the spectrum of the operator B *lie outside some interval* $(0, m)$, $m>0$. *If*

$$(F(u)-F(v),\ u-v)>\frac{1}{m}\| u-v\|^2,$$

then equation (19.6) has at most one solution.

We now show how to reduce Hammerstein equations in certain vector spaces to equivalent equations with monotone operators in Hilbert space. We first remark that monotonicity of the operator $F:E\to E'$ does not necessarily imply monotonicity of the Hammerstein operator $\Gamma:E\to E$. This is readily verified if we let F be a Nemytskii operator $h:L^p\to L^q$ $(p^{-1}+q^{-1}=1,\ p>1)$ generated by a function $g(u, x)$ which is nondecreasing in u (see Example 5.4), and B the integral operator with kernel $K(x, y)=\alpha(x)\alpha(y)$.

The idea underlying the reduction of equation (19.6) to an equation with monotone operators is the same as in subsection 6.6. Let E be a real embeddable (see Definition 15.1) reflexive locally convex space, E' its strong dual, $F:E\to E'$ a monotone operator and $B:E'\to E$ a positive bounded selfadjoint linear operator. We have shown that in this case equations (19.6) and (19.7) are equivalent. However, since F is monotone, it follows that for any $y_1,\ y_2\in H$

$$(TF(Ay_1)-TF(Ay_2),\ y_1-y_2)=$$
$$=\langle F(Ay_1)-F(Ay_2),\ T'y_1-T'y_2\rangle=$$
$$=\langle F(Ay_1)-F(Ay_2),\ Ay_1-Ay_2\rangle\geqslant 0.$$

Thus, we have

Lemma 19.1. If F *is a monotone operator in* E, *then the right-hand side of (19.7) is a monotone operator in the Hilbert space* H.

We now consider a more general case. As in subsection 15.6, let E be an embeddable reflexive locally convex space, H the corresponding Hilbert space, and B a regular (see Definition 15.3) bounded operator from E' to E. Then, by Theorem (15.2), equality (15.9) holds, and equation (19.6) is equivalent (see subsection 6.6) to

$$VF(Ay)-(P_1-P_2)y=0 \quad (y\in H), \tag{19.10}$$

where A and V are as in Theorem 15.2. Since $V' = A$, it follows that $VF(Ay)$ is a monotone operator in H if $F: E \to E'$ is a monotone operator. Indeed,

$$(VF(Ay_1) - VF(Ay_2),\ y_1 - y_2) = \\ = \langle F(Ay_1) - F(Ay_2),\ Ay_1 - Ay_2 \rangle \geqslant 0.$$

Suppose now that the spectrum of the operator B_H (see subsection 15.6) lies outside an interval $(0, m), m > 0$, and that the operator $F: E \to E'$ satisfies the strong monotonicity condition

$$\langle F(x) - F(y),\ x - y \rangle \geqslant \frac{1+\alpha}{m} \| x - y \|_H^2 \qquad (\forall x, y \in E), \qquad (19.11)$$

where $0 < \alpha < 1$. Consider the mapping

$$\Phi(u) = VF(Ay) - (P_1 - P_2)y \quad (\forall y \in H)$$

from H to H. We have

$$(\Phi(u) - \Phi(v),\ u - v) = \langle F(Au) - F(Av),\ Au - Av \rangle + \\ + \| P_2(u - v) \|_H^2 - \| P_1(u - v) \|_H^2 \geqslant \\ \geqslant \frac{1+\alpha}{m} \| A(u - v) \|_H^2 + \| P_2(u - v) \|_H^2 - \| P_1(u - v) \|_H^2.$$

However, since (17.7) holds in this case too,

$$\| A(u - v) \|_H^2 \geqslant m \| P_1(u - v) \|^2.$$

Consequently,

$$(\Phi(u) - \Phi(v),\ u - v) \geqslant \alpha \| u - v \|^2.$$

We have thus proved

Lemma 19.2. Let the hypotheses of Theorem 15.2 and inequality (19.11) hold. Then equation (19.6) is equivalent to equation (19.10), and the left-hand side of the latter is a strongly monotone operator in the corresponding Hilbert space.

19.3. Equation in certain Banach spaces*

Theorem 19.4. Let the Banach space E, Hilbert space H and operator B from E^* to E satisfy the hypotheses of Theorem 16.3,

* Vainberg and Lavrent'ev /1/.

and let $F: E \to E^*$ *be a hemicontinuous operator such that*

$$\langle F(x+h) - F(x),\ h\rangle \leqslant 0 \quad (\forall x,\ h \in E).$$

Then, equation (19.6) has a unique solution, which lies in E.

Proof. Consider the mapping

$$\psi(u) = u - TF(T^*u) \quad (\forall u \in H),$$

which, by Theorem 16.3 and the hypotheses of our theorem, is defined in H. But for any $u,\ v \in H$,

$$\begin{aligned}(\psi(u) - \psi(v),\ u - v) &= \\ &= (u - v,\ u - v) - (TF(T^*u) - TF(T^*v),\ u - v) = \\ &= \|u - v\|^2 - \langle F(T^*u) - F(T^*v),\ T^*u - T^*v\rangle \geqslant \|u - v\|^2,\end{aligned}$$

so that ψ is a strongly monotone operator in H, whence, by Theorem 18.5, the equation

$$u - TF(T^*u) = 0 \tag{19.12}$$

has a unique solution u_0 in H. Applying the operator T^* to both sides of (19.12) and setting $T^*u_0 = x_0$, we see via Theorem 16.3, that x_0 is a solution of equation (19.6). Since equations (19.6) and (19.12) are, by Lemma 6.1, equivalent, x_0 is the unique solution of (19.6).

Remark 19.1. Theorem 19.1 remains valid if we replace the inequality $\langle F(x+h) - F(x), h\rangle \leqslant 0$ by

$$\langle F(x+h) - F(x),\ h\rangle \leqslant \gamma(\|h\|_E)\|h\|_E,$$

where

$$\sup_{t \in \sigma} \frac{\gamma(t)}{r} \leqslant \frac{1}{M} - \alpha(r),\ \gamma(t) > 0,\ \sigma = [0,\ Mr],$$

and M is a constant such that

$$\begin{gathered}\|T^*u\| \leqslant M\|u\|_H \quad (u \in H),\\ \alpha(r) > 0,\quad r\alpha(r) \to +\infty \quad \text{as} \quad r \to +\infty.\end{gathered}$$

Indeed, we then have, for $u,\ v \in H\ (u \neq v)$ and $r = \|u - v\|$,

$$\begin{aligned}(\psi(u) - \psi(v),\ u - v) &\geqslant \\ &\geqslant \|u - v\|^2 - \gamma(\|T^*(u - v)\|_E)\ \|T^*(u - v)\|_E \geqslant \\ &\geqslant \|u - v\|^2 - M\|u - v\|\gamma(\|T^*(u - v)\|_E) = \\ &= \|u - v\|^2(1 - M\gamma(\|T^*(u - v)\|_E)\|u - v\|^{-1}) \geqslant \\ &\geqslant r^2\left(1 - M\sup_{t \in \sigma}\frac{\gamma(t)}{r}\right) \geqslant r^2 M\alpha(r), \qquad \sigma \in [0,\ Mr].\end{aligned}$$

Consequently,

$$\lim_{r\to\infty}\frac{1}{r}(\psi(u)-\psi(v),\ u-v)=+\infty,$$

i. e., the operator $\psi(u)$ is strictly monotone and coercive in the Hilbert space H, so that, by Theorem 18.2, equation (19.12) has a unique solution. Hence, by Lemma 6.1, equation (19.6) has a unique solution in E. Observe that Remark 19.1 holds if $\gamma(t)=at$, where $aM^2<1$.

Remark 19.2. If we replace the inequality $\langle F(x+h)-F(x), h\rangle\leqslant 0$ in the assumptions of Theorem 19.4 by the inequalities

$$\langle F(x+h)-F(x),\ h\rangle\leqslant\gamma_1(\|h\|_E)\|h\|_E, \tag{19.13}$$

$$\langle F(x),\ x\rangle\leqslant\gamma_2(\|x\|_E)\|x\|_E, \tag{19.14}$$

where

$$\sup_{t\in\sigma}\frac{\gamma_1(t)}{r}\leqslant\frac{1}{M},\quad \sup_{t\in\sigma}\frac{\gamma_2(t)}{r}\leqslant\frac{1}{M}-\alpha(r),$$
$$\gamma_1(t)\geqslant 0,\quad \gamma_2(t)\geqslant 0,$$

$\sigma=[0, Mr]$, M is as in Remark 19.1, and $r\alpha(r)\to+\infty$ as $r\to+\infty$, then equation (19.6) has at least one solution.

Indeed, as in the previous remark, we infer from (19.13) that

$$(\psi(u)-\psi(v),\ u-v)\geqslant 0,$$

so that $\psi(u)$ is a monotone operator in H. It follows from (19.14) that

$$\begin{aligned}(\psi(u),\ u)&=(u,\ u)-(TF(T^*u),\ u)=\\ &=(u,\ u)-\langle F(T^*,\ u),\ T^*u\rangle\geqslant(u,\ u)-\gamma_2(\|T^*u\|_E)\|T^*u\|_E\geqslant\\ &\geqslant(u,\ u)-\gamma_2(\|T^*u\|_E)M\|u\|_H=\\ &=(u,\ u)[1-M\gamma_2(\|T^*u\|_E)\|u\|_H^{-1}]\geqslant\\ &\geqslant(u,\ u)\left[1-M\sup_{t\in\sigma}\frac{\gamma_2(t)}{\|u\|_H}\right]\geqslant(u,\ u)M\alpha(\|u\|_H),\\ &\sigma=[0,\ M\|u\|_H],\end{aligned}$$

so that $\psi(u)$ is a coercive operator in H. It therefore follows from Theorem 18.2 that the equation $\psi(u)=0$ has a solution $u_0\in H$. Applying the operator T^* to the equality $u_0-TF(T^*u_0)=0$, and noting that (by Theorem 16.3) $T^*T=B$, we obtain

$$x_0 = BF(x_0),$$

where $x_0 = T^* u_0$. This proves our assertion.

Note also that if

$$\sup_{t \in \sigma} \frac{\gamma_1(t)}{r} < \frac{1}{M}, \qquad \sigma = [0, Mr],$$

then $\psi(u)$ is a strictly monotone operator and the solution is unique.

We now consider the indefinite case, employing the notation and propositions of subsection 16.4.

Theorem 19.5. Suppose that:

1. The Banach space E, Hilbert space H and operator $B: E^* \to E$ satisfy the hypotheses of Theorem 16.7 (or 16.8).

2. The spectrum of the operator B_H lies outside some interval $(0, m)$, where m is a positive number.

3. F is a continuous operator from E into E^* such that

$$\langle F(x+h) - F(x), h \rangle \geqslant \frac{1+\alpha}{m} \| h \|_H^2 \; (\forall x, h \in V^*[D]), \qquad (19.15)$$

where $D = \{x \in H_0 : Ax \in E\}$ is dense in H and $0 < \alpha \leqslant 1$.

Then equation (19.6) has a unique solution in E.

Proof. Let E_t be a resolution of the identity for the operator B_H. Consider the subspaces $H_1, H_2 \subset H$ and projections P_1, P_2 as in the proof of Theorem 19.2, and set

$$\Phi(u) = VF(V^* u) - (P_1 - P_2) u \quad (u \in H).$$

By the hypotheses of our theorem and Theorem 16.7 (or 16.8), this operator is defined in H.

We claim that $V^* x = Ax$ for any $x \in D$. Indeed, if $x, y \in D$, then

$$(x, Ay) = (Ax, y) = (Vx, y) = (x, V^* y).$$

Hence, since D is dense in H, it follows that $Ay = V^* y$. Let u, v be arbitrary vectors in D. Then

$$\begin{aligned}
&(\Phi(u) - \Phi(v), u - v) = \\
&= (V[F(V^* u) - F(V^* v)], u - v) - ((P_1 - P_2)(u - v), u - v) = \\
&\quad = \langle F(V^* u) - F(V^* v), V^* u - V^* v \rangle + \| P_2(u - v) \|_H^2 - \\
&\qquad - \| P_1(u - v) \|_H^2 \geqslant \frac{1+\alpha}{m} \| V^*(u - v) \|_H^2 + \\
&\qquad\qquad + \| P_2(u - v) \|_H^2 - \| P_1(u - v) \|_H^2.
\end{aligned}$$

Now, by (17.7), $\| V^*(u-v)\|_H^2 \geqslant m\| P_1(u-v)\|_H^2$, whence

$$(\Phi(u)-\Phi(v),\ u-v) \geqslant \alpha \| u-v\|_H^2,$$

so that $\Phi(u)$ is strongly monotone on the dense subset D of H. Since Φ is continuous in H (because F is continuous), the last inequality holds throughout H and thus the operator Φ is strongly monotone in H. Therefore, by Theorem 18.5, the equation $\Phi(u)=0$ has a unique solution $u_0 \in H$. Applying the operator W^* to

$$VF(V^*u_0) = (P_1 - P_2)\, u_0$$

and taking into account that (see Theorem 16.7 or 16.8 and subsection 6.6) $W^*V = B$ and $W^*(P_1 - P_2) = V^*$, we find that

$$BF(x_0) = x_0,$$

where $x_0 = V^*u_0 \in E$. Since equations (19.6) and (19.10) are by subsection 6.6 equivalent, x_0 is the unique solution of equation (19.6). Q. E. D.

Remark 19.3. If $E \subset H \subset E^*$, we may set $D = E$ in (19.15). Moreover, F need only be hemicontinuous and not necessarily continuous, since (Theorem 1.2) a monotone hemicontinuous operator is demicontinuous.

19.4. Equations in locally convex spaces*

Here we shall use the notation and propositions of subsections 15.5 and 15.6 to prove existence theorems for equations (19.6) of Hammerstein type in locally convex spaces. The two propositions presented will suffice to illustrate the method.

Theorem 19.6. Suppose that:

1. *E is a real reflexive locally convex embeddable space and H the corresponding Hilbert space.*
2. *B is a positive bounded selfadjoint operator from E' to E.*
3. *$F: E \to E'$ is a hemicontinuous operator such that*

$$\langle F(x+h) - F(x),\ h\rangle \leqslant \alpha \| h\|_H^2$$

for any $x, h \in E$, where $\alpha \| B_H \| < 1$.

Then equation (19.6) has a unique solution, which lies in E.

* Vainberg /16/.

Proof. By Theorem 15.1, we have $B=T'T$, where $T'=A=B_H^{1/2}$, B_H is the restriction of B to H, and $T:E'\to H$ is a bounded extension of A. Since, as we have shown, equations (19.6) and (19.7) are equivalent, it suffices to show that equation (19.7) has a unique solution. To check this, we consider the operator

$$\Phi(u)=u-TF(Au),\quad u\in H,$$

which is defined in H and is hemicontinuous, by hypothesis. We have

$$\begin{aligned}(\Phi(u)-\Phi(v),\ u-v)= \\ =\|u-v\|^2-(T[F(Au)-F(Av)],u-v)= \\ =\|u-v\|^2-\langle F(Au)-F(Av),\ Au-Av\rangle\geqslant \\ \geqslant\|u-v\|^2-\alpha\|A(u-v)\|^2\geqslant\|u-v\|^2- \\ -\alpha\|B_H\|\|u-v\|^2=c\|u-v\|^2,\end{aligned}$$

where $c=1-\alpha\|B_H\|>0$.

The operator Φ is thus monotone and hemicontinuous, so that, by Theorem 18.5, there exists a unique vector $u_0\in H$ such that $u_0=TF(Au_0)$. Applying the operator T' and setting $Au_0=x_0\in E$, we obtain $x_0=BF(x_0)$. Q. E. D.

Theorem 19.7. Suppose that:

1. E is a real reflexive locally convex embeddable space and H the corresponding Hilbert space.

2. B is a regular (see Definition 15.3) bounded operator from E' to E.

3. The spectrum of B_H lies outside some interval $(0,m)$, where m is a positive number.

4. $F:E\to E'$ is a hemicontinuous operator satisfying inequality (19.11).

Then equation (19.6) has a unique solution, which lies in E.

Proof. By Theorem 15.2, $B=V'W=W'V$, where $V'=A$ and $V:E'\to H$ is a bounded extension of the operator A. Consider the operator

$$\Phi(u)=VF(Au)-(P_1-P_2)u\quad(u\in H),$$

which, by the hypothesis of the theorem and Lemma 19.2, is hemicontinuous and monotone in H. It follows by Theorem 18.5 that equation (19.10) has a unique solution $u_0\in H$, i. e.,

$$VF(Au_0)=(P_1-P_2)u_0.$$

Now, applying W' and noting that $W'V = B$ and $W'(P_1 - P_2) = V' = A$, we obtain

$$BF(x_0) = x_0,$$

where $x_0 = Au_0 \in E$. Equations (19.6) and (19.10) are equivalent by Lemma 19.2, so this solution is unique. Q. E. D.

19.5. Equations in Hilbert space

The theorems proved above for equations of Hammerstein type in Banach and locally convex spaces have several implications concerning such equations in Hilbert space. We cite one such proposition, as a corollary of Theorem 19.1.

*Theorem 19.8.** *Let F be a bounded monotone hemicontinuous operator in a real Hilbert space H such that*

$$(F(u), u) \geqslant 0 \quad \text{if} \quad \|u\| > \lambda > 0 \quad (\lambda = \text{const})$$

and let $B: H \to H$ be a positive bounded linear operator. Then the equation

$$u + BF(u) = 0 \quad (u \in H)$$

has a solution. If B is strictly positive or F strictly monotone, this solution is unique.

To prove this, we need only observe that the Hilbert space H has property (π) (see Definition 19.1), so that the theorem follows from Theorem 19.1. In this case $A = I$, where I is the identity operator in H.

19.6. Equations with semimonotone operators

Here we consider the equation $x + BF(x) = 0$ in a Banach space E with property (π) (see Definition 19.1), on the assumption that the operator $F: E \to E^*$ is semimonotone (see Definition 18.3).

Theorem 19.9. Suppose that:

1. *E is a Banach space with property (π).*
2. *$F: E \to E^*$ is a bounded hemicontinuous semimonotone operator such that $\langle F(x), x\rangle \geqslant 0$ if $\|x\| > \lambda > 0$ ($\lambda =$ const).*
3. *B is a positive bounded linear operator from E^* to E.*

* Dolph and Minty /1/.

Then the equation

$$x + BF(x) = 0$$

has a solution lying in E.

Proof. By property (π), there exists an operator $A: E^* \to E$ with the properties indicated in Definition 19.1. Consider the sequence

$$\Phi_n z = A\frac{z}{n} + B^* z + BFB^* z \quad (\forall z \in E^*; \quad n = 1, 2, 3, \ldots).$$

For each n, the operator $\Phi_n : E^* \to E$ is hemicontinuous and semimonotone. Hemicontinuity follows from the hemicontinuity of F and A and the continuity of B. Since F is semimonotone, there exists a strongly continuous operator $G: E \to E^*$ such that $(F + G)$ is monotone. Since G is strongly continuous, the operator $BGB^*: E^* \to E$ is also strongly continuous. We claim that the operator

$$T_n = \Phi_n + BGB^*$$

is monotone. Indeed,

$$\begin{aligned}\langle T_n x - T_n y, x - y\rangle = \left\langle A\frac{x}{n} - A\frac{y}{n}, x - y\right\rangle + \\ + \langle B(x - y), x - y\rangle + \langle (FB^* x + GB^* x) - \\ - (FB^* y + GB^* y), B^*(x - y)\rangle \geqslant 0,\end{aligned}$$

since all the terms are nonnegative. Since the operator A is coercive, so is Φ_n, and hence, by Theorem 18.4, there exists a solution to the equation $\Phi_n z = 0$, i. e.,

$$A\frac{z_n}{n} + B^* z_n + BFB^* z_n = 0 \qquad (n = 1, 2, 3, \ldots). \tag{19.16}$$

These equalities coincide with (19.1), so that, proceeding as in the proof of Theorem 19.1, we again get (19.4) and (19.5). To prove that $y_0 = F(x_0)$, we adopt a slightly different method. Set $z_{n_k} = u_k$, $B^* z_{n_k} = x_k$, $FB^* z_{n_k} = y_k$. Then (19.4) yields

$$x_k \rightharpoonup x_0 \in E, \quad y_k \rightharpoonup y_0 \in E^* \tag{19.17}$$

as $k \to \infty$. Since the operator $T = F + G$ is monotone,

$$\langle Tx - Tx_k, x - x_k\rangle \geqslant 0$$

or

$$\langle Tx, x-x_k\rangle - \langle Tx_k, x\rangle \geqslant \langle Tx_k, -x_k\rangle = \\ = \langle Fx_k, -x_k\rangle + \langle Gx_k, -x_k\rangle.$$

But, using (19.16) and our notation, we have

$$-x_k = -B^*z_{n_k} = A\frac{u_k}{n_k} + BFB^*z_{n_k} = A\frac{u_k}{n_k} + By_k,$$

whence

$$\langle Tx, x-x_k\rangle - \langle Tx_k, x\rangle \geqslant \left\langle Fx_k, A\frac{u_k}{n_k}\right\rangle + \\ + \langle y_k, By_k\rangle + \left\langle Gx_k, A\frac{u_k}{n_k}\right\rangle + \langle Gx_k, By_k\rangle. \tag{19.18}$$

However, it follows from (19.17) that as $k\to+\infty$,

1) $\langle Tx, x-x_k\rangle \to \langle Tx, x-x_0\rangle$,

2) $\langle Tx_k, x\rangle = [\langle y_k, x\rangle + \langle Gx_k, x\rangle] \to \langle y_0, x\rangle + \langle Gx_0, x\rangle$,

3) $\left\langle Fx_k, A\frac{u_k}{n_k}\right\rangle = \left\langle y_k, A\frac{u_k}{n_k}\right\rangle \to 0$ (see the proof of Theorem 19.1),

4) $\varliminf_{k\to\infty} \langle y_k, By_k\rangle \geqslant \langle y_0, By_0\rangle$ (see Example 8.6),

5) $\left\langle Gx_k, A\frac{u_k}{n_k}\right\rangle \to \langle Gx_0, A0\rangle = 0$,

6) $Gx_k \to Gx_0$, $\| By_k \| \leqslant N = \text{const}$,

$$|\langle Gx_k - Gx_0, By_k\rangle| \leqslant N\| Gx_k - Gx_0\| \to 0.$$

Consequently,

$$\langle Gx_k, By_k\rangle = [\langle Gx_0, By_k\rangle + \langle Gx_k - Gx_0, By_k\rangle] \to \langle Gx_0, By_0\rangle.$$

If we let $k\to\infty$ in (19.18), these relations together with (19.5) give

$$\langle Tx, x-x_0\rangle - \langle y_0, x\rangle - \langle Gx_0, x\rangle \geqslant \langle y_0, x_0\rangle - \langle Gx_0, x_0\rangle,$$

or

$$\langle Tx - (y_0 + Gx_0), x - x_0\rangle \geqslant 0.$$

Thus, by Lemma 18.1,

$$y_0 + Gx_0 = Tx_0 = Fx_0 + Gx_0,$$

i. e., $y_0 = Fx_0$. Finally, we substitute this into (19.5) to obtain $x_0 + BF(x_0) = 0$. Q. E. D.

§ 20. EQUATIONS OF HAMMERSTEIN TYPE WITH UNBOUNDED OPERATORS

20.1. The case of a Banach space

Here E will be a real embeddable reflexive Banach space (see Definition 15.2) and H the corresponding Hilbert space. In this space E we consider an equation of Hammerstein type

$$x + BF(x) = 0, \tag{20.1}$$

where $F: E \to E^*$ is a semimonotone operator and B is a linear operator from $E^*$to E, not necessarily bounded. We assume,however, that the following condition is satisfied.

Condition 20.1. The linear operator B from E^* to E satisfies the following conditions:

1. It may be represented as $B = CA$, where A is a bounded linear operator from E^* to H with dense domain* and C a linear operator from H to E with dense domain having a bounded right inverse C^{-1}, i. e., $CC^{-1}x = x$ for any $x \in E$.

2. The operators C^{-1} and $C^{-1}A^*$ are positive and

$$\|C^{-1}x\|_H^2 \leqslant b(C^{-1}x, x) \quad (\forall x \in E),$$

$$\|x\|_E^2 \leqslant m^2 + a\|C^{-1}x\|_H^2 \quad (\text{if } \|A^*x\| \leqslant m),$$

where a, b, m are constants.

Theorem 20.1. Suppose that:

1. E is a real reflexive embeddable Banach space, H the corresponding Hilbert space.

2. F is a bounded hemicontinuous semimonotone operator from E to E^* such that $\langle F(u), u\rangle \geqslant 0$ if $\|u\| > \lambda$, where λ is a positive number.

* The extension of the operator A to all of E^* is also denoted by A, so that in what follows we shall verify whether the vector Ax lies in the domain $D(C)$ of the operator C.

3. *The positive linear operator* B *satisfies Condition 20.1.*

4. *The operator* $\omega_n = \frac{1}{n}C^{-1} + AFA^*$ *is coercive for any* n.

Then equation (20.1) has a solution lying in E.

Proof. Consider the sequence of operators

$$\Psi_n z = \frac{1}{n}C^{-1}z + C^{-1}A^*z + AFA^*z$$
$$(z \in E;\ n = 1,\ 2,\ 3,\ \ldots),$$

which map E into H. The hypotheses of the theorem imply that Ψ_n is hemicontinuous. It is also coercive, since $\Psi_n = \omega_n + C^{-1}A^*$, where ω_n is coercive and

$$(C^{-1}A^*z,\ z) \geqslant 0$$

by assumption. We claim that the operator Ψ_n is semimonotone. Indeed, since F is semimonotone, there exists a strongly continuous operator G such that $F + G = T$ is monotone. We shall prove that the operator

$$T_n = \Psi_n + AGA^*$$

is monotone. Indeed, for any $y_1,\ y_2 \in E$,

$$(T_n y_1 - T_n y_2,\ y_1 - y_2) = \frac{1}{n}(C^{-1}(y_1 - y_2),\ y_1 - y_2) +$$
$$+ (C^{-1}A^*(y_1 - y_2),\ y_1 - y_2) + \langle (FA^*y_1 + GA^*y_1) -$$
$$- (FA^*y_2 + GA^*y_2),\ A^*(y - y_2)\rangle,$$

and since each term on the right is nonnegative, T_n is a monotone operator. Thus Ψ_n is a semimonotone operator, since AGA^* is a strongly continuous operator mapping H into itself. Since Ψ_n is hemicontinuous, monotone and coercive, it follows from Theorem 18.4 that there is a solution $z_n \in E$ to the equation $\Psi_n z = 0$, i. e.,

$$\frac{1}{n}C^{-1}z_n + C^{-1}A^*z_n + AFA^*z_n = 0 \qquad (n = 1,\ 2,\ 3,\ \ldots). \qquad (20.2)$$

We claim that the sequence $\{C^{-1}z_n/n\}$ converges to zero. In fact, by (20.2),

$$\frac{1}{n}(C^{-1}z_n,\ z_n) + (C^{-1}A^*z_n,\ z_n) + (AFA^*z_n,\ z_n) = 0.$$

But $(C^{-1}A^*z_n,\ z_n) \geqslant 0$ and $(C^{-1}z_n,\ z_n) > 0$ for $z_n \neq 0$, so that, for $z_n \neq 0$,

$$(AFA^*z_n,\ z_n) = \langle FA^*z_n,\ A^*z_n\rangle < 0.$$

The hypotheses of the theorem now yield $\| A^*z_n \| \leqslant \lambda$ $(n=1, 2, 3, \ldots)$, and since F is bounded, it follows that $\| FA^*z_n \| \leqslant l$. Consequently,

$$\frac{1}{n}(C^{-1}z_n, z_n) \leqslant \frac{1}{n}(C^{-1}z_n, z_n) + (C^{-1}A^*z_n, z_n) =$$
$$= -(AFA^*z_n, z_n) = -\langle FA^*z_n, A^*z_n \rangle \leqslant l\lambda.$$

Since $\|C^{-1}z_n\|^2 \leqslant b(C^{-1}z_n, z_n)$, we have

$$\frac{1}{n}\|C^{-1}z_n\|^2 \leqslant \frac{b}{n}(C^{-1}z_n, z_n) \leqslant bl\lambda,$$

or

$$\left\| \frac{C^{-1}z_n}{n} \right\|^2 \leqslant \frac{bl\lambda}{n},$$

and so

$$\lim_{n\to\infty} n^{-1}C^{-1}z_n = 0. \qquad (20.3)$$

Now, since the sequences $\{A^*z_n\}$ and $\{FA^*z_n\}$ in E and E^*, respectively, are bounded and the space E is reflexive, there exist subsequences such that as $k \to \infty$,

$$A^*z_{n_k} \longrightarrow x_0 \in E, \quad FA^*z_{n_k} \longrightarrow y_0 \in E^*. \qquad (20.4)$$

Substituting n_k for n in (20.2) and letting $k \to \infty$, we have from (20.3) and (20.4) that

$$C^{-1}x_0 + Ay_0 = 0. \qquad (20.5)$$

It remains to show that $y_0 = F(x_0)$. To verify this, we use the fact that the operator $F + G = T$ is monotone; for the sake of brevity, we set

$$x_k = A^*z_{n_k}, \quad y_k = FA^*z_{n_k} = Fx_k.$$

We may then write

$$\langle Tx - Tx_k, \quad x - x_k \rangle \geqslant 0,$$

or

$$\langle Tx, x - x_k \rangle - \langle Tx_k, x \rangle \geqslant \langle Fx_k, -x_k \rangle + \langle Gx_k, -x_k \rangle.$$

But it follows from (20.2) that

$$\frac{1}{n_k} C^{-1} z_{n_k} + C^{-1} x_k + A y_k = 0,$$

whence Ay_k belongs to the domain $D(C)$ of the operator C.

Applying the operator C to the last equality, we obtain

$$\frac{1}{n_k} z_{n_k} + x_k + B y_k = 0. \tag{20.6}$$

Hence, by combining the previ ous relations and the fact that $Tx_k = Fx_k + Gx_k = y_k + Gx_k$, we obtain

$$\begin{aligned} \langle Tx, x - x_k \rangle - \langle y_k, x \rangle - \langle Gx_k, x \rangle \geqslant \\ \geqslant \left\langle y_k, \frac{1}{n_k} z_{n_k} \right\rangle + \langle y_k, By_k \rangle - \langle Gx_k, x_k \rangle. \end{aligned} \tag{20.7}$$

As $k \to +\infty$, however, we have

1) $\langle Tx, x - x_k \rangle \to \langle Tx, x - x_0 \rangle$;
2) $\langle y_k, x \rangle \to \langle y_0, x \rangle$; $\langle Gx_k, x \rangle \to \langle Gx_0, x \rangle$;
3) $\left\langle y_k, \frac{1}{n_k} z_{n_k} \right\rangle \to 0$ since, by assumption, if, $\| A^* z_{n_k} \| \leqslant \lambda$, then

$$\left\| \frac{1}{n_k} z_{n_k} \right\|_E^2 \leqslant \left(\frac{\lambda^2}{n_k^2} + a \left\| \frac{1}{n_k} C^{-1} z_{n_k} \right\|^2 \right) \to 0, \quad \| y_k \| \leqslant l;$$

4) $\langle By_k, y_k \rangle = \langle B(y_k - y_0), y_k - y_0 \rangle + \langle By_0, y_k - y_0 \rangle + \langle By_k, y_0 \rangle$,

because (20.5) implies that $y_0 \in D(B)$. Hence, condition (20.6) and the fact that the operator B is positive, yield

$$\langle By_k, y_k \rangle \geqslant \langle By_0, y_k - y_0 \rangle - \left\langle \frac{1}{n_k} z_{n_k}, y_0 \right\rangle - \langle x_k, y_0 \rangle.$$

But as $k \to \infty$ we have $\langle x_k, y_0 \rangle \to \langle x_0, y_0 \rangle$,

$$\langle By_0, y_k - y_0 \rangle \to 0, \quad \left\langle \frac{1}{n_k} z_{n_k}, y_0 \right\rangle \to 0,$$

and since $\langle By_k, y_k \rangle \geqslant 0$, it follows that

$$\lim_{k \to \infty} \langle By_k, y_k \rangle \geqslant - \langle x_0, y_0 \rangle.$$

Finally, since G is strongly continuous,

5) $\langle Gx_k, x_k\rangle \to \langle Gx_0, x_0\rangle$.

Passing to the lower limit as $k\to\infty$ in (20.7) and using these relations we see that

$$\langle Tx, x-x_0\rangle - \langle y_0, x\rangle - \langle Gx_0, x\rangle \geqslant -\langle x_0, y_0\rangle - \langle Gx_0, x_0\rangle,$$

or

$$\langle Tx-(y_0+Gx_0), x-x_0\rangle \geqslant 0.$$

Hence, by Lemma 18.1,

$$y_0+Gx_0=Tx_0=F(x_0)+Gx_0,$$

or

$$y_0=F(x_0).$$

Substituting this into (20.5), we have

$$C^{-1}x_0+AF(x_0)=0.$$

Since this implies that $AF(x_0)\in D(C)$, we may apply the operator C to the last equality and, using the fact that $CA=B$, we obtain $x_0+BF(x_0)=0$, i. e., $x_0\in E$. Q. E. D.

Remark 20.1. If, in addition, we assume that either the operator B is positive and strictly monotone, or F is monotone and B strictly positive, then equation (20.1) has a unique solution. For the proof, see the end of subsection 19.1.

20.2. The case of a Hilbert space

We now consider equation (20.1) in a real Hilbert space, where the restrictions become more intuitive, and present a proposition due to Kositskii. As opposed to subsection 19.5 (see Theorem 19.8) here we do not assume that the operator B is bounded.

Theorem 20.2. Suppose that:*

1. F is a semimonotone bounded hemicontinuous and coercive operator mapping a real Hilbert space H into itself.

* Kositskii /1,2/.

2. *B is a closed linear operator from H to H with dense domain $D(B)$, such that for any $x \in D(B)$ and $y \in D(B^*)$*

$$(Bx,\ x) \geqslant 0, \quad (B^*y,\ y) \geqslant 0.$$

Then equation (20.1) has a solution in H.

It suffices to show that the hypotheses of this theorem imply those of Theorem 20.1. Now any real Hilbert space may be regarded as an embeddable Banach space, so we have only to prove that the remaining hypotheses of Theorem 20.1 are also valid. By the theorem on the canonical representation of closed linear operators,* the operator B can be expressed as KU, where U is a partially isometric operator and K a positive selfadjoint operator.

Let E_t be a resolution of the identity for the operator K, so that (see Akhiezer and Glazman /1/)

$$K = \int_{-\infty}^{+\infty} t\, dE_t.$$

Consider the real functions of a real variable

$$\alpha(t) = \begin{cases} 1, & t \leqslant 1 \\ t, & t \geqslant 1 \end{cases}; \quad \beta(t) = \begin{cases} t, & t \leqslant 1, \\ 1, & t > 1, \end{cases}$$

with product $\alpha(t)\beta(t) = t$, and denote the corresponding functions of the operator K by C and Γ. Then

$$C = \alpha(K) = \int_{-\infty}^{+\infty} \alpha(t)\, dE_t; \quad \Gamma = \beta(K) = \int_{-\infty}^{+\infty} \beta(t)\, dE_t;$$

$$C\Gamma = \int_{-\infty}^{+\infty} t\, dE_t = K,$$

the operators C and Γ are selfadjoint, commute, and $\|\Gamma\| \leqslant 1$. By the spectral mapping theorem (Dunford and Schwartz /2/), the spectrum $\sigma(C)$ of C is $\alpha(\sigma(K)$, where $\sigma(K)$ is the spectrum of K, and thus $0 \notin \sigma(C)$, i. e., C is invertible. In fact, by standard theorems on functions of an operator, its inverse is

$$C^{-1} = \int_{-\infty}^{+\infty} \frac{1}{\alpha(t)}\, dE_t,$$

* Dunford and Schwartz /2/, Theorem 12.77.

so that

$$\|C^{-1}\| \leqslant \max \frac{1}{\alpha(t)} = 1.$$

Now set $A = \Gamma U$. Then $B = CA$, where A is bounded and C has a bounded inverse. We have thus proved part 1 of Condition 20.1, and proceed to verify part 2. That the operator C^{-1} is positive follows from its representation (it is selfadjoint and positive). Then

$$(C^{-1}A^*x, x) = (C^{-1}U^*\Gamma x, x) = (U^*\Gamma CC^{-1}x, C^{-1}x) = \\ = (U^*K^*C^{-1}x, C^{-1}x) = (B^*C^{-1}x, C^{-1}x) \geqslant 0.$$

Now, since C^{-1} is selfadjoint and positive and $\|C^{-1/2}\| \leqslant 1$, we have

$$\|C^{-1}x\|^2 = \|C^{-1/2}C^{-1/2}x\|^2 \leqslant \|C^{-1/2}\|^2 \|C^{-1/2}x\|^2 \leqslant \\ \leqslant \|C^{-1/2}x\|^2 = (C^{-1/2}x, C^{-1/2}x) = (C^{-1}x, x).$$

Finally, if $\|\Gamma x\| \leqslant m$ then, since U is partially isometric,

$$\|A^*x\| = \|U^*\Gamma x\| = \|\Gamma x\|,$$

so that

$$m^2 \geqslant \|\Gamma x\|^2 = \int_{-\infty}^{+\infty} \beta^2(t)\, d(E_t x, x) \geqslant \int_1^{+\infty} \beta^2(t)\, d(E_t x, x) = \\ = \int_1^{+\infty} d(E_t x, x) = \|(I - E_1)x\|^2$$

and

$$\|C^{-1}x\|^2 = \int_{-\infty}^{+\infty} \frac{1}{\alpha^2(t)}\, d(E_t x, x) \geqslant \int_{-\infty}^{1} \frac{1}{\alpha^2(t)}\, d(E_t x, x) = \\ = \int_{-\infty}^{1} d(E_t x, x) = \|E_1 x\|^2.$$

Hence, if $\|A^*x\| \leqslant m$, then

$$\|x\|^2 = (I - E_1)x\|^2 + \|E_1 x\|^2 \leqslant m^2 + \|C^{-1}x\|^2.$$

Condition 20.1 therefore holds.

We now proceed to the other hypotheses of Theorem 20.1. Since F is coercive, there exists a positive λ such that if $\|x\| > \lambda$ then $(Fx, x) > 0$, and thus condition 2 of Theorem 20.1 holds. It remains to show that condition 4 of Theorem 20.1 is satisfied. By our previous arguments,

$$\|C^{-1}x\|^2 \geqslant \|E_1 x\|^2, \ \|A^* x\|^2 = \|\Gamma x\|^2 \geqslant \|Px\|^2, \tag{20.8}$$

where $P = I - E_1$, so that for any $x \in H$

$$\|x\|^2 = \|E_1 x\|^2 + \|Px\|^2. \tag{20.9}$$

Since F is by hypothesis coercive, there exists an increasing function $\gamma(t)$, tending to infinity as $t \to +\infty$, such that

$$(F(x), \ x) \geqslant \|x\| \gamma(\|x\|),$$

and for $\|x\| \geqslant \lambda$ we have $\gamma(\|x\|) > 0$.

Now consider the sphere $S = \{x \in H : \|x\| = \sqrt{2}\lambda\}$ in H. Let $S_1 \subset S$ denote the set of vectors for which $\|Px\|^2 \geqslant 2^{-1}\|x\|^2$. Then for all vectors in $S_2 = S \setminus S_1$, we have

$$\|E_1 x\|^2 > 2^{-1}\|x\|^2.$$

Hence, for any ray tx $(t \geqslant 1)$

$$\|Ptx\|^2 \geqslant 2^{-1}\|tx\|^2 \qquad (x \in S_1)$$
$$\|E_1 tx\|^2 > 2^{-1}\|tx\|^2 \qquad (x \in S_2).$$

Since for $x \in S_1$ we have

$$\|Px\| \geqslant \frac{1}{\sqrt{2}}\|x\| = \lambda,$$

it follows from (20.8) that $\|A^* x\| \geqslant \lambda$, and thus

$$\gamma(\|A^* tx\|) > 0 \qquad (t \geqslant 1, \ x \in S_1).$$

Hence, if $x \in S_1$ and $t \geqslant 1$,

$$(AFA^* tx, \ tx) = (FA^* tx, \ A^* tx) \geqslant \gamma(\|A^* tx\|) \|A^* tx\| \geqslant$$
$$\geqslant \gamma(\|Ptx\|) \|Ptx\| \geqslant \frac{1}{\sqrt{2}} t\|x\| \gamma\left(\frac{1}{\sqrt{2}} t\|x\|\right),$$

i. e., for any $x \in S_1$,

$$\lim_{t\to+\infty}\frac{(AFA^*tx,\, tx)}{t\,\|x\|}=+\infty.$$

The operator ω_n is thus coercive on the set tS_1 $(t\geqslant 1)$, since the term $n^{-1}(C^{-1}x, x)$ is positive for $x\neq 0$. We now show that the operator ω_n is coercive on the set tS_2 $(t\geqslant 1)$. Indeed, we have seen that $(C^{-1}x, x)\geqslant\|C^{-1}x\|^2$. Thus, for $x\in S_2$, it follows from (20.8) that

$$\frac{1}{n}(C^{-1}tx,\ tx)\geqslant\frac{t^2}{n}\|E_1x\|^2>\frac{t^2}{2n}\|x\|^2.$$

Consequently, $n^{-1}(C^{-1}x, x)$ is coercive on tS_2 $(t\geqslant 1)$, and, since (AFA^*x, x) is either bounded or tends to $+\infty$ as $\|x\|\to\infty$, it follows that ω_n is coercive on tS_2 $(t\geqslant 1)$ for any fixed n. Thus, condition 4 of Theorem 20.1 is satisfied, and the proof of Theorem 20.2 is complete.

20.3. Equations reducible to equations of Hammerstein type*

Here we consider equations of the form

$$Bx+F(x)=0, \tag{20.10}$$

where F is a nonlinear operator and B a linear operator, both from a real reflexive Banach space E to E^*. Such equations are either reducible to equations of Hammerstein type or else existence theorems for them may be proved by the same method as for equations of Hammerstein type.

Theorem 20.3. Suppose that:

1. E is a Banach space with property (π).

2. $F:E\to E^*$ is a bounded hemicontinuous and semimonotone operator such that

$$\langle F(x),\ x\rangle\geqslant 0 \quad \text{if} \quad \|x\|>\lambda>0 \quad (\lambda=\text{const}).$$

3. B is a positive linear operator with dense domain $D(B)\subset E$ and range $R(B)=E^*$, having a bounded right inverse.

Then equation (20.10) has a solution in E.

Proof. Since B is a positive operator, its inverse B^{-1} is also positive. Hence, by Theorem 19.9, the equation

$$x+B^{-1}F(x)=0$$

* Equations of this type in Banach spaces have been studied by Zeragia /1,2/, Browder /7,8/, Kositskii /1,2/ and others.

has a solution $x_0 \in E$. Since $B^{-1}F(x_0) \in D(B)$, it follows from the last equality that $x_0 \in D(B)$, so that, applying B, we obtain

$$Bx_0 + F(x_0) = 0.$$

Q. E. D.

Remark 20.2. The theorem remains valid for the equation $Bx + F(x) = y$, where y is any vector in E^*, since the operator $F_y = F - y$ has the same property as F.

We now consider a more general case. To do this for the case of embeddable spaces (Definition 15.2), we introduce

Condition 20.2. The linear operator B from E to E^* satisfies the following conditions:

1. It may be expressed as $B = CA$, where A is a bounded linear operator from E to H $(E \subset H \subset E^*)$ with dense domain,* C is a linear operator from H to E^* with dense domain, having a bounded right inverse C^{-1}, i. e., $CC^{-1}x = x$ for any $x \in E^*$.

2. The operators C^{-1} and AV, where $V = (C^{-1})^*$, are positive, for any $x \in E^*$

$$\|C^{-1}x\|_H \leqslant b(C^{-1}x, x),$$

and if $\|Vx\|_E \leqslant m$ for certain vectors $x \in E$, then for such vectors

$$\|x\|_H^2 \leqslant m^2 + a\|C^{-1}x\|_H^2,$$

where a, b, m are constants.

Theorem 20.4. Suppose that:

1. E is a real reflexive embeddable Banach space and H the corresponding Hilbert space.

2. F is a hemicontinuous bounded semimonotone operator from E to E^* such that $\langle F(x), x\rangle \geqslant 0$ if $\|x\| > \lambda$, where λ is a positive number.

3. B is a positive linear operator satisfying Condition 20.2.

4. The operator $\omega_n = n^{-1}C^{-1} + C^{-1}FV$ is coercive for any n.

Then equation (20.10) has a solution lying in E.

Proof. Consider the sequence of operators

$$\Phi_n x = \frac{1}{n} C^{-1}x + AVx + C^{-1}FVx$$

$$(x \in E;\ n = 1,\ 2,\ 3,\ \dots),$$

* The continuous extension of A to all of E is also denoted by A.

mapping E into H. It follows from our hypotheses that each operator Φ_n is hemicontinuous and coercive. Since F is semimonotone, there exists a strongly continuous operator $G: E \to E^*$ such that the operator $F + G = T$ is monotone. Since G is strongly continuous, the operator $C^{-1}GV$, mapping H into itself, is strongly continuous. Consider the operator

$$T_n = \Phi_n + C^{-1}GV$$

as a mapping of all of E into H. Then, for any $y_1, y_2 \in E$,

$$\begin{aligned}(T_n y_1 - T_n y_2,\ y_1 - y_2) &= \\ = \frac{1}{n}(C^{-1}(y_1 - y_2),\ y_1 - y_2) &+ (AV(y_1 - y_2),\ y_1 - y_2) + \\ + \langle (FVy_1 + GVy_1) &- (FVy_2 + GVy_2),\ Vy_1 - Vy_2\rangle,\end{aligned}$$

and since each term on the right is positive (because the operator $F + G = T$ is monotone), it follows that the operator T_n is monotone. The operator Φ_n is therefore semimonotone. Thus, for any fixed n, the operator Φ_n satisfies the hypotheses of Theorem 18.4, so that there exists a solution $z_n \in E$ to the equation $\Phi_n z = 0$, i. e.,

$$\frac{1}{n}C^{-1}z_n + AVz_n + C^{-1}FVz_n = 0 \quad (n = 1, 2, 3, \ldots). \qquad (20.11)$$

One now proves along lines similar to the proof of Theorem 20.1 that the sequence $n^{-1}C^{-1}z_n$ converges to zero. Indeed, by (20.11),

$$\frac{1}{n}(C^{-1}z_n,\ z_n) + (AVz_n,\ z_n) + \langle FVz_n,\ Vz_n\rangle = 0.$$

But, for $z_n \neq 0$,

$$\frac{1}{n}(C^{-1}z_n,\ z_n) > 0; \quad (AVz_n,\ z_n) \geqslant 0,$$

whence

$$\langle FVz_n,\ Vz_n\rangle < 0.$$

Condition 2 of the theorem now implies that

$$\| Vz_n \| \leqslant \lambda \qquad (n = 1,\ 2,\ 3,\ \ldots),$$

and since F is bounded, it follows that

$$\| FVz_n \| \leqslant l \qquad (n = 1,\ 2,\ 3,\ \ldots).$$

Consequently,

$$\frac{1}{n}(C^{-1}z_n, z_n) \leqslant \frac{1}{n}(C^{-1}z_n, z_n) + (AVz_n, z_n) = \\ = -\langle FVz_n, Vz_n\rangle \leqslant l\lambda.$$

Condition 3 of the theorem now yields

$$\frac{1}{n}\|C^{-1}z_n\|^2 \leqslant \frac{b}{n}(C^{-1}z_n, z_n) \leqslant bl\lambda,$$

or

$$\left\|\frac{C^{-1}z_n}{n}\right\|^2 \leqslant \frac{1}{n}\, bl\lambda,$$

and thus

$$\lim_{n\to\infty} n^{-1}C^{-1}z_n = 0. \tag{20.12}$$

Now, since the subsequences $\{Vz_n\}$ and $\{FVz_n\}$ in the spaces E and E^*, respectively, are bounded, the reflexivity of E implies the existence of two subsequences such that

$$Vz_{n_k} \rightharpoonup x_0 \in E, \quad FVz_{n_k} \rightharpoonup y_0 \in E^*$$

as $k \to +\infty$. Hence, invoking (20.12) and going to the limit in (20.11) via the sequence $n = n_k$, we obtain

$$Ax_0 + C^{-1}y_0 = 0. \tag{20.13}$$

We now show that $y_0 = F(x_0)$. Set

$$x_k = Vz_{n_k}, \quad y_k = FVz_{n_k} = Fx_k.$$

Noting that the operator $T = F + G$ is monotone, we may write

$$\langle Tx - Tx_k, x - x_k\rangle \geqslant 0$$

or

$$\langle Tx, x - x_k\rangle - \langle Tx_k, x\rangle \geqslant \langle -y_k, x_k\rangle - \langle Gx_k, x_k\rangle.$$

By (20.11), it follows that

$$\frac{1}{n_k}C^{-1}z_{n_k} + Ax_k + C^{-1}y_k = 0,$$

whence $Ax_k \in D(C)$, so that application of the operator C to the last equality gives

$$\frac{1}{n_k} z_{n_k} + Bx_k + y_k = 0. \tag{20.14}$$

Hence, combining the above results and noting that $Fx_k = y_k$, we have

$$\begin{aligned} \langle Tx, x - x_k\rangle - \langle y_k, x\rangle - \langle Gx_k, x\rangle \geqslant \\ \geqslant \frac{1}{n_k}(z_{n_k}, x_k) + \langle Bx_k, x_k\rangle - \langle Gx_k, x_k\rangle. \end{aligned} \tag{20.15}$$

But, as $k \to \infty$, we have

1) $\langle Tx, x - x_k\rangle \to \langle Tx, x - x_0\rangle$;
2) $\langle y_k, x\rangle \to \langle y_0, x\rangle$;
3) $\langle Gx_k, x\rangle \to \langle Gx_0, x\rangle$.

Now, since $\|Vz_{n_k}\| \leqslant \lambda$, it follows from Condition 20.2 and (20.12) that

$$\left\|\frac{z_{n_k}}{n_k}\right\|_H^2 \leqslant \left(\frac{\lambda^2}{n_k^2} + a\left\|\frac{C^{-1}z_{n_k}}{n_k}\right\|_H^2\right) \to 0.$$

Recalling that the space E is embeddable, we see that $\|x\|_H \leqslant \alpha\|x\|_E$ (α= const) for any $x \in E$. By the previous results,

4) $\left|\left(\frac{z_{n_k}}{n_k}, x_k\right)\right| \leqslant \left\|\frac{z_{n_k}}{n_k}\right\|_H \|x_k\|_H \leqslant \left\|\frac{z_{n_k}}{n_k}\right\| \alpha\lambda \to 0.$

We now claim that

5) $\varliminf_{k\to\infty} \langle Bx_k, x_k\rangle \geqslant -\langle y_0, x_0\rangle$. Indeed,

$$\begin{aligned} \langle Bx_k, x_k\rangle = \langle Bx_0, x_k - x_0\rangle + \langle B(x_k - x_0), x_k - x_0\rangle + \\ + \langle Bx_k, x_0\rangle, \end{aligned}$$

whence, as B is positive, we get

$$\langle Bx_k, x_k\rangle \geqslant \langle Bx_0, x_k - x_0\rangle + \langle Bx_k, x_0\rangle.$$

By (20.14), therefore,

$$\langle Bx_k, x_k\rangle \geqslant \langle Bx_0, x_k - x_0\rangle - \left\langle \frac{1}{n_k} z_{n_k}, x_0\right\rangle - \langle y_k, x_0\rangle,$$

and since $\left\langle \frac{1}{n_k} z_{n_k}, x_0\right\rangle \to 0$ as $k \to \infty$, it follows that

$$\varliminf_{k\to\infty} \langle Bx_k, x_k\rangle \geqslant -\langle y_0, x_0\rangle.$$

Finally, since the operator G is strongly continuous, it follows that

6) $\langle Gx_k, x_k\rangle \to \langle Gx_0, x_0\rangle$

as $k\to\infty$. Going to the lower limit in (20.15) as $k\to\infty$, we get

$$\langle Tx, x-x_0\rangle - \langle y_0, x\rangle - \langle Gx_0, x\rangle \geqslant -\langle y_0, x_0\rangle - \langle Gx_0, x_0\rangle,$$

or

$$\langle Tx-(y_0+Gx_0), x-x_0\rangle \geqslant 0.$$

Hence, by Lemma 18.1,

$$y_0+Gx_0 = Tx_0 = F(x_0)+Gx_0, \text{ or } y_0 = F(x_0).$$

Substituting this into (20.13), we obtain

$$Ax_0 + C^{-1}F(x_0) = 0.$$

This means that $Ax_0 \in D(C)$, and, after application of C, we finally have

$$Bx_0 + F(x_0) = 0.$$

Q. E. D.

Observe that Remark 20.2 also applies to this theorem.

For Hilbert spaces, we have

Theorem 20.5. Suppose that:*

1. F is an operator mapping a real Hilbert space H into itself, which is bounded, hemicontinuous, semimonotone and coercive.

2. B is a closed linear operator from H to H with dense domain $D(B)$, $D(B)\subset D(B^*)$, such that for any $y\in D(B^*)$,

$$(B^*y, y)\geqslant 0.$$

Then the equation $Bx+F(x)=y$ has a solution for any $y\in H$.

This theorem is derived from Theorem 20.4 in the same way as Theorem 20.2 was derived from Theorem 20.1. The next and last theorem is due to Browder.**

Theorem 20.6. Suppose that:

1. B is a closed linear operator from a real reflexive Banach space E to E^*, with dense domain $D(B)$, and its adjoint B^* coincides with the closure of its restriction to $D(B)\cap D(B^*)$.

* See Kositskii /1, 2/ for a direct proof of this theorem.

** Browder /8/, Theorem 1.

2. F is a hemicontinuous and bounded operator mapping E into E^*.

3. The operator $B+F=T$ is coercive and semimonotone.
Then the range of T is all of E^*.

We conclude this chapter with the following remarks.

1. Since the monotonicity of F does not necessarily imply that of the operator BF, it was necessary to establish the theorems in subsection 19.2 concerning the equivalence of equations (19.6) and (19.7), and also of equations (19.6) and (19.10), with suitable, superposition operators $TF(A)$ and $VF(A)$. These superposition operators are monotone in the corresponding Hilbert space. Thus, in §19 we used the idea of replacing a Hammerstein operator by a suitable superposition operator, just as in § 17.

2. As in previous chapters, we have considered here only those properties of monotone mappings which were used in the proofs of the fundamental propositions. Other properties of monotone mappings and properties of monotone sets are not considered here, though some of them, such as the surjectivity of mappings of maximal monotone sets (Minty /1/), are of interest in the study of nonlinear equations. The interested reader may consult the publications of Browder, Browder and Figueiredo, Brezis, Brezis and Sibony, Gossez, Kato, Kachurovskii, Kirpotina, Levitin and Polyak, Lions, Minty, Petryshyn, Polyak.

Chapter VII

NONLINEAR EQUATIONS WITH ACCRETIVE OPERATORS AND GALERKIN-PETROV APPROXIMATIONS

In the first two sections of this chapter we study nonlinear equations with accretive operators in certain Banach spaces. In a Hilbert space, nonlinear accretive operators coincide with nonlinear monotone operators, studied in the previous chapter.

Nonlinear equations with accretive operators in Banach spaces were first studied by the author /12, 13/. Such equations with more general accretive operators were later studied by Browder and his associates. Since nonlinear accretive operators in Banach spaces are defined in terms of semi-inner products and duality mappings, various properties of duality mappings are studied in §§ 21, 22. Approximate methods for solution of nonlinear equations are also studied in this chapter. In § 21 we consider steepest descent methods for nonlinear equations with accretive operators, and in § 23 approximations of the Galerkin-Petrov type for nonlinear equations with accretive and monotone operators.

§ 21. NONLINEAR EQUATIONS WITH ACCRETIVE OPERATORS SATISFYING A LIPSCHITZ CONDITION

21.1. Duality mapping and semi-inner product

Definition 21.1. A mapping $U\colon E \to E^*$, where E is a Banach (or normed) space, is a *duality mapping** if for any $x \in E$

$$\| U(x)\| = \| x\|, \ \langle Ux, x\rangle = \| U(x)\| \| x\| = \| x\|^2. \qquad (21.1)$$

If the norm of E is Gâteaux-differentiable, i. e., the unit sphere is smooth (see subsections 2.6, 2.7 and Corollary 2.1), it follows from Lemma 2.5) that a duality mapping in such spaces is positive-homogeneous and has the form

$$U(x) = \| x\| \operatorname{grad} \| x\|, \ x \neq 0; \ U(0) = 0. \qquad (21.2)$$

* Duality mappings were first considered by the author (for spaces with Gâteaux-differentiable norm) in his 1959 paper /11/ and in /13/, in connection with the theory of nonlinear equations and the generalization of the steepest descent method to Banach spaces. A detailed account of duality mappings was later given by Browder /29, 32, 38/ and others.

Hence we see, via (2.6), that in the L-spaces $L^p(G)$, with $p>1$ and G a measurable set (of finite or infinite measure) in a finite-dimensional euclidean space,

$$U(x)=\|x\|^{2-p}|x(u)|^{p-2}x(u), \qquad u\in G. \tag{21.2'}$$

In the sequence spaces l^p (see subsection 2.6), $p>1$, we have

$$U(x)=\|x\|^{2-p}z, \tag{21.3}$$

where

$$x=(x_1,\ x_2,\ x_3,\ \ldots),\ z= \\ =(|x_1|^{p-2}x_1,\ |x_2|^{p-2}x_2,\ \ldots)\in l^{p/(p-1)}.$$

A duality mapping may be constructed in any Banach (or normed) space E, in the following manner. By the Hahn-Banach theorem,* for any $x\in E$ there exists at least one bounded linear functional $y_x\in E^*$ such that $\|y_x\|=1$ and $\langle y_x,\ x\rangle=\|x\|$. Taking one such functional and setting $U(x)=\|x\|y_x$ and $U(-x)=-\|x\|y_x$, we get

$$\|U(x)\|=\|x\|,\ \langle Ux,\ x\rangle=\|Ux\|\|x\|=\|x\|^2.$$

Generally speaking, a duality mapping is multi-valued,** i. e., to each $x\in E$ there corresponds a set u_x:

$$u_x\subset v_x=\{v\in E^*:\langle v,\ x\rangle=\|x\|^2,\ \|v\|\leqslant\|x\|\}.$$

It turns out that this set v_x is convex. For if $v_1,\ v_2\in v_x$, then $v=(\lambda v_1+(1-\lambda)v_2)\in v_x$ $(0\leqslant\lambda\leqslant1)$, since $\langle v,x\rangle=\langle\lambda v_1+(1-\lambda)v_2,x\rangle=$ $=\|x\|^2$;

$$\|\lambda v_1+(1-\lambda)v_2\|\leqslant\|x\|.$$

By the definition of v_x, we have $v=0$, if $x=0$; if $x\neq0$, it follows from $\|x\|^2=\langle v,x\rangle\leqslant\|v\|\|x\|$ that $\|x\|\leqslant\|v\|$, i. e., $v_x=u_x$, where

$$u_x=\{u\in E^*:\langle u,\ x\rangle=\|x\|^2;\ \|u\|=\|x\|\}.$$

Hence the set u_x is convex. Since $u_x\subset S_r^*=\{u\in E^*:\|u\|=\|x\|=r\}$, it follows that if E^* is strictly convex, any duality mapping is single-valued. Recall that a space X is strictly convex if $\|x+y\|=\|x\|+\|y\|$

* Hille and Phillips /1/, Theorem 2.7.4.

** See Browder /32/, where multi-valued duality mappings are considered.

holds only for collinear x and y.* A sphere $\|x\| = r$ in a strictly convex space certainly cannot contain any interval. Thus, if E^* is strictly convex, u_x consists of only one point. We have thus proved

Lemma 21.1. A duality mapping $U: E \to E^*$ is single-valued if the dual E^* of E is strictly convex.

Lemma 21.2. Any duality mapping $U: E \to E^*$ is coercive and monotone, and if E is strictly convex U is a strictly monotone mapping.

Proof. Coerciveness follows from the definition of U. Since

$$\langle Ux, y\rangle \leqslant \|x\|\|y\|, \ \langle Uy, x\rangle \leqslant \|y\|\|x\|$$

for any $x, y \in E$, it follows that

$$\langle Ux - Uy, x - y\rangle \geqslant (\|x\| - \|y\|)^2 \geqslant 0.$$

Finally, if E is strictly convex, then** $\sup_{x \in S} \langle v, x\rangle$ is attained on the unit sphere $S = \{x \in E : \|x\| = 1\}$ for at most one $x \in S$, and since $\langle Ux, x\rangle = \|Ux\|\|x\|$ we have that for $x \neq y$ and $\|x\| = \|y\|$

$$\left\langle U\frac{x}{\|x\|}, \frac{y}{\|y\|}\right\rangle < \left\| U\frac{x}{\|x\|}\right\| \left\|\frac{y}{\|y\|}\right\| = 1,$$

i. e., $\langle Ux, y\rangle < \|x\|\|y\|$. Hence,

$$\langle Ux - Uy, x - y\rangle > (\|x\| - \|y\|)^2 = 0.$$

If $x \neq y$ and $\|x\| \neq \|y\|$, the previous results give

$$\langle Ux - Uy, x - y\rangle \geqslant (\|x\| - \|y\|)^2 > 0.$$

Consequently, a duality mapping in a strictly convex space is a strictly monotone operator Q. E. D. †

Lemma 21.3. † *If E is a real reflexive Banach space and its dual E^* is strictly convex, any duality mapping $U: E \to E^*$ is demi-continuous.*

Proof. Since U is homogeneous, it suffices to consider U on the unit sphere $S = \{x \in E \ \|x\| = 1\}$. Let $x_k \to x_0$ $(x_k, x_0 \in S)$ as $k \to \infty$. Since E is reflexive, the sequence $\{Ux_k\}, \|Ux_k\| = 1$, contains a subsequence $Ux_{n_k} \rightharpoonup v \in E^*$ and, by weak convergence,

* See Dunford and Schwartz /1/.

** Dunford and Schwartz /1/, V, 11.7.

† Browder /15/.

$$\| v \| \leqslant \varliminf_{k\to\infty} \| Ux_{n_k} \| = \lim_{k\to\infty} \| Ux_{n_k} \| = 1.$$

Next,

$$\langle v, x_0 \rangle = \lim_{k\to\infty} \langle Ux_{n_k}, x_{n_k} \rangle = \lim_{k\to\infty} \| x_{n_k} \|^2 = 1,$$

$\| v \| \geqslant \langle v, x_0 \rangle = 1$. Hence, $\| v \| = 1$ and

$$\langle v, x_0 \rangle = \| v \| \| x_0 \|,$$

i. e., v is a value of the duality mapping at the vector x_0. Since E^* is strictly convex, it follows from Lemma 21.1 that the duality mapping is single-valued, and so $Ux_0 = v$. Thus, $Ux_{n_k} \rightharpoonup Ux_0$ as $k \to \infty$. Since this is true for any weakly convergent subsequence $\{Ux_{k_i}\}$, it follows that $Ux_n \rightharpoonup Ux_0$ as $n \to \infty$. Q. E. D.

Remark 21.1. If E is reflexive and E^* strictly convex, Lemma 21.3 shows that a duality mapping is hemicontinuous.

By Theorem 18.2, Lemmas 21.2 and 21.3 imply

Lemma 21.4. If E is a reflexive Banach space and E^* is strictly convex, then the mapping $U : E \to E^*$, where $Ux = \|x\| y_x$, is surjective, and it maps the unit sphere $\|x\| = 1$ of E onto the unit sphere $\|y\| = 1$ of E^*.*

All the above propositions concerning duality mappings remain valid if the duality mapping is normalized as follows.** Let $\mu(t)$ be a continuous increasing real function defined on , such that $\mu(0) = 0$ and $\mu(t) \to +\infty$ as $t \to +\infty$. Then $U : E \to E^*$ is said to be a duality mapping with gauge function μ if

$$\| U(x) \| = \mu(\| x \|), \qquad \langle U(x), x \rangle = \| x \| \mu(\| x \|).$$

In particular, when $\mu(t) = t$ we have the usual duality mappings. If the unit sphere in E is smooth (compare Example 5.7), the duality mapping is a potential operator with potential $\nu(\|x\|)$, where $\nu'(t) = \mu(t)$.

Given a duality mapping $U : E \to E^*$, we now introduce a semi-inner product in E. Let x, y, z, be any vectors in E. Set

$$\langle x, U(z) \rangle = [x, z].$$

* Browder /1/; see also Lin'kov /1/, where a similar remark is made for spaces with Gâteaux-differentiable norm, i. e., spaces in which $y_x = \operatorname{grad} \|x\|$.

** Browder and Figueiredo /1/.

Then

1. $[\lambda x, z] = \lambda [x, z]$.
2. $[x+y, z] = [x, z] + [y, z]$.
3. $[x, x] = \| x \|^2 > 0$ for $x \neq 0$.
4. $| [x, z] |^2 \leqslant [x, x][z, z]$.

Definition 21.2. A functional $[x, z]$, satisfying conditions 1 — 4 is called a semi-inner product. *

Given a semi-inner product, we may introduce the notion of a dissipative operator ** in a general Banach space.

Definition 21.3. A linear operator A with domain and range in a normed space E with semi-inner product $[x, z]$ is said to be dissipative if $\mathrm{Re}[Ax, x] \leqslant 0$.

21.2. Nonlinear accretive operators

Let E be a real Banach space with semi-inner product

$$[x, z] = \langle x, U(z) \rangle,$$

where U is a duality mapping from E to E^*, and let X be some set in E.

Definition 21.4. An operator $F(x)$ from X to E is said to be accretive (or U-monotone) if for any $x, x+h \in X$,

$$[F(x+h) - F(x), h] \geqslant 0, \tag{21.4}$$

and strictly accretive if equality can hold in (21.4) only when $h = 0$. If E is a complex normed space, the same definition applies, but with (21. 4) replaced by the inequality

$$\mathrm{Re}[F(x+h) - F(x), h] \geqslant 0. \tag{21.4'}$$

Remark 21.2. Let $X \subset E$ be an open convex set containing zero, and suppose $F: X \to E$ is a mapping with a derivative $F'(x)$ existing, at least in the weak sense, at each point $x \in X$, i. e., for any $z \in E^*$, the limit

$$\langle F'(x) h, z \rangle = \lim_{t \to 0} \left\langle \frac{F(x+th) - F(x)}{t}, z \right\rangle$$

exists. We then have the following equivalent definition of an accretive operator.

* Lumer /1/.

** Lumer and Phillips /1/.

Definition 21.5. A Gâteaux-differentiable operator $F(x)$ from X to E is said to be *accretive* if

$$[F'(x)h,\ h]\geqslant 0 \tag{21.5}$$

for any $x,\ x+h\in E$.

Indeed, by Lemma 2.2, inequality (21.5) implies

$$[F(x+h)-F(x),\ h]=[F'(x+th)h,\ h]\geqslant 0,$$

and the positive homogeneity of $U(x)$ implies, by (21.4), that for $t>0$

$$\left[\frac{F(x+th)-F(x)}{t},\ h\right]\geqslant 0.$$

Hence, letting $t\to+0$, we obtain inequality (21.5).

Definition 21.6. An operator $F(x)$ from E to E is said to be *strongly accretive* if for any $x,\ x+h\in D_r=\{x\in E:\|x\|\leqslant r\}$,

$$[F(x+h)-F(x),\ h]\geqslant m(r)\|h\|^2, \tag{21.6}$$

where $m(t)$ is a positive decreasing function defined for $t\geqslant 0$ such that

$$\lim_{t\to+\infty} tm(t)=+\infty.$$

An example of a strongly accretive operator is the mapping

$$\Phi=(I+F):E\to E,$$

where I is the identity operator and F a contraction, since if $\|F(x+h)-F(x)\|\leqslant k\|h\|$ then

$$[\Phi(x+h)-\Phi(x),\ h]\geqslant\|h\|^2(1-k).$$

If $k=1$, the mapping $\Phi:E\to E$ is accretive.

Note that if $E=H$, where H is a real Hilbert space, the definitions of accretive, strictly and strongly accretive operators are equivalent, respectively, to those of monotone, strictly and strongly monotone operators, for if $E=H$ the inner product in H is a semi-inner product.

Henceforth we shall let D_r denote the ball $\|x\|\leqslant r$ in the space considered.

21.3. Accretive operators satisfying a Lipschitz condition*

Here we assume that E is a Banach space whose norm is uniformly Fréchet-differentiable on the sphere $\|x\|=1$. By Lemma 2.5, for any α the remainder of the Fréchet differential satisfies $\omega(\alpha x, \alpha h)=\alpha\omega(x, h)$, since

$$\omega(x, h)=\|x+h\|-\|x\|-\langle \operatorname{grad}\|x\|, h\rangle.$$

Hence $\|x\|$ is uniformly differentiable in the annulus

$$0<r_1\leqslant\|x\|\leqslant r_2.$$

It is clear that in such spaces a semi-inner product and a duality mapping as defined by (21.2) always exist.

Theorem 21.1. Let $F(x)$ be a strongly accretive operator, defined in a real reflexive Banach space E with uniformly Fréchet-differentiable norm, which satisfies in each ball D_r a Lipschitz condition

$$\|F(x+h)-F(x)\|\leqslant N(r)\|h\|;\ x,\ x+h\in D_r,$$

where $N(r)$ is a real continuous function, defined for $r\geqslant 0$, with arbitrary rate of increase. Then the mapping $F: E\rightarrow E$ is a homeomorphism.

Proof. Let y_0 be a fixed vector in E. We consider the mapping $\Phi(x)=F(x)-y_0$, which satisfies the hypotheses of the theorem, and prove the existence of a vector x_0 for which $\Phi(x_0)=0$. Setting

$$\alpha=\inf\|\Phi(x)\|, \tag{21.7}$$

we show that $\alpha=0$. Since

$$[x, z]=\langle U(z), x\rangle=\langle\|z\|\operatorname{grad}\|z\|, x\rangle,$$

it follows from (21.6) that

$$\|\Phi(x)-\Phi(0)\|\|x\|\geqslant\langle Ux, \Phi(x)-\Phi(0)\rangle\geqslant m(\|x\|)\|x\|^2,$$

and thus

$$\|\Phi(x)\|\geqslant\|\Phi(x)-\Phi(0)\|-\|\Phi(0)\|\geqslant$$
$$\geqslant m(\|x\|)\|x\|-\|\Phi(0)\|,$$

* Vainberg /13/.

i. e.,

$$\lim_{\|x\|\to\infty} \|\Phi(x)\| = +\infty.$$

Hence, there exists a ball

$$D_{r_0-1} = \{x \in E : \|x\| \leqslant r_0 - 1\}$$

such that the inequality

$$\|\Phi(x)\| \leqslant \lambda x \qquad (\lambda > 1),$$

implies that, if we assume $\alpha > 0$, then

$$\|x\| \leqslant r_0 - 1. \tag{21.8}$$

Set $m(r_0) = m_0$, $N(r_0) = N_0$, and consider the annulus

$$K_0 = \{y \in E : \alpha \leqslant \|y\| \leqslant \lambda\alpha\}.$$

Since the functional $\varphi(x) = \|x\|$ is uniformly differentiable in the annulus K_0, it follows that for the given number $m_0/4N_0$ there exists $\varepsilon > 0$ such that, for $\|\eta\| \leqslant \varepsilon N_0$,

$$\|y + \eta\| = \|y\| + \langle \operatorname{grad}\|y\|, \eta\rangle + \omega(y, \eta), \; y \in K_0,$$

where

$$|\omega(y, \eta)| < \frac{m}{4N_0}\|\eta\|. \tag{21.9}$$

Let us assume that $\varepsilon < \min\{1, 2\lambda\alpha/m_0\}$. By (21.7) and (21.8), there exists a vector $z_0 \in D_{r_0-1}$ such that

$$\alpha \leqslant \Phi(z_0) < \alpha + \frac{m_0\varepsilon}{2\lambda}. \tag{21.10}$$

Now set

$$h = -\varepsilon\frac{\Phi(z_0)}{\|\Phi(z_0)\|} \tag{21.11}$$

and, using Lemma 2.5, write

$$\|\Phi(z_0+h)\|=\|\Phi(z_0)+[\Phi(z_0+h)-\Phi(z_0)]\|=$$
$$=\|\Phi(z_0)\|+\langle \operatorname{grad}\|\Phi(z_0)\|,\ \Phi(z_0+h)-\Phi(z_0)\rangle+\omega(y,\eta)=$$
$$=\|\Phi(z_0)\|-\frac{1}{\varepsilon}\left\langle \varepsilon\operatorname{grad}\left\|-\varepsilon\frac{\Phi(z_0)}{\|\Phi(z_0)\|}\right\|,\ \Phi(z_0+h)-\Phi(z_0)\right\rangle+$$
$$+\omega(y,h)=\|\Phi(z_0)\|-\frac{1}{\varepsilon}\langle Uh,\ \Phi(z_0+h)-\Phi(z_0)\rangle+$$
$$+\omega(y,\eta),$$

where

$$y=\Phi(z_0),\quad \eta=\Phi(z_0+h)-\Phi(z_0).$$

Hence, by the Lipschitz inequalities, (21.6), (21.9) and (21.10), we obtain

$$\|\Phi(z_0+h)\|<$$
$$<\alpha+\frac{m_0\varepsilon}{2\lambda}-m_0\varepsilon+\frac{m_0}{4N_0}\|\Phi(z_0+h)-\Phi(z_0)\|\leqslant$$
$$\leqslant\alpha+\frac{m_0\varepsilon}{2\lambda}-m_0\varepsilon+\frac{m_0}{4N_0}N_0\varepsilon=\alpha-\frac{m_0\varepsilon}{2}\left(\frac{3}{2}-\frac{1}{\lambda}\right)<\alpha.$$

This inequality contradicts (21.7). It follows that $\inf\|\Phi(x)\|=0$ and therefore we can select a minimizing sequence $\{x_n\}$ with $\|x_n\|\leqslant R_0$ and

$$\lim_{n\to\infty}\Phi(x_n)=0. \tag{21.12}$$

Now, condition (21.6) implies that

$$2R_0\|\Phi(x_m)-\Phi(x_n)\|\geqslant$$
$$\geqslant\langle U(x_m-x_n),\ \Phi(x_m)-\Phi(x_n)\rangle\, m(R_0)\|x_m-x_n\|^2,$$

so that, since $\{\Phi(x_n)\}$ is a fundamental sequence, so is $\{x_n\}$; hence, since the space E is complete, $x_n\to x_0$ as $n\to\infty$. Thus, by (21.12), we have $\Phi(x_0)=0$, whence $F(x_0)=y_0$, i. e., the mapping $F:E\to E$ is surjective. The vector x_0 satisfying $F(x)=y_0$ is unique, since if x_1 and x_2 are two solutions of the equation $F(x)=y_0$ then, by (21.6),

$$0=\langle U(x_2-x_1),\ F(x_2)-F(x_1)\rangle\geqslant$$
$$\geqslant m(R)\|x_2-x_1\|^2,\quad x_1,\ x_2\in D_R,$$

whence $x_1=x_2$. Thus, the mapping $F:E\to E$ is bijective. Finally, let us demonstrate that the inverse F^{-1}, which exists by virtue of the injectivity of F, is continuous. Consider any two vectors $x,\ y\in E$ and a ball D_R containing both of them. Then, by (21.6),

$$2R\|F(x)-F(y)\| \geqslant \langle U(x-y),\ F(x)-F(y)\rangle \geqslant \\ \geqslant m(R)\|x-y\|^2$$

or, if we set $F(x)=u,\ F(y)=v$,

$$\|F^{-1}(u)-F^{-1}(v)\| \leqslant 2R[m(R)]^{-1}\|u-v\|,$$

so that the inverse mapping is continuous. The mapping $F: E \to E$ is therefore a homeomorphism. Q. E. D.

Theorem 21.2. Let $F(x)$ be a strictly accretive operator, defined in a real Banach space E with uniformly Fréchet-differentiable norm, which satisfies (21.6) in every ball D_r with a positive decreasing function $m(t)$ defined for $t \geqslant 0$, as well as a Lipschitz inequality

$$\|F(x+h)-F(x)\| \leqslant N(r)\|h\| \quad (x,\ x+h \in D_r); \tag{21.13}$$

moreover, let the functional $f(x)=\|F(x)\|$ be increasing (see Definition 10.2). Then the equation $F(x)=0$ has a unique solution in E.

The proof is similar to that of the preceding theorem. In fact, setting $\alpha = \inf\|F(x)\|$ and using the fact that $f(x)$ is increasing, we may assert the existence of $\lambda > 1$ for which inequalities (21.8), (21.10) hold, together with other inequalities, leading to (21.12), with Φ replaced by F. As before, it follows from (21.12) that there exists a unique solution x_0 to the equation $F(x)=0$.

Remark 21.3. Under the hypotheses of Theorems 21.1 and 21.2, the functional $\|F(x)\|$ is increasing. Hence there exists a sphere

$$S_R = \{x \in E : \|x\| = R > 0\}$$

such that $\|F(x)\| > \|F(0)\| \geqslant \alpha$ for any $x \in S_R$. The equation thus has a solution provided inequalities (21.6) and (21.13) hold for every $r \leqslant R$. We cite one proposition of this kind, due to Shragin /14/.

Theorem 21.3. Let $F(x)$ be a strictly accretive operator defined in a ball D_R of a real Banach space with uniformly Fréchet-differentiable norm, and suppose that in any ball D_r, $r \leqslant R$, $F(x)$ satisfies inequalities (21.6) and (21.13). Suppose, furthermore, that there exists $\gamma > 0$ such that the set

$$\{x \in D_R : \|F(x)\| \leqslant \gamma\}$$

is nonempty and

$$\sup \{\|x\| : \|F(x)\| \leqslant \gamma\} = \beta < R.$$

Then there exists a unique solution to the equation $F(x) = 0$.

Proof. Set $m(R) = m$, $N(R) = N$ and

$$\alpha = \inf \{\|F(x)\| : x \in D_R\} \leqslant \gamma,$$

where $m > 0$ by assumption. Let $\{x_n\}$ be a minimizing sequence (i.e., $\|F(x_n)\| \to \alpha$) lying in D_R. Then, as in the proof of Theorem 21.1, one shows that $x_n \to z_0 \in D_R$ as $n \to \infty$, so, by the continuity of $F(x)$, we have $\|F(z_0)\| = \alpha$.

Let $\alpha > 0$. Proceeding just as in the proof of Theorem 21.1, and assuming in addition that $\varepsilon < R - \beta$, we get $\|z + h\| \leqslant \beta - \varepsilon \leqslant R$, since $\|z\| \leqslant \beta$, and

$$\| F(z + h) \| \leqslant \alpha - m\varepsilon + \frac{1}{4} m\varepsilon < \alpha,$$

contradicting the assumption. Hence $\alpha = 0$ and so $F(z_0) = 0$. Q. E. D.

21.4. Steepest-descent approximations in Hilbert space and in L-spaces

Here we prove the convergence of steepest-descent approximations to solutions of equations with accretive operators in a real Hilbert space H and real L-spaces L^p $(p > 1)$. The norms of these spaces are uniformly Fréchet-differentiable. Indeed, consider the unit sphere $S = \{x \in H : \|x\| = 1\}$ in H. Since grad $\|x\| = x/\|x\|$ (see Example 2.2), it follows that

$$\| x + h \| - \| x \| = \left(\frac{x}{\|x\|}, \ h\right) + \omega(x, h),$$

whence, for any $x \in S$,

$$| \omega(x, h) | \leqslant \frac{2 \| h \|^2}{1 + \| x + h \|},$$

so that the sphere S is uniformly Fréchet-differentiable. Since any L-space with $p > 1$ is uniformly convex,* its norm is uniformly Fréchet-differentiable.**

Theorem 21.4.† *Let* $F(x)$ *be an operator defined in a real Hilbert space* H, *such that in every ball* D_r

* See Clarkson /1/.

** See Shmul'yan /2/ and Day /1/, VII, § 2(8).

† Vainberg /13/, Theorem 3.1.

$$\| F(x+h) - F(x) \| \leqslant N(r) \| h \| \quad (x,\ x+h \in D_r),$$
$$(h,\ F(x+h) - F(x)) \geqslant m(r) \| h \|^2 \quad (x,\ x+h \in D_r),$$

where $m(t)$ is a positive decreasing function defined for $t \geqslant 0$ and $tm(t) \to +\infty$ as $t \to +\infty$. Then the equation $F(x)=0$ has a unique solution x_0 in H and the iterative process

$$x_{n+1} = x_n - \varepsilon F(x_n), \qquad n = 1,\ 2,\ 3,\ \dots, \tag{21.14}$$

where $x_1 = 0$, converges to x_0 provided $\varepsilon_1 \leqslant \varepsilon \leqslant \varepsilon_2$, where ε_1 and ε_2 are the roots (which are positive) of the equation

$$N(r_0)\varepsilon^2 - 2m(r_0)\varepsilon + 1 - q^2 = 0,$$

in which

$$\left(1 - \frac{m^2(r_0)}{N^2(r_0)}\right)^{1/2} < q < 1.$$

Here $r_0 = 2R$, $R \geqslant \| x_0 \|$, where R is given by

$$R = \sup\{\| x_0 \| : \| x_0 \| m(\| x_0 \|) \leqslant \| F(0) \|\}. \tag{21.15}$$

The following error estimate holds:

$$\| x_{n+1} - x_0 \| \leqslant q^n R.$$

Proof. By Theorem 21.1, our assumptions imply that the equation $F(x)=0$ has a unique solution x_0 in H. Substituting x_0 into the inequality

$$\| x \| \| F(x) - F(0) \| \geqslant (x,\ F(x) - F(0)) \geqslant m(\| x \|) \| x \|^2,$$

we obtain (21.15), from which we have the estimate $\| x_0 \| \leqslant R$. Now, in any real Hilbert space, for any $y,\ \eta \in H$,

$$\| y + \eta \|^2 = \| y \|^2 + 2(y,\ \eta) + \| \eta \|^2.$$

Thus, if we set $y = x_n - x_0$, $\eta = -\varepsilon F(x_n)$ and suppose that $\| x_n - x_0 \| \leqslant R$, it follows from the hypotheses of the theorem that

$$\begin{aligned} \| x_{n+1} - x_0 \|^2 &= \| x_n - x_0 - \varepsilon F(x_n) \|^2 = \\ &= \| x_n - x_0 \|^2 - 2\varepsilon (x_n - x_0,\ F(x_n)) + \| \varepsilon F(x_n) \|^2 = \\ &= \| x_n - x_0 \|^2 - 2\varepsilon (x_n - x_0,\ F(x_n) - F(x_0)) + \\ &+ \varepsilon^2 \| F(x_n) - F(x_0) \|^2 \leqslant \| x_n - x_0 \|^2 (1 - 2\varepsilon m(r_0) + \\ &+ N^2(r_0)\varepsilon^2) \leqslant q^2 \| x_n - x_0 \|^2. \end{aligned}$$

It is assumed here that

$$1 - 2\varepsilon m(r_0) + N^2(r_0)\varepsilon^2 \leqslant q^2 < 1.$$

This inequality holds only if

$$q > \left(1 - \frac{m^2(r_0)}{N^2(r_0)}\right)^{1/2}; \qquad \varepsilon_1 \leqslant \varepsilon \leqslant \varepsilon_2.$$

Since by assumption $\| x_1 - x_0 \| = \| x_0 \| < R$, it follows that

$$\| x_2 - x_0 \| \leqslant q \| x_1 - x_0 \| \leqslant qR < R,$$

so that $x_2 \in D_{r_0}$ and, by induction, $x_n \in D_{r_0}$. Therefore

$$\| x_{n+1} - x_0 \| \leqslant q^n \| x_1 - x_0 \| \leqslant q^n R.$$

This inequality proves the theorem.

If we assume in the above theorem that $m = \text{const}$ and $N = \text{const}$, the existence of a solution to the equation $F(x) = 0$ follows from the contracting-mapping principle. Indeed, set

$$\Phi(x) = x - \lambda F(x),$$

where $0 < \lambda < 2mN^{-2}$. Then for any vectors $x, y \in H$,

$$\begin{aligned} &\| \Phi(y) - \Phi(x) \|^2 = \| y - x - \lambda [F(y) - F(x)] \|^2 = \\ &= \| y - x \|^2 - 2\lambda (y - x, F(y) - F(x)) + \lambda^2 \| F(y) - F(x) \|^2 \leqslant \\ &\leqslant \| y - x \|^2 (1 - 2\lambda m + \lambda^2 N^2). \end{aligned}$$

Since $m < N$, it follows that $\lambda^2 N^2 - 2\lambda m + 1 > 0$ for any real λ and thus, if $0 < \lambda < 2mN^{-2}$,

$$1 - 2\lambda m + \lambda^2 N^2 = q^2 < 1.$$

Consequently,

$$\| \Phi(y) - \Phi(x) \| \leqslant q \| y - x \|, \quad 0 < q < 1, \tag{21.16}$$

so that the usual successive approximations $x_{n+1} = \Phi(x_n)$ converge (for any initial approximation) to a solution of the equation

$$x = \Phi(x) \equiv x - \lambda F(x),$$

i. e., to a solution of the equation $F(x) = 0$. Clearly, inequality (21.16) will hold in the ball D_{r_0} if

$$0 < \lambda < 2m(r_0)N^{-2}(r_0).$$

Hence, if $\Phi(x)$ maps the ball D_{r_0} into itself, the solution x_0 can be found by the usual method of successive approximations.

Piontkovskii /1/ has pointed out that if the initial approximation x_1 in Theorem 21.4 is chosen arbitrarily and R is taken sufficiently large (so that the ball D_{r_0} contains the vectors x_0 and x_1), the assertion of the theorem remains valid, and, moreover, if

$$0 < 1 - 2m(r_0) + N^2(r_0) \leqslant q < s,$$

we can set $\varepsilon = 1$ in (21.4).

We now turn to the real L-spaces $L^p(G)$, $p > 1$. Since the norms in these spaces are uniformly Fréchet-differentiable, Theorems 21.1 — 21.3 are valid. By (21.2), if $h \neq 0$, inequality (21.6) becomes

$$\int_G |h(u)|^{p-2} h(u)[F(x(u) + h(u)) - F(x(u))]\,du \geqslant$$
$$\geqslant m(r)\|h\|^p. \qquad (21.17)$$

It turns out that certain analogs of Theorem 21.4 may be proved in these spaces. We shall cite one such theorem, due to T. A. Blinova.

Theorem 21.5. Let $F: L^p \to L^p$ *be an operator satisfying (21.13) and (21.17), where* $m(t)$ *is a positive decreasing function such that* $tm(t) \to +\infty$ *as* $t \to +\infty$. *Then there exists an interval* $[\varepsilon_1, \varepsilon_2]$, *depending on* p, *such that for* $\varepsilon \in [\varepsilon_1, \varepsilon_2]$ *the iterative process (21.14) with* $x_1 = 0$ *converges to the solution* x_0 *of the equation* $F(x) = 0$ *and*

$$\|x_{n+1} - x_0\| \leqslant q^n R,$$

where $q < 1$ *and* R *is given by (21.15).*

Proof. The existence of a unique solution x_0 follows from Theorem 21.1; we prove the convergence of (21.4) for $p \neq 2$, since the case $p = 2$ is covered by Theorem 21.4. We first observe that when determining the gradient of the norm in L^p, $1 < p < 2$ (see subsection 2.6), we started with a function $\varphi(\xi)$ and obtained the inequality

$$\|x + h\|^p \leqslant \|x\|^p + p\int_G |x(u)|^{p-2} x(u)h(u)\,du + C_2\|h\|^p. \qquad (21.18)$$

A similar inequality can be obtained for $p > 2$:

$$\|x+h\|^p \leqslant \|x\|^p + p\int_G |x(u)|^{p-2} x(u) h(u)\,du + C_2\|h\|^p + \\ + C_2\int_G h^2(u)|x(u)|^{p-2}\,du, \tag{21.19}$$

by starting from the function $\psi(\xi)$ considered in subsection 2.6, with $0 < C_2 = \text{const}$. Proceeding now to estimate the difference

$$\|x_{n+1} - x_0\|^p = \|x_n - x_0 - \varepsilon F(x_n)\|^p,$$

we consider two cases. If $1 < p < 2$, we use inequality (21.18) to show that

$$\|x_{n+1} - x_0\|^p \leqslant \\ \leqslant \|x_n - x_0\|^p - \varepsilon p\int_G F(x_n)|x_n - x_0|^{p-2}(x_n - x_0)\,du + \\ + C_2\|\varepsilon F(x_n)\|^p = \|x_n - x_0\|^p - \\ - \varepsilon p\int_G |x_n - x_0|^{p-2}(x_n - x_0)[F(x_n) - F(x_0)]\,du + \\ + C_2\varepsilon^p\|F(x_n) - F(x_0)\|^p \leqslant \\ \leqslant \|x_n - x_0\|^p - \varepsilon p m(r_0)\|x_n - x_0\|^p + C_2\varepsilon^p N^p(r_0)\|x_n - x_0\|^p,$$

where $r_0 = 2R$. Consequently,

$$\|x_{n+1} - x_0\|^p \leqslant \\ \leqslant \|x_n - x_0\|^p(1 - \varepsilon p m(r_0) + C_2\varepsilon^p N^p(r_0)) \equiv \|x_n - x_0\|^p \alpha(\varepsilon).$$

Since $\alpha(0) = 1$, $\alpha'(0) = -pm(r_0) < 0$ and $\alpha''(\varepsilon) > 0$ for $\varepsilon > 0$, there exists a positive number $q < 1$ such that the equation $\alpha(\varepsilon) = q^p$ has two positive roots $\varepsilon_1, \varepsilon_2$. Thus, if $\varepsilon_1 \leqslant \varepsilon \leqslant \varepsilon_2$, then $0 < \alpha(\varepsilon) \leqslant q^p$. Hence, by our previous results,

$$\|x_{n+1} - x_0\| \leqslant q\|x_n - x_0\|, \quad 0 < q < 1,$$

and the fact that $x_n \in D_{r_0}$ for any n may be proved as before (see the proof of Theorem 21.4). Consequently,

$$\|x_{n+1} - x_0\| \leqslant q^n\|x_1 - x_0\| \leqslant q^n R.$$

If $p > 2$, we use inequality (21.9) and show that

$$\|x_{n+1}-x_0\|^p=\|x_n-x_0-\varepsilon F(x_n)\|^p\leqslant$$
$$\leqslant\|x_n-x_0\|^p-\varepsilon p\int_G F(x_n)\,|x_n-x_0|^{p-2}(x_n-x_0)\,du+$$
$$+C_2\varepsilon^p\|F(x_n)\|^p+C_2\varepsilon^2\int_G [F(x_n)]^2|x_n-x_0|^{p-2}\,du.$$

But

$$\|F(x_n)\|=\|F(x_n)-F(x_0)\|\leqslant N(r_0)\|x_n-x_0\|,$$
$$\int_G [F(x_n)-F(x_0)]^2|x_n-x_0|^{p-2}\,du\leqslant$$
$$\leqslant\left(\int_G |F(x_n)-F(x_0)|^p\,du\right)^{2/p}\left(\int_G |x_n-x_0|^p\,du\right)^{(p-2)/p}=$$
$$=\|F(x_n)-F(x_0)\|^2\|x_n-x_0\|^{p-2}.$$

Hence, by the preceding results and (21.17),

$$\|x_{n+1}-x_0\|\leqslant\|x_n-x_0\|^p\beta(\varepsilon);$$
$$\beta(\varepsilon)=1-\varepsilon pm(r_0)+C_2\varepsilon^pN^p(r_0)+C_2\varepsilon^2N^2(r_0).$$

Since $\beta(0)=1$, $\beta'(0)<0$ and $\beta(\varepsilon)$ is a convex function (because $\beta''(\varepsilon)>0$), there exists a positive number $q<1$ for which the equation $\beta(\varepsilon)=q^p$ has two positive roots ε_1, ε_2. Thus $0<\beta(\varepsilon)\leqslant q^p$ for $\varepsilon_1\leqslant\varepsilon\leqslant\varepsilon_2$ and, by the above estimates, we have

$$\|x_{n+1}-x_0\|\leqslant q\|x_n-x_0\|,\qquad 0<q<1.$$

We have here used the fact that $x_n\in D_{r_0}$, which, as we have seen, is true for any $p>1$. Thus, in this case again we have $\|x_{n+1}-x_0\|\leqslant q^nR$. Q. E. D.

21.5. Steepest-descent approximations in sequence spaces

We shall consider the real sequence spaces l^p only for $1<p\neq 2$, for l^2 is a Hilbert space and steepest-descent approximations in Hilbert space have already been discussed (subsection 21.4). The spaces l^p are uniformly convex (see Clarkson /1/), and so their norms are uniformly Fréchet-differentiable.* Thus Theorems 21.1 — 21.3 are true for the spaces l^p, $p>1$. We wish to establish

* See Shmul'yan, /2/ and Day, /1/, VII, § 2(8).

a certain analog of Theorem 21.4 for these spaces. To do this, we need several inequalities. In subsection 2.6, we used the function

$$\varphi(\xi)=(|1+\xi|^p-1-p\xi)/|\xi|^p \qquad 1<p<2,$$

to prove the inequality

$$|1+\xi|^2-1-p\xi\leqslant C|\xi|^p, \qquad C>0.$$

If we set $\xi=\eta_i/x_i$ $(x_i\neq 0)$, then

$$|x_i+\eta_i|^p-|x_i|^p-p\eta_i|x_i|^{p-1}\operatorname{sign} x_i\leqslant C|\eta_i|^p.$$

This inequality clearly holds for $x_i=0$ as well. Now consider any two vectors x and η in l^p,

$$x=(x_1, x_2, x_3, \ldots); \quad \eta=(\eta_1, \eta_2, \eta_3, \ldots).$$

Summing the above inequality over i, we get

$$\|x+\eta\|^p-\|x\|^p-p\sum_{i=1}^{\infty}\eta_i|x_i|^{p-2}x_i\leqslant C\|\eta\|^p. \tag{21.20}$$

Now, as above (see subsection 2.6), using the function

$$\psi(\xi)=(|1+\xi|^p-1-p\xi)(|\xi|^p+\xi^2)^{-1}, \quad p>2,$$

we arrive at the inequality

$$\|x+\eta\|^p\leqslant\|x\|^p+p\sum_{i=1}^{\infty}\eta_i|x_i|^{p-2}x_i+$$
$$+C\|\eta\|^p+C\sum_{i=1}^{\infty}\eta_i^2|x_i|^{p-2}, \tag{21.21}$$

where $C>0$, $x\in l^p$, $\eta\in l^p$. Furthermore, inequality (21.6) now becomes

$$\sum_{i=1}^{\infty}|\eta_i|^{p-2}\eta_i[F_i(x+\eta)-F_i(x)]\geqslant m(r)\|\eta\|^p, \tag{21.22}$$

where $(F_1, F_2, F_3, \ldots)=F: l^p\to l^p$.

Theorem 21.6. Let $F: l^p\to l^p$ *be an operator satisfying (21.13) and (21.22), where* $m(t)$ *is a positive decreasing function such that* $tm(t)\to+\infty$ *as* $t\to+\infty$. *Then there exists an interval* $[\varepsilon_1, \varepsilon_2]$, *depending on* p, *such that for any positive* $\varepsilon\in[\varepsilon_1, \varepsilon_2]$, *the iterative process*

$$x^{(n+1)} = x^{(n)} - \varepsilon F(x^{(n)}), \qquad n = 1, 2, 3, \ldots, \tag{21.23}$$

with $x^{(1)} = 0$, *converges to the unique solution* $x^{(0)}$ *of the equation* $F(x) = 0$, *and, moreover, we have the estimate* $\|x^{(n+1)} - x^{(0)}\| \leqslant q^n R$, *where* $q < 1$, *and* R *is given by (21.15).*

Proof. By virtue of (21.23), we need only estimate the difference

$$\| x^{(n+1)} - x_0 \|^p = \| x^{(n)} - x^{(0)} - \varepsilon F(x^{(n)}) \|^p, \tag{21.24}$$

where x_0 is the unique solution of the equation $F(x) = 0$ whose existence follows from Theorem 21.1.

For $1 < p < 2$, we use (21.20) to estimate this difference, writing

$$\| x^{(n+1)} - x^{(0)} \|^p \leqslant \| x^{(n)} - x^{(0)} \|^p + \\ + C\| \varepsilon F(x^{(n)}) \|^p - p\varepsilon \sum_{i=1}^{\infty} F_i(x^{(n)}) \,| x_i^{(n)} - x_i^{(0)} |^{p-2} (x_i^{(n)} - x_i^{(0)}).$$

Since $F(x^{(0)}) = 0$, inequality (21.13) implies that

$$\sum_{i=1}^{\infty} F_i(x^{(n)}) \,| x_i^{(n)} - x_i^{(0)} |^{p-2} (x_i^{(n)} - x_i^{(0)}) \leqslant \\ \leqslant \| F(x^{(n)}) \| \, \| x^{(n)} - x^{(0)} \|^{p-1} = \\ = \| F(x^{(n)}) - F(x^{(0)}) \| \, \| x^{(n)} - x^{(0)} \|^{p-1} \leqslant N(r) \| x^{(n)} - x^{(0)} \|^p.$$

Hence, by (21.22), we obtain

$$\| x^{(n+1)} - x^{(0)} \|^p \leqslant \| x^{(n)} - x^{(0)} \|^p (1 - p\varepsilon m(r) + C\varepsilon^p N^p(r)),$$

and, as in the proof of Theorem 21.5, the assertion of the theorem now follows.

If $p > 2$, we use (21.21) to estimate the difference (21.24), writing

$$\| x^{(n+1)} - x^{(0)} \|^p \leqslant \\ \leqslant \| x^{(n)} - x^{(0)} \|^p - p\varepsilon \sum_{i=1}^{\infty} F_i(x^{(n)}) \,| x_i^{(n)} - x_i^{(0)} |^{p-2} (x_i^{(n)} - x_i^{(0)}) + \\ + C\varepsilon^p \| F(x^{(n)}) \|^p + C\varepsilon^2 \sum_{i=1}^{\infty} (F(x_i^{(n)}))^2 \,| x_i^{(n)} - x_i^{(0)} |^{p-2}.$$

But

$$\sum_{i=1}^{\infty} (F(x_i^{(n)}))^2 \,| x_i^{(n)} - x_i^{(0)} |^{p-2} \leqslant \| F(x^{(n)}) \|^2 \| x^{(n)} - x^{(0)} \|^{p-2} = \\ = \| F(x^{(n)}) \|^2 \| x^{(n)} - x^{(0)} \|^{p-2},$$

and, as in the case $1 < p < 2$, we thus find that

$$\| x^{(n+1)} - x^{(0)} \|^p \leqslant$$
$$\leqslant \| x^{(n)} - x^{(0)} \|^p (1 - \varepsilon pm(r) + C\varepsilon^p N^p(r) + C\varepsilon^2 N^2(r)).$$

As in the proof of Theorem 21.5, this completes the proof for $p > 2$.

§ 22. EQUATIONS WITH GENERAL ACCRETIVE OPERATORS

22.1. Some types of continuity of duality mappings

If the norm of a space E is Fréchet-differentiable then, as we have seen, the only duality mapping is $Ux = \|x\| \operatorname{grad} \|x\|$ which, by Lemma 2.6, is continuous from E to E^*. In other cases, this mapping need not be continuous. Lemma 21.3 gives a sufficient condition for U to be continuous relative to the strong topology of E and the weak topology of E^*. We now present a sufficient condition for uniform continuity.

Lemma 22.1. Let E be a real reflexive Banach space whose dual is uniformly convex. Then the duality mapping $U: E \to E^*$ is uniformly continuous on every bounded set in E.

Proof. Since E^* is uniformly convex,* its norm is uniformly Gâteaux-differentiable in any annulus $0 < r_1 \leqslant \|x\| \leqslant r_2$ in the space E. Hence, by Lemma 21.1, the duality mapping is uniquely determined and it is of the form (21.1), i.e.,

$$Ux = \| x \| \operatorname{grad} \| x \| = \frac{1}{2} \operatorname{grad} \| x \|^2.$$

Since the functional $\varphi(x) = \frac{1}{2}\| x \|^2$ is uniformly continuous and uniformly differentiable on every bounded subset of E, its gradient Ux is uniformly continuous (see VM, Theorem 4.2). Q. E. D.

We now proceed to an investigation of the weak continuity of duality mappings. We have seen that in a real Hilbert space the only duality mapping is $Ux = x$, which transforms every weakly convergent sequence into a weakly convergent one. In the sequence spaces l^p, $p > 1$, the weak convergence of a sequence $\{x^{(n)}\}$ to $x^{(0)}$, where $x^{(n)} = (x_1^{(n)}, x_2^{(n)}, x_3^{(n)}, \ldots)$, means that i $x_i^{(n)} \to x_i^{(0)}$ as $n \to \infty$ for each i. Hence, by (21.3), a duality mapping transforms every weakly convergent sequence in l^p into a weakly convergent sequence in

* See Shmul'yan /2/ and Day /1/, VII, § 2(8).

the dual l^q $(p^{-1}+q^{-1}=1)$. By contrast, in the L-spaces L^p, $1<p\neq 2$ duality mappings need not be weakly continuous. This may be seen in the following example.

Example 22.1.* Consider the real space $L^p[0,1]$, $1<p<\infty$, $p\neq 2$. In this space (see (21.2')), the only duality mapping is

$$Ux=\|x\|^{2-p}|x(t)|^{p-2}x(t), \quad 0\leqslant t\leqslant 1.$$

Let $u_0(t)$ be a periodic function with period 1 such that

$$\int_0^1 u_0(t)\,dt=0, \quad \int_0^1 |u_0(t)|^p\,dt=1,$$

$$\int_0^1 |u_0(t)|^{p-2}u_0(t)\,dt=c_0\neq 0.$$

For each positive integer k, set

$$u_k(t)=u_0(kt), \quad t\in[0,1].$$

Then

$$\|u_k\|^p=\int_0^1 |u_k(t)|^p\,dt=\frac{1}{k}\int_0^k |u_0(\tau)|^p\,d\tau=1.$$

Then, for any $c\in[0,1]$,

$$\int_0^c u_k(t)\,dt=\int_0^c u_0(kt)\,dt=\frac{1}{k}\int_0^{kc} u_0(\tau)\,d\tau=$$

$$=\frac{1}{k}\int_0^{kc-[kc]} u_0(\tau)\,d\tau\to 0$$

as $k\to\infty$. Thus, by the criterion for weak convergence in $L^p[0,1]$, the subsequence of unit vectors $u_k(t)$ converges weakly to 0 as $k\to\infty$. On the other hand, for the subsequence $v_k(t)=Uu_k(t)=|u_k(t)|^{p-2}u_k(t)$ we have

$$\int_0^1 |u_k(t)|^{p-2}u_k(t)\,dt=\frac{1}{k}\int_0^k |u_0(\tau)|^{p-2}u_0(t)\,dt=c_0\neq 0,$$

so that $v_k(t)\not\rightharpoonup 0$ as $k\to\infty$.

* Browder and Figueiredo /1/.

22.2. Auxiliary propositions*

Lemma 22.2. *Suppose that:*
1. E *is a real reflexive space with strictly convex dual* E^*.
2. U *is a duality mapping from* E *to* E^*.
3. F *is a hemicontinuous operator from* E *to* E.

Suppose, moreover, that for fixed $u_0, v_0 \in E$ *and any* $u \in E$

$$\langle U(u-u_0),\ F(u)-v_0\rangle \geqslant 0;$$

then $F(u_0)=v_0$.

Proof. Setting $u_t=u_0+tv$, where v is an arbitrary vector in E, we get

$$0 \leqslant \langle U(u_t-u_0),\ F(u_t)-v_0\rangle = \langle U(tv),\ F(u_t)-v_0\rangle$$

or, since U is a duality mapping,

$$\langle Uv,\ F(u_t)-v_0\rangle \geqslant 0.$$

Hence, letting $t \to +0$, we get

$$\langle Uv,\ F(u_0)-v_0\rangle \geqslant 0.$$

Replacing v by $(-v)$ in the preceding inequalities and taking into account that $U(-v)=-Uv$, we get

$$\langle Uv,\ F(u_0)-v_0\rangle \leqslant 0.$$

Consequently, $\langle Uv, F(u_0)-v_0\rangle = 0$ for any $v \in E$. Lemma 21.4 now entails that $F(u_0)=v_0$. Q. E. D.

Lemma 22.3. *Let* E_n *be a finite-dimensional Banach space,* U *a duality mapping from* E_n *to* E_n, *and* T *a continuous mapping from* E_n *to* E_n. *Then, if*

$$\langle Ux,\ Tx\rangle \geqslant 0, \tag{22.1}$$

on some sphere $\|x\| = r > 0$, *there exists a vector* $x_0 \in D_r = \{x \in E_n : \|x\| \leqslant r\}$ *such that* $Tx_0 = 0$.

Proof. Let $G = I - T$, where I is the identity operator in E_n. We claim that G has a fixed point in D_r; this will imply the assertion

* In this and the following subsections, we follow Browder and Figueiredo /1/.

of the lemma. To prove this, we consider a mapping V defined by

$$Vx = \begin{cases} Gx, & \text{if} \quad \| Gx \| < r, \\ \frac{Gx}{\| Gx \|} r, & \text{if} \quad \| Gx \| \geqslant r. \end{cases}$$

Since V is a continuous mapping of D_r into itself, it follows from Brouwer's theorem* that V has a fixed point x_0 in D_r. If $\|x_0\| < r$, then $x_0 = Vx_0 = Gx_0 = x_0 - Tx_0$, i. e., $Tx_0 = 0$. If $\|x_0\| = r$, then $x_0 = = rGx_0/\|Gx_0\|$, where $\|Gx_0\| \geqslant r$ so that

$$Gx_0 = \lambda x_0, \qquad \lambda = \| Gx_0 \| / r \geqslant 1.$$

Now the assumption that $\lambda > 1$ leads to a contradiction for, by (22.1),

$$\langle Ux_0, Gx_0 \rangle = \langle Ux_0, x_0 \rangle - \langle Ux_0, Tx_0 \rangle \leqslant \langle Ux_0, x_0 \rangle,$$

and since $\lambda > 1$, it follows that

$$\langle Ux_0, Gx_0 \rangle = \lambda \langle Ux_0, Gx_0 \rangle > \langle Ux_0, x_0 \rangle.$$

Consequently, $\lambda = 1$, and so $Tx_0 = 0$. Q. E. D.

Definition 22.1. We shall say that a Banach space E possesses property $(\pi)_1$ if there exist a directed family of finite-dimensional subspaces $\{E_\alpha\}$, ordered by inclusion, and a corresponding family of projections P_α from E onto E_α, such that $\bigcup_\alpha E_\alpha$ is dense in E and $\|P_\alpha\| = 1$ for each α (see Definition 18.4).

Examples of spaces with property $(\pi)_1$ are Hilbert spaces and sequence spaces l^p, $p \geqslant 1$. For l^p, the family $\{E_\alpha\}$ consists of the subspaces l^p_n of sequences $(x_1, x_2, \ldots, x_n, 0, 0, 0, \ldots,) = x_{(n)}$, and the projections are $P_n x = x_n$, where $x = (x_1, x_2, \ldots, x_n, x_{n+1}, \ldots) \in l^p$. Other examples of spaces with property $(\pi)_1$ are L-spaces $L^p(m)$ with σ-finite measure m, $p > 1$. Let the functions $f(x) \in L^p$ be defined on a set A. Let s_i be disjoint sets such that $\bigcup_{i=1}^n s_i = A$, and denote their characteristic functions by $f_1(x), f_2(x), \ldots, f_n(x)$. Since any function $f(x) \in L^p$ is the limit of a sequence of simple functions,** the linear span of the characteristic functions $(f_1, f_2, \ldots, f_n)$ is a finite-dimensional subspace of $L^p(m)$. Setting

$$P_n f = \sum_{i=1}^{n} \left(\frac{1}{nm(s_i)} \right)^{1/q} \int_{s_i} f(x)\, dx, \quad \frac{1}{p} + \frac{1}{q} = 1,$$

* See Brouwer /1/ or Aleksandrov /1/.

** See Kolmogorov and Fomin /1/.

and using Hölder's inequality, we get

$$\| P_n f \| \leqslant \left(\int_A | f(x) |^p \, dx \right)^{1/p} = \| f \|,$$

i. e., $\| P_n \| = 1$.

Lemma 22.4. Suppose that:

1. E is a real Banach space with property $(\pi)_1$, $\{E_\alpha\}$ and $\{P_\alpha\}$ being the corresponding families of finite-dimensional subspaces and projections.

2. The dual E^* is strictly convex.

3. U is a duality mapping from E to E^*.

Then for any $x \in E_\alpha$,

$$P_\alpha^* U x = U x,$$

where P_α^* is the adjoint of P_α.

Proof. The mapping $P_\alpha^* U x$ satisfies the conditions for a duality mapping, since

$$\langle P_\alpha^* U x, x \rangle = \langle U x, P_\alpha x \rangle = \langle U x, x \rangle = \| U x \| \| x \|,$$

and, since $\| P_\alpha^* \| = \| P_\alpha \| = 1$,

$$\| P_\alpha^* U x \| \leqslant \| P_\alpha^* \| \| U x \| = \| U x \| = \| x \|.$$

Hence (see the proof of Lemma 21.1), $\| P_\alpha^* U x \| = \| x \|$. Thus $P_\alpha^* U x$ is indeed a duality mapping. Since the dual is strictly convex, it follows from Lemma 21.2 that the duality mapping is single-valued, and so $P_\alpha^* U x = U x$.

Lemma 22.5. Suppose that:

1. E is a real reflexive Banach space with property $(\pi)_1$, and $\{P_\alpha\}$ being the corresponding families of finite-dimensional subspaces and projections.

2. The dual E^* is strictly convex.

3. U is a duality mapping from E to E^* and T a demicontinuous operator such that $\langle U x, T x \rangle \geqslant 0$ for $\| x \| = r > 0$.

Then for any α the equation $P_\alpha T(x) = 0$ has a solution in $D_r \cap E_\alpha$, where

$$D_r = \{ x \in E : \| x \| \leqslant r, \ r > 0 \}.$$

Proof. Since P_α is the projection of E onto E_α, it follows* (see Dunford and Schwartz, /1/, VI. 3. 3 and VI. 9.19) that P_α^* is

* See also the proof of Lemma 18.6.

the projection of E^* onto E_α^*, where the dimension of E_α^* is α, so that we may identify E_α and E_α^*. $U_\alpha x = P_\alpha^* Ux$ is a duality mapping from E_α to E_α^* and, by Lemma 22.4, for any $x \in E_\alpha$ we have $U_\alpha x = Ux$. Next, since T transforms any convergent sequence into a weakly convergent sequence and the projection P_α transforms any weakly convergent sequence into a weakly convergent one, it follows that $P_\alpha T$ transforms any convergent sequence in E_α into a weakly convergent sequence in E_α. But in finite-dimensional spaces weak convergence is equivalent to strong convergence, so that $P_\alpha T$ is continuous in E_α. Hence, in view of the inequality

$$\langle U_\alpha x, P_\alpha T(x)\rangle = \langle Ux, P_\alpha T(x)\rangle = \\ = \langle P_\alpha^* Ux, T(x)\rangle = \langle Ux, T(x)\rangle \geqslant 0,$$

valid for any $x \in E_\alpha$, $\|x\| = r$, Lemma 22.3 implies that the equation $P_\alpha T(x) = 0$ has a solution, which lies in $D_r \cap E_\alpha$. Q. E. D.

22.3. Existence theorems

Theorem 22.1. Suppose that:

1. E is a real reflexive Banach space having property $(\pi)_1$, with strictly convex dual.

2. The duality mapping $U: E \to E^*$ is continuous and sequentially weakly continuous.

3. F is an accretive (U-monotone), demicontinuous operator from E to E.

4. There exists a ball $D_r = \{x \in E: \|x\| \leqslant r, r > 0\}$ on whose surface $\langle Ux, F(x)\rangle \geqslant 0$.

Then there exists $x_0 \in D_r$ such that $F(x_0) = 0$.

Proof. Let $\{E_\alpha\}$ be the family of finite-dimensional subspaces of E and $\{P_\alpha\}$ the corresponding projections, which exist by hypothesis. Set $F_\alpha = P_\alpha F$. By Lemma 22.5, the equation $P_\alpha F_\alpha(x) = 0$ has a solution $x_\alpha \in D_r \cap E_\alpha$. Let $V_\gamma = \bigcup\{x_\alpha\}$, where the union extends over all α such that $E_\alpha \supset E_\gamma$. Since D_r is a ball in a reflexive Banach space, it is weakly compact, so that there exists a vector $x_0 \in D_r$ in the intersection of the weak closures of the sets V_γ. We claim that $F(x_0) = 0$. To show this, we take any vector $v \in \bigcup_\gamma E_\gamma$ and choose β such that $v \in E_\beta$. For any x_α, $\alpha \geqslant \beta$, and the corresponding V_β, the fact that F is accretive implies that

$$\langle U(v - x_\alpha), Fv - Fx_\alpha\rangle \geqslant 0. \qquad (22.2)$$

Since v and x_a are in E_a, we have

$$\langle U(v - x_a), Fx_a\rangle = \langle P_a^* U(v - x_a), Fx_a\rangle = \\ = \langle U(v - x_a), P_a F x_a\rangle = 0$$

since $P_a F(x_a) = 0$, and, by Lemma 22.4,

$$P_a^* U(v - x_a) = U(v - x_a).$$

Consequently, by (22.2), $\langle U(v - x_a), Fv\rangle \geqslant 0$. Consider the real-valued functional $h(u) = \langle U(v - u), Fv\rangle$ on D_r. Since U is weakly continuous, $h(u)$ is continuous in the weak topology on D_r. Since h is nonnegative on V_γ, it follows that it is also nonnegative on the weak closure of V_γ, whence $h(x_0) \geqslant 0$ i. e.,

$$\langle U(v - x_0), Fv\rangle \geqslant 0 \quad \left(\forall v \in \bigcup_\gamma E_\gamma\right).$$

Let v be an arbitrary vector in E. Since $\bigcup_\gamma E_\gamma$ is dense in E, there exists a sequence $\{v_k\} \subset \bigcup_\gamma E_\gamma$ such that $v_k \to v$. Using the continuity of U and the fact that F transforms any convergent sequence into a weakly convergent one, we find that

$$\lim_{k \to \infty} \langle U(v_k - x_0), Fv_k\rangle = \langle U(v - x_0), Fv\rangle. \tag{22.3}$$

Since $\langle U(v_k - x_0), Fv_k\rangle \geqslant 0$, it follows from (22.3) that $\langle U(v - x_0), Fv\rangle \geqslant 0$. By Lemma 22.2, this inequality implies that $F(x_0) = 0$. Q. E. D.

Now (22.2) remains valid if the accretive operator F is continuous, since by Lemma 21.3 the mapping U is demicontinuous; hence, if F is continuous, we may drop the assumption of Theorem 22.1 that U is continuous. We thus have

Theorem 22.2. Suppose that:

1. E is a real reflexive Banach space having property $(\pi)_1$, with strictly convex dual.

2. The duality mapping $U: E \to E^*$ is sequentially weakly continuous.

3. $F: E \to E$ is a continuous accretive operator.

4. On the surface of some ball $D_r = \{x \in E : \|x\| \leqslant r, r > 0\}$,

$$\langle Ux, F(x)\rangle \geqslant 0.$$

Then there exists a vector $x_0 \in D_r$ such that $F(x_0) = 0$.

Theorem 22.3. Suppose that:

1. E is a real reflexive Banach space having property $(\pi)_1$, with strictly convex dual E^*.

2. The duality mapping $U: E \to E^*$ is sequentially weakly continuous.

3. F is an accretive operator from E to E.

4. Either F is continuous, or U is continuous and F demicontinuous.

5.
$$\langle Ux, F(x)\rangle \geqslant \|x\| \gamma(\|x\|), \tag{22.4}$$
where $\gamma(t)$ is a real function such that $\gamma(t) \to +\infty$ as $t \to +\infty$.

Then the mapping $F: E \to E$ is surjective.

Proof. Let y be a fixed vector in E. The mapping $T_y = F - y$ is also an accretive operator from E to E, and, for any $r > 0$ such that $\gamma(r) \geqslant \|y\|$, we have

$$\langle Ux, T_y x\rangle = \langle Ux, Fx\rangle - \langle Ux, y\rangle \geqslant \|x\|(\gamma(r) - \|y\|) \geqslant 0.$$

Hence, by conditions 1–4 of the theorem and by Theorems 22.1 and 22.2, there exists a vector $x_0 \in E$ such that $T_y(x_0) = 0$, i. e., $F(x_0) = y$. Thus F maps E onto itself. Q. E. D.

Remark 22.1. Under the hypotheses of Theorems 22.1 and 22.2, if F is strictly accretive, we may also assert that the equation $F(x) = 0$ has a unique solution, and the mapping $F: E \to E$ in Theorem 22.3 is bijective.

We present yet another proposition based on the uniform continuity of the duality mapping.

Theorem 22.4. Let E be a real reflexive Banach space with uniformly convex dual and F a continuous bounded accretive operator with a bounded inverse defined on the range of F, such that

$$\langle U(x-y), F(x) - F(y)\rangle \geqslant \|x-y\| \gamma(\|x-y\|) \quad (\forall x, y \in E),$$

where $\gamma(t)$ is an increasing continuous function, defined for $t \geqslant 0$, such that $\gamma(0) = 0$.

Then the mapping $F: E \to E$ is surjective.

By Lemma 22.1, this theorem is a corollary of Theorem 2 in Browder /38/.

§23. GALERKIN APPROXIMATIONS

We shall now invoke Galerkin's method and some of its modifications to find approximate solutions of nonlinear equations with accretive and monotone operators.* Galerkin's method finds extensive application in the theory of differential and integral equations. For example, it was used by Vishik** to prove his fundamental theorems on the solvability of boundary-value problems for quasilinear strongly elliptic differential equations. Browder's proof of Theorem 18.4 (Browder /5/) is essentially based on Galerkin's method. Leray and Lions /1/ use the method to prove certain results of Vishik by the method of monotone operators.

23.1. Galerkin approximations and Galerkin systems

Let E be a separable normed space with a basis and F a mapping of E into E^*. Galerkin's method of approximating solutions to the equation $F(x) = 0$ is based on the following idea. One first takes a basis in E, i.e., a linearly independent system of vectors $\varphi_1, \varphi_2, \varphi_3, \ldots$, such that any vector $x \in E$ may be expressed uniquely in the form

$$x = \sum_{k=1}^{\infty} \alpha_k(x) \varphi_k.$$

Using this basis, one constructs a sequence of finite-dimensional subspaces $\{E_n\}$, where E_n is the n-dimensional subspace spanned by $\varphi_1, \varphi_2, \ldots, \varphi_n$.

The n-th Galerkin approximation to the solution of the equation $F(x) = 0$ is defined as a vector $x_n \in E_n$,

$$x_n = \sum_{k=1}^{n} a_k \varphi_k, \tag{23.1}$$

satisfying the system of equations

$$\left\langle F\left(\sum_{k=1}^{n} a_k \varphi_k\right), \varphi_i \right\rangle = 0 \qquad (i = 1, 2, \ldots, n). \tag{23.2}$$

This system of equations is known as a Galerkin system. Although system (23.2) is identical to system (12.3), the problem of its solvability is to be solved otherwise. In § 12 we used the properties of a functional, whereas our examination of system (23.2) will use the properties of the mapping $F(x)$.

* Concerning Galerkin's method, see Mikhlin /2/, /3/, and also Krasnosel'skii, Vainikko et al /1/.

** See, for example, Vishik /5/.

23.2. Relation to projective methods

Let P_n be the projection of E onto E_n and P_n^* be its adjoint, which (see Dunford and Schwartz /1/, VI. 3. 3 and VI. 9. 19) projects E^* onto an n-dimensional subspace E_n^*.

The projective method for approximating solutions to the equation $F(x) = 0$, where $F: E \to E^*$, is based on replacing the original equation by an equation in a finite-dimensional space, namely,

$$P_n^* F(P_n x) = 0, \tag{23.3}$$

and the solution of this equation is called an approximate solution of the initial equation. We claim that equation (23.3) is equivalent to system (23.2). Indeed, if h is any vector in E^*, equation (23.3) is equivalent to the equation

$$0 = \langle P_n^* F(P_n x), h\rangle = \langle F(P_n x), P_n h\rangle =$$
$$= \left\langle F\left(\sum_{k=1}^{n} a_k \varphi_k\right), \sum_{i=1}^{n} \beta_i \varphi_i\right\rangle = \sum_{i=1}^{n} \beta_i \left\langle F\left(\sum_{k=1}^{n} a_k \varphi_k\right), \varphi_i\right\rangle,$$

which, in view of the arbitrary choice of β_i, is equivalent to system (23.3). Thus, Galerkin's method, as applied here to solution of the equation $F(x) = 0$, coincides with the projective method.

Furthermore, if P is a mapping from E_x to E_y, where E_x and E_y are normed spaces, the projective method for the equation $P(x) = 0$ proceeds as follows. One takes two sequences of subspaces, $\{E_x^{(n)}\}$ and $\{E_y^{(n)}\}$, whose unions are dense in E_x and E_y, respectively, and considers the sequences of projections $\{P_n\}$ and $\{Q_n\}$, where $P_n E_x = E_x^{(n)}$, $Q_n E_y = E_y^{(n)}$. The equation $P(x) = 0$ is then replaced by the equation

$$Q_n P(P_n x) = 0, \tag{23.4}$$

whose solution is again called an approximate solution of the initial equation.

If $E_x = E_y = H$, where H is a Hilbert space, then the projective method is referred to as the Bubnov-Galerkin method; in the general case, the projective method is known as the Galerkin-Petrov method.

To conclude this subsection, we observe that, by (23.4), if $F(x)$ is an accretive operator in a normed space E or a mapping from E to E, the Galerkin approximations for the equation $F(x) = 0$ may be found via the finite-dimensional equation

$$P_n F(P_n x) = 0. \tag{23.5}$$

23.3. Solvability of Galerkin systems

Lemma 23.1. Let* $F: E \to E^*$ *be a hemicontinuous operator, where* E *is a real separable normed space, and suppose that, on the sphere* $\|x\| = r > 0$,

$$\langle F(x), x\rangle > 0 \quad (\|x\| = r). \tag{23.6}$$

Then the Galerkin system (23.2) is solvable for any n *and the Galerkin approximations* x_n *satisfy the estimate* $\|x_n\| < r$.

Proof. Since (23.2) and (23.3) are equivalent, it suffices to prove that equation (23.3) is solvable. Set $P_n x = z \in E_n$. Then, by Remark 1.4, the mapping $\Phi_n(z) = P_n^* F(z)$ is continuous from E_n to E_n^* ($E_n^* = P_n^* E^*$), so that its domain and range are both in an n-dimensional space. Identifying E_n and E_n^*, we may view $\Phi_n(z)$ as a continuous mapping in n-space. Now, since for $\|z\| = r$

$$\langle \Phi_n(z), z\rangle = \langle P_n^* F(z), z\rangle = \langle F(z), P_n z\rangle = \langle F(z), z\rangle > 0,$$

it follows that $\|\Phi_n(z)\| > 0$ on the sphere $\|z\| = r$. Hence, reasoning as in the proof of Lemma 18.2, we find that the equation $\Phi_n(z) = 0$ has a solution in the ball $\|z\| < r$. Q. E. D.

Lemma 23.2. Let $F: E \to E$ *be a hemicontinuous mapping, where* E *is a real separable normed space with property* $(\pi)_1$, E^* *is strictly convex, and suppose that on the sphere* $\|x\| = r > 0$

$$\langle Ux, F(x)\rangle \geqslant 0 \quad (\|x\| = r), \tag{23.7}$$

where U *is a duality mapping from* E *to* E^*. *Then the Galerkin system (23.5) is solvable for any* n *and the Galerkin approximations* x_n *satisfy the estimate* $\|x_n\| \leqslant r$.

Proof. Set $P_n x = z \in E_n$ for any $x \in E$ and consider the mapping $\Phi_n(z) = P_n F(z)$ from E_n to E_n. By Remark 1.4, $\Phi_n(z)$ is continuous. Now, since by Lemma 22.4 $P_n^* Uz = Uz$, it follows that for $\|z\| = r$

$$\langle Uz, \Phi_n(z)\rangle = \langle Uz, P_n F(z)\rangle = \langle P^* Uz, F(z)\rangle = \\ = \langle Uz, F(z)\rangle \geqslant 0.$$

Hence, by Lemma 22.3, the equation $\Phi_n(z) = 0$ has a solution in the ball $\|z\| \leqslant r$ in E_n. Q. E. D.

Remark 23.1. If F is either strictly monotone or strictly accretive, then, under the hypotheses of Lemmas 23.1 and 23.2, respectively, one can state that the Galerkin system has a unique solution for any n.

* Compare Kachurovskii /4/, Lemma 4.1.

23.4. Some properties of Galerkin approximations

Let x_n denote the Galerkin approximations of the equation $F(x) = 0$, where F is a mapping from a normed space E to E^* or from E to E. In the first case, the x_n are to be found from system (23.2) or from equation (23.3), in the second, from equation (23.5).

Lemma 23.3.* *Suppose the Galerkin approximations x_n of system (23.2) exist for any n, and let $\|x_n\| \leqslant r$. Then, if $F(x)$ is a bounded operator, the sequence $\{F(x_n)\}$ converges E-weakly to zero.*

Proof. Since $\varphi_1, \varphi_2, \varphi_3, \ldots$ is a basis in E, any vector $h \in E$ is expressible as $h = h_n + h^{(n)}$, where $h^{(n)} = \sum_{k=n+1}^{\infty} \alpha_k(h)\varphi_k \to 0$ as $n \to \infty$. Hence, by (23.3),

$$\begin{aligned}\langle F(x_n), h\rangle &= \langle F(x_n), h_n\rangle + \langle F(x_n), h^{(n)}\rangle = \\ &= \langle F(x_n), P_n h_n\rangle + \langle F(x_n), h^{(n)}\rangle = \\ &= \langle P_n^* F(x_n), h_n\rangle + \langle F(x_n), h^{(n)}\rangle = \langle F(x_n), h^{(n)}\rangle.\end{aligned}$$

Hence,

$$|\langle F(x_n), h\rangle| \leqslant \|F(x_n)\| \|h^{(n)}\| \to 0 \quad \text{as} \quad n \to \infty.$$

Thus $F(x_n) \underset{E}{\rightarrow} 0$. Q. E. D.

Let E be a separable normed space with basis $\varphi_1, \varphi_2, \varphi_3, \ldots$; let $\{E_n\}$ be the sequence of subspaces considered in subsection 23.1, $\{P_n\}$, the sequence of projections ($P_n E = E_n$), and $\{P_n^*\}$ the sequence of their adjoints. As remarked in subsection 23.2, P_n^* projects E^* onto an n-dimensional subspace E_n^*.

Let h be an arbitrary vector in E^*, $h_n = P_n^* h$ and set $h^{(n)} = h - h_n$. The sequence $\{E_n^*\}$ is said to be ultimately dense in E^* if, for any $h \in E^*$, $h^{(n)} \to 0$ as $n \to \infty$.

Lemma 23.4. *Suppose that the Galerkin approximations $x_n = P_n x_{(n)}$ of equation (23.5) exist for any n and that $\|x_n\| \leqslant r =$ const. If $F: E \to E$ is a bounded operator and the sequence $\{E_n^*\}$ is ultimately dense in E^*, then the sequence $\{F(x_n)\}$ converges weakly to zero.*

Proof. Let h be any vector in E^*. Then $h = h_n + h^{(n)} = P_n^* h_n + h^{(n)}$ and, by (23.5), we have

$$\langle F(x_n), h\rangle = \langle P_n F(x_n), h_n\rangle + \langle F(x_n), h^{(n)}\rangle = \langle F(x_n), h^{(n)}\rangle,$$

where

$$|\langle F(x_n), h\rangle| \leqslant \|F(x_n)\| \|h^{(n)}\| \to 0 \quad \text{as} \quad n \to \infty.$$

* Krasnosel'skii, Vainikko et al. /1/, p. 300.

23.5. Weak convergence of Galerkin approximations

Theorem 23.1. Let E be a real reflexive separable Banach space and $F:E\to E^*$ a semimonotone bounded mapping. If the Galerkin approximations x_n exist for any n and $\|x_n\|\leqslant r=$ const, then the weak limit of any weakly convergent subsequence of $\{x_n\}$ is a solution of the equation $F(x)=0$.

Proof. Since F is semimonotone, there exists a strongly continuous operator $C:E\to E^*$ such that $F+C=T$ is monotone. Let x be any vector in E. Then

$$\begin{aligned}\langle F(x),\ x-x_n\rangle= \\ =\langle Tx-Tx_n,\ x-x_n\rangle-\langle Cx-Cx_n,\ x-x_n\rangle+ \\ +\langle F(x_n),\ x\rangle-\langle F(x_n),\ x_n\rangle\geqslant \\ \geqslant-\langle Cx-Cx_n,\ x-x_n\rangle+\langle F(x_n),\ x\rangle,\end{aligned}$$

since

$$\langle Tx-Tx_n,\ x-x_n\rangle\geqslant 0,$$
$$\langle F(x_n),\ x_n\rangle=\langle F(x_n),\ P_nx_n\rangle=\langle P_n^*F(x_n),\ x_n\rangle=0.$$

Consider a subsequence such that $x_{n_k}\to x_0$ as $k\to\infty$. Then, letting $k\to\infty$ in the inequality

$$\langle F(x),\ x-x_{n_k}\rangle\geqslant-\langle Cx-Cx_{n_k},\ x-x_{n_k}\rangle+\langle F(x_{n_k}),\ x\rangle$$

we see, by Lemma 23.3 and the choice of the operator C, that

$$\langle F(x),\ x-x_0\rangle\geqslant-\langle Cx-Cx_0,\ x-x_0\rangle$$

or $\langle Tx-Cx_0,\ x-x_0\rangle\geqslant 0$. Hence, by Lemma 18.1, $Cx_0=Tx_0=F(x_0)+Cx_0$, i. e., $F(x_0)=0$. Q. E. D.

Remark 23.2. In Theorem 23.1, if the equation $F(x)=0$ has a unique solution x_0 then, by Lemma 16.1, the Galerkin approximations x_n converge weakly to x_0. The proof of the next theorem is similar to that of Theorem 22.1.

Theorem 23.2. Suppose that:

1. E is a real reflexive Banach space having property $(\pi)_1$, with strictly convex dual.

2. The duality mapping $U:E\to E^*$ is continuous and sequentially weakly continuous.

3. $F:E\to E$ is a demicontinuous accretive operator.

If the Galerkin approximations x_n of the solution to the equation $F(x)=0$ exist for any n and $\|x\|\leqslant r\,(r>0)$, then the weak limit of any subsequence $\{x_{n_k}\}$ is a solution of the equation $F(x)=0$.

Proof. Let $\{E_n\}$ and $\{P_n\}$ be the corresponding families of finite-dimensional subspaces of E and projections onto these subspaces (see condition 1 of the theorem). By assumption, $P_nF(x_n) = 0$ and $P_nx_n = x_n$. Since $\|x_n\| \leqslant r = \text{const}$ and E is reflexive, one can select a subsequence of $\{x_n\}$ such that $x_{n_k} \rightharpoonup x_0$, where $\|x_0\| \leqslant r$. We claim that $F(x_0) = 0$. Indeed, take any vector $v \in \bigcup_m E_m$ and choose $m = n_k = \alpha$ such that $v \in E_\alpha$. Since F is accretive, we may write

$$\langle U(v - x_\alpha),\ Fv - Fx_\alpha\rangle \geqslant 0,$$

and since $v - x_\alpha \in E_\alpha$, it follows from Lemma 22.4 and the equality $P_\alpha F(x_\alpha) = 0$ that

$$\langle U(v - x_\alpha),\ Fx_\alpha\rangle = \langle P_\alpha^* U(v - x_\alpha),\ Fx_\alpha\rangle = \\ = \langle U(v - x_\alpha)\, P_\alpha F x_\alpha\rangle = 0.$$

Hence

$$\langle U(v - x_\alpha),\ Fv\rangle \geqslant 0.$$

Letting $\alpha = n_k \to \infty$ and noting that if $x_\alpha \rightharpoonup x_0$ then $U(v - x_\alpha) \rightharpoonup U(v - x_0)$, and if $n > \alpha$ the $v \in E_n$, we get

$$\langle U(v - x_0),\ Fv\rangle \geqslant 0 \qquad \Big(\forall v \in \bigcup_m E_m\Big). \tag{23.8}$$

Since $\bigcup_m E_m$ is dense in E, it follows that for any $x \in E$ there exists a sequence $\{v_n\} \subset \bigcup_m E_m$ such that $v_n \to x$. Now set $v = v_n$ in (23.8) and observe that $Fv_n \rightharpoonup Fx$ and $U(v_n - x_0) \to U(x - x_0)$; thus, letting $n \to \infty$, we deduce from (23.8) that $\langle U(x - x_0), Fx\rangle \geqslant 0$. Hence, by Lemma 22.2, $F(x_0) = 0$. Q. E. D.

Remark 23.3. If the operator F in Theorem 23.2 is strictly accretive, the equation $F(x) = 0$ has a unique solution x_0, so that, by Lemma 16.1, the Galerkin approximations x_n converge weakly to x_0 as $n \to \infty$.

23.6. Strong convergence of Galerkin approximations

Theorem 23.3. Suppose that:

1. *E is a real reflexive Banach space with basis $\{\varphi_k\}$.*
2. *$F: E \to E^*$ is a hemicontinuous uniformly monotone bounded operator such that $\langle F(x), x\rangle > 0$ if $\|x\| \geqslant r > 0$.*

Then the Galerkin approximations x_n exist for any n and converge to the unique solution of the equation $F(x) = 0$.

Proof. Theorem 18.1 and the strict monotonicity of F imply that the equation $F(x) = 0$ has a unique solution x_0, $\|x_0\| \leqslant r$. By Lemma 23.1, the Galerkin approximations x_n exist for any n and $\|x_n\| \leqslant r$. Since the operator F is uniformly monotone by assumption, we have

$$I_n \equiv \langle F(x_n) - F(x_0),\ x_n - x_0\rangle \geqslant \gamma(\|x_n - x_0\|), \qquad (23.9)$$

where $\gamma(t)$ is a continuous increasing function, defined for $t \geqslant 0$, $\gamma(0) = 0$. But since $F(x_0) = 0$ and the Galerkin approximations satisfy the relation $P_n^* F(x_n) = 0$, we have $\langle F(x_n),\ x_n\rangle = \langle F(x_n),\ P_n x_n\rangle =$ $= \langle P_n^* F(x_n),\ x_n\rangle = 0$, and thus

$$I_n = \langle F(x_n),\ x_n\rangle - \langle F(x_n),\ x_0\rangle = -\langle F(x_n),\ x_0\rangle.$$

By Lemma 22.3, then, $I_n \to 0$ as $n \to \infty$, and since $\gamma(t)$ has a continuous increasing inverse γ^{-1}, vanishing at zero, it follows from (23.9) that $\|x_n - x_0\| \leqslant \gamma^{-1}(I_n) \to 0$ as $n \to \infty$. Q. E. D.

23.7. Galerkin approximations for equations of Hammerstein type

Here we consider a real reflexive Banach space possessing property (π) (see Definition 19.1), with basis $\{\varphi_k\}$, such that (see subsection 23.4) the sequence $\{E_n^*\}$ is ultimately dense in E^*. We shall refer to these, in ad hoc fashion, as nice spaces. The Galerkin approximations for an equation of Hammerstein type $x + BF(x) = 0$, where $F: E \to E^*$ is a nonlinear operator and $B: E^* \to E$ a linear operator, are solutions of the equation $P_n x + P_n B P_n^* F(P_n x) = 0$ or

$$z + P_n B P_n^* F(z) = 0 \qquad (z \in E_n). \qquad (23.10)$$

Theorem 23.4. Let F be a uniformly monotone hemicontinuous bounded operator from a nice space E to E^*, such that

$$\langle F(x),\ x\rangle \geqslant 0 \quad \text{if} \quad \|x\| > M > 0 \quad (M = \text{const}),$$

and B a bounded positive linear operator from E^* to E. Then the Galerkin approximations x_n exist for any n and converge to the unique solution of the equation.

Proof.* By Theorem 19.1 and the uniform monotonicity of F, our equation has a unique solution x_0 and, as is clear from the proof of Theorem 19.1, $\|x_0\| \leqslant M$.

* The proof uses certain ideas from A. M. Vainberg /1/ and Kositskii /1, 2/.

Setting $B_n = P_nBP_n^*$ and $F_n = P_n^*F$, we can rewrite equation (23.10) as

$$z + B_nF_n(z) = 0 \quad (z \in E_n). \tag{23.11}$$

Here $F_n : E_n \to E_n^*$, $B_n : E_n^* \to E_n$; F_n is uniformly monotone, since $\langle F_nz_1 - F_nz_2, z_1 - z_2\rangle = \langle Fz_1 - Fz_2, P_n(z_1 - z_2)\rangle = \langle Fz_1 - Fz_2, z_1 - z_2\rangle$; B_n is a positive operator from E_n^* to E_n, because

$$\langle y, B_ny\rangle = \langle y, P_nBP_n^*y\rangle = \langle P_n^*y, BP_n^*y\rangle = \langle y, By\rangle \geqslant 0$$

for any $y \in E_n^*$, and $\langle F_nz, z\rangle = \langle Fz, P_nz\rangle = \langle Fz, z\rangle \geqslant 0$ if $\|z\| > M > 0$. Thus, again by Theorem 19.1 and the uniform (hence also strong) monotonicity of F_n, equation (23.11) has a unique solution x_n for any n and $\|x_n\| \leqslant M$. Thus the Galerkin approximations x_n exist for any n and are uniformly bounded.

To prove that the approximations x_n converge, we subtract the equalities $x_n + BFx_0 = 0$ and $x_0 + B_nF_nx_n = 0$, to obtain

$$x_n - x_0 + B_n(Fx_n - Fx_0) + (B_nFx_0 - BFx_0) = 0.$$

Now applying the functional $Fx_n - Fx_0$, we get

$$\langle Fx_n - Fx_0, BFx_0 - B_nFx_0\rangle = \\ = \langle Fx_n - Fx_0, x_n - x_0\rangle + \langle Fx_n - Fx_0, B_n(Fx_n - Fx_0)\rangle.$$

Since F is uniformly monotone and B_n is positive, we have

$$I_n \equiv \langle Fx_n - Fx_0, BFx_0 - B_nFx_0\rangle \geqslant \gamma(\|x_n - x_0\|),$$

where $\gamma(t)$ is a continuous increasing function of $t \in [0, +\infty)$, $\gamma(0) = 0$. Since $\gamma(t)$ has a continuous increasing inverse γ^{-1}, the last inequality gives

$$\|x_n - x_0\| \leqslant \gamma^{-1}(I_n). \tag{23.12}$$

We now write

$$BFx_0 - B_nFx_0 = (BFx_0 - P_nBFx_0) + P_nB(Fx_0 - P_n^*Fx_0).$$

But,

$$\|BFx_0 - P_nBFx_0\| \to 0$$

as $n \to \infty$, since the space E has a basis, and $\| Fx_0 - P_n^* Fx_0 \| \to 0$, since the sequence $\{E_n^*\}$ is ultimately dense in E. Hence,

$$\| BFx_0 - B_n Fx_0 \| \to 0,$$

as $n \to \infty$, and thus, since $\| Fx_n - Fx_0 \|$ is uniformly bounded, we have $I_n \to 0$. By (23.12), therefore, $\| x_n - x_0 \| \to 0$, as $n \to \infty$. Q. E. D.

23.8. Galerkin approximations in the case of unbounded operators

We now consider an equation of Hammerstein type $x + BF(x) = 0$ in a real Hilbert space H with orthogonal basis $\varphi_1, \varphi_2, \varphi_3, \ldots$, assuming that B is a closed (not necessarily bounded) operator from H to H with dense domain $D(B)$, and $F: H \to H$.

For such equations, the equations of the Galerkin system are of the form

$$P_n x + P_n B P_n F(P_n x) = 0 \qquad (P_n x = z \in H_n), \tag{23.13}$$

where P_n is the orthogonal projection of H onto H_n, the subspace spanned by the vectors $\varphi_1, \varphi_2, \ldots, \varphi_n$. It is assumed here that $\varphi_k \in D(B)$. The theorem quoted below is due to Kositskii /1, 2/.

Theorem 23.5. Suppose that:

1. B is a closed linear operator from H to H with dense domain $D(B)$, for each basis vector we have $\varphi_k \in D(B)$, $(Bx, x) \geqslant 0$ for any $x \in D(B)$ and $(B^* y, y) \geqslant 0$ for any $y \in D(B^*)$.

2. F is a strongly monotone demicontinuous bounded operator from H to H. Then equation (23.13) has a unique solution $x_n \in H_n$ and the Galerkin approximations x_n converge to the unique solution x_0 of the equation $x + BF(x) = 0$.

Chapter VIII

APPLICATIONS

This chapter consists of three sections, in which the methods developed above are applied to the investigation of nonlinear integral and differential equations.

In § 24 we prove existence and uniqueness theorems for nonlinear Hammerstein integral equations $u = Bhu$, both in function spaces and in vector-function spaces. These theorems differ from those established in VM in an important respect: In VM it was assumed throughout that the Nemytskii operator h is potential and the linear integral operator B selfadjoint, and these assumptions were essential. Many theorems of VM also assume that the operator B is completely continuous. Here all the theorems are free of this last assumption while in some theorems we do not even require the operator B to be bounded. Neither do we assume throughout that the operator B is selfadjoint and the operator h potential. It is worth noting that in function spaces h is a potential operator if it is either bounded or has both domain and range in the space in question; in spaces of vector-functions, however, h is a potential operator if and only if condition (5.2) holds.

In § 25 we show how to apply general theorems on equations with monotone operators to boundary-value problems for partial differential equations of elliptic or parabolic type.

The work of Vishik /1 – 6/ on the solvability of certain important classes of boundary-value problems for elliptic and parabolic partial differential equations in Sobolev spaces is well known. Essentially, Vishik's method is based on finite-dimensional approximation of the equation by modified Galerkin approximations and proving the convergence of the Galerkin approximations to a solution of the equation. These investigations were developed further by Browder /1 – 7, 11, 16, 18/, Leray and Lions /1/, Dubinskii /1/ and others.

Browder, and later Leray and Lions, showed that Vishik's results can be obtained by the method of monotone operators. In § 25 we present Browder's approach.

In the last section we employ monotonicity properties to investigate the initial-value problem for nonlinear differential equations in Hilbert spaces and certain Banach spaces. In so doing we include some results due to Browder, Kato and Lions.

§ 24. NONLINEAR HAMMERSTEIN INTEGRAL EQUATIONS

Here we shall prove existence and uniqueness theorems for nonlinear Hammerstein equations in various function spaces.

24.1. Equations with positive operators

We consider Hammerstein equations $u+Bhu=0$ and $u-Bhu=0$, assuming that the linear integral operator B,

$$Bv=\int_G K(x,y)\,v(y)\,dy, \tag{24.1}$$

is positive, i. e.,

$$\langle Bu,u\rangle=\int_G\int_G K(x,y)\,u(x)\,u(y)\,dx\,dy\geqslant 0.$$

Theorem 24.1. Let $g(u,x)$ be a real function, continuous and nondecreasing in $u\in(-\infty,+\infty)$ for almost all fixed $x\in G$, where G is a measurable set of positive (finite or infinite) Lebesgue measure in euclidean n-space, and suppose that g generates a bounded Nemytskii operator*

$$hu=g(u(x),x),$$

from $L^p(G)$ to $L^q(G)$ $(p>1,\ p^{-1}+q^{-1}=1)$, i. e. (see VM, Theorem 19.1),

$$\| g(u,x)|\leqslant a(x)+b|u|^{p-1},\quad a(x)\in L^q,\quad b>0.$$

Suppose, moreover, that $\langle hu,u\rangle\geqslant 0$ for $\|u\|>M>0$ and let $B:L^q\to L^p$ be a positive integral operator.

Then the equation $u+Bhu=0$ has a solution $u_0\in L^p$. This solution u_0 is unique if either B is a strictly positive operator or h is a strictly monotone operator, i. e. (see Example 5.4), the function $g(u,x)$ is increasing in u for almost every $x\in G$.

Proof. Since the L-spaces are uniformly convex, * it follows from a theorem of Shmul'yan** that the norm in these spaces is uniformly Fréchet-differentiable, so that the duality mapping $U(x)=\|x\|\operatorname{grad}\|x\|$ (see § 21) has all the properties of the operator A in Definition 19.1; in other words, every L-space has property (π). Our theorem therefore follows from Theorem 19.1, since by a theorem of Banach † the fact that the operator B maps L^p into L^q implies that it is bounded.

* Clarkson /1/.

** Shmul'yan /2/.

† Banach /1/, p.75.

Note that in this theorem we assume neither that mes $G < +\infty$, $p \geqslant 2$, nor that the operator $B: L^q \to L^p$ is selfadjoint. However, if we assume either of these conditions, the assumption that the function $g(u, x)$ is monotone in u may be weakened.

To illustrate, let $p \geqslant 2$ and mes $G < +\infty$. Then $L^p(G) = L^p$ is embeddable, the corresponding Hilbert space being $H = L^2(G)$. Since in this case $L^p \subset L^2$, we agree to let $\|u\|$ denote the norm of a vector $u \in L^p$, the norm of this vector in L^2 being denoted by $\|u\|_2$. Let $\|v\|_q$ denote the norm of a vector $v \in L^q = (L^p)^*$. Now consider the operator $B: L^q \to L^p$ $(p^{-1} + q^{-1} = 1)$ defined by (24.1), and assume that it is positive and selfadjoint. Then, by Theorem 15.1 and Remark 15.2 we have $B = T^*T$, where $T^* = A$ is the positive square root of the restriction B_H of B to L^2. Since L^p is a reflexive space, (see Dunford and Schwartz /1/, Lemma VI. 2. 6) $T = T^{**} = A^*$ and so

$$B = AA^*, \qquad (24.2)$$

where $A: L^2 \to L^p$, $A^*: L^q \to L^2$, and $A = B_H^{1/2}$. Consider any vector $v \in L^q$. Then

$$\| A^*v \|_2^2 = (A^*v,\ A^*v) = \langle Bv,\ v \rangle \leqslant \| B \| \| v \|_q^2,$$

so that $\|A^*\|^2 \leqslant \|B\|$ and, by the theorem on adjoint operators, $\|A\|^2 \leqslant \|B\|$.

Now consider the Hammerstein equation $u - Bhu = 0$, where h is a Nemytskii operator. By Lemma 6.1, this equation is equivalent to the equation

$$\psi(z) \equiv z - A^*hAz = 0 \quad (z \in L^2). \qquad (24.3)$$

Since a Nemytskii operator $h: L^p \to L^q$ $(p > 1,\ p^{-1} + q^{-1} = 1)$ is potential (see Example 5.1 and VM §19) and its potential has the form

$$f(u) = \int_G dx \int_0^{u(x)} g(v,\ x)\, dv,$$

it follows that

$$\psi(z) = \operatorname{grad} \varphi(z); \quad \varphi(z) = \frac{1}{2}(z,\ z) - f(Az).$$

Existence and uniqueness theorems for equation (24.3) may now be established either by the variational method, i. e., by investigating the functional $\varphi(z)$, or by the method of monotone operators,

i. e., by using the monotonicity and other properties of the operator $\psi(z)$. We present a few propositions illustrating each approach.

Theorem 24.2. Let $h: L^p \to L^q$ $(p \geqslant 2, p^{-1}+q^{-1}=1)$ *be a Nemytskii operator such that*

$$\langle hu - hv, u - v\rangle \leqslant a\| u - v\|^2 \quad (a > 0,\ a\| B\| < 1),$$

and B *a positive bounded selfadjoint linear integral operator from* L^q *to* L^p. *Then the Hammerstein equation* $u - Bhu = 0$ *has a unique solution in* L^p.

Proof. Since the Hammerstein equation is equivalent to equation (24.3), it suffices to show that (24.3) has a unique solution. In fact, for any vectors $x, y \in L^2$

$$\begin{aligned}
&(\psi(x) - \psi(y),\ x - y) = \\
&\quad = (x - y,\ x - y) - (A^*hAx - A^*hAy,\ x - y) = \\
&\quad = (x - y,\ x - y) - \langle hAx - hAy,\ Ax - Ay\rangle \geqslant \\
&\quad \geqslant (x - y,\ x - y) - a\| A(x - y)\|^2 \geqslant \\
&\quad \geqslant (x - y,\ x - y) - a\| B\|(x - y,\ x - y) = c\| x - y\|_2^2 \\
&\qquad\qquad (c = 1 - a\| B\| > 0),
\end{aligned}$$

i. e., ψ is a strictly monotone operator. Hence, by Theorem 18.5, the equation $\psi(z) = 0$ has a unique solution $z_0 \in H$, i.e., $z_0 - A^*hAz_0 = 0$. Applying the operator A and setting $Az_0 = u_0$, we get $u_0 - Bhu_0 = 0$. Q. E. D.

Remark 24.1. If it is assumed that $a > 0$ but $a\|B\| = 1$, we can only state that $\psi(z)$ is monotone in L^2. If we then assume, furthermore, that $\langle hu, u\rangle \leqslant a_1\|u\|^2$, where $a_1\|B\| < 1$, then the mapping ψ is also coercive, since as before,

$$(\psi(x),\ x) \geqslant (x,\ x)(1 - a_1\| B\|).$$

Since ψ is monotone and coercive, it follows from Theorem 18.2 that equation (24.3) has a solution, and so the Hammerstein equation also has a solution.

Theorem 24.3. Suppose that B *is an operator satisfying the assumptions of Theorem 24.2, and the Nemytskii operator*

$$h: L^p \to L^q \quad (p \geqslant 2,\ p^{-1} + q^{-1} = 1)$$

and its generating function $g(u, x)$ *satisfy the inequalities*

$$\langle hu - hv,\ u - v\rangle \leqslant a\| u - v\|^2 \quad (a > 0,\ a\| B\| \leqslant 1), \tag{24.4}$$

$$2\int_0^u g(v, x)\,dv \leqslant a_1u^2 + b(x)|u|^\alpha + c(x), \tag{24.5}$$

where $a_1\|B\|<1$, $0<\alpha<2$, $0\leqslant b(x)\in L^\gamma$, $\gamma=2(2-\alpha)^{-1}$, $0\leqslant c(x)\in L$. *Then the equation* $u-Bhu=0$ *has a solution.*

Proof. As in the proof of Theorem 24.2, we find that

$$(\psi(x)-\psi(y),\ x-y)\geqslant 0 \quad (\forall x,\ y\in L^2),$$

whence, by Theorem 8.4, the potential of the operator $\psi(x)$, i. e., the functional

$$\varphi(x)=\frac{1}{2}(x,\ x)-f(Ax), \quad x\in L^2, \quad A=B_H^{1/2}$$

is weakly lower semicontinuous in L^2. Repeating the arguments of VM, pp. 268—269, we find that

$$\varphi(x)\geqslant\frac{1}{2}(x,\ x)^{\alpha/2}\left[(1-a_1\|B_H\|)(x,\ x)^{(2-\alpha)/2}-\beta\right]-C,$$

where

$$C=\int_G c(t)\,dt, \quad \beta=\left(\int_G (b(t))^\gamma\,dt\right)^{1/\gamma}.$$

It follows that there is a sphere $\|x\|=r$ in L^2 on which $\varphi(x)>\varphi(0)$, i. e., the functional $\varphi(x)$ has the m-property. Hence, by Theorem 9.4, there exists a vector $x_0\in L^2$ ($\|x_0\|<r$) such that $\psi(x_0)=0$. Consequently, $x_0-A^*hAx_0=0$. Applying the operator A and setting $Ax_0=u_0$, we get $u_0=Bhu_0$. Q. E. D.

A sufficient condition for inequality (24.5) to hold is, for example, $ug(u,x)\leqslant a_1u^2$ where $a_1\|B_H\|<1$, for this implies (24.5) with $b(x)\equiv 0$ and $c(x)\equiv 0$. We then have

$$(\psi(x)-\psi(0),\ x)\geqslant\|x\|[(1-a_1\|B_H\|)\|x\|-\psi(0)],$$

whence it follows that the operator $\psi(x)$ is coercive.

24.2. Equations with indefinite operators

In this subsection we consider Hammerstein equations $u-Bhu=0$ in the spaces $L^p=L^p(G)$, where $p\geqslant 2$ and mes $G<\infty$, assuming that the integral operator $B: L^q\to L^p\ (p^{-1}+q^{-1}=1)$ defined by (24.1) is selfadjoint and indefinite. Since $L^p\subset L^2\subset L^q$, it follows that L^p is an embeddable reflexive space. As in subsection 6.6, we set

$$B_H^+ = \frac{1}{2}(|B_H| + B_H), \qquad B_H^- = \frac{1}{2}(|B_H| - B_H),$$
$$A = (B_H^+)^{1/2} + (B_H^-)^{1/2}, \qquad C = (B_H^+)^{1/2} - (B_H^-)^{1/2},$$

where B_H is the restriction of the integral operator B to L^2. Suppose, as well, that B is a regular operator (see Definition 15.3). This is true, in particular, if the subspace of L^2 generated by either the positive or negative spectrum of B_H is finite-dimensional.

By Theorem 15.2, these conditions imply that

$$B = V^*W = W^*V,$$

where V and W are extensions of the operators A and C, respectively, bounded from L^q to L^2, $V^* = A$ and $W^* = C$. Since L^q is a reflexive space, it follows from a standard theorem (Dunford and Schwartz /1/, VI. 2. 6) that $V = V^{**} = A^*$ and $W = W^{**} = C^*$, so that

$$B = AC^* = CA^*. \tag{24.6}$$

Let P_1 be the orthogonal projection of L^2 onto the subspace $H_1 \subset L^2$ generated by the positive spectrum of B_H and P_2 the orthogonal projection of L^2 onto $L^2 \ominus H_1 = H_2$. Then (see subsection 6.6) the Hammerstein equation under consideration is equivalent to the equation

$$A^*hAy - (P_1 - P_2)y = 0 \quad (y \in L^2). \tag{24.7}$$

Theorem 24.4. *Let B be a bounded selfadjoint regular indefinite integral operator from L^q to L^p ($p \geqslant 2$, $p^{-1} + q^{-1} = 1$), such that the spectrum of its restriction B_H lies outside some interval $(0, m)$, $m > 0$. Furthermore, let h be a Nemytskii operator from L^p to L^q, satisfying the following strong monotonicity condition:*

$$\langle hu - hv, u - v \rangle \geqslant \frac{1+\alpha}{m} \| u - v \|_2^2 \quad (\forall u, v \in L^p),$$

where $\alpha > 0$. Then the Hammerstein equation has a unique solution in L^p.

Proof. Consider the mapping

$$\psi(y) = A^*hAy - (P_1 - P_2)y \quad (y \in H).$$

It is continuous, since h is defined from L^p into L^q (see VM, §19). Now, we have

$$(\psi(x)-\psi(y),\ x-y)=$$
$$=\langle hAx-hAy,\ Ax-Ay\rangle+\|P_2(x-y)\|^2-\|P_1(x-y)\|^2\geqslant$$
$$\geqslant\frac{1+\alpha}{m}\|A(x-y)\|_2^2-\|P_1(x-y)\|^2+\|P_2(x-y)\|^2.$$

But by inequality (17.7), which holds in this case, we have

$$(\psi(x)-\psi(y),\ x-y)\geqslant\alpha\|x-y\|^2.$$

Since $\psi(x)$ is continuous and strongly monotone, Theorem 18.5 guarantees that equation (24.7) has a unique solution $x_0\in L^2$, i. e.,

$$A^*hAx_0-(P_1-P_2)x_0=0.$$

Applying the operator C and taking into account that $C(P_1-P_2)=A$, we find via (24.6) that $Bhu_0-u_0=0$, where $u_0=Ax_0$. Q. E. D.

Theorem 24.5. Let B be a bounded selfadjoint integral operator from L^q to L^p $(p\geqslant 2,\ p^{-1}+q^{-1}=1)$, whose restriction B_H to L^2 is a quasinegative operator. Let the generating function $g(u,x)$ of the continuous Nemytskii operator $h: L^p\to L^q$ be nondecreasing in $u\in(-\infty,+\infty)$ for almost every $x\in G$ and such that

$$\int_0^u g(v,x)\,dv\geqslant\frac{1}{m}u^2+b(x)|u|^{\alpha}+c(x)$$

for almost every $x\in G$, where m is the infimum of the positive spectrum of the operator B_H, $0<\alpha<2$, $0\geqslant b(x)\in L^{\gamma}$, $\gamma=2(2-\alpha)^{-1}$, $0\geqslant c(x)\in L$. Then the Hammerstein equation $u-Bhu=0$ has a solution.

Proof. Since the positive spectrum of B_H (see VM) consists of a finite number of eigenvalues of finite multiplicity, the operator B is regular (Definition 15.3), so that (24.6) is true.

Consider the functional

$$\varphi(v)=2f(Av)+\|P_2v\|^2-\|P_1v\|^2,$$
$$f(u)=\int_G dx\int_0^{u(x)} g(v,x)\,dv,$$

on L^2. Since grad $f(u)=hu=g(u(x),x)$ our hypothesis implies that this operator is monotone from L^p to L^q, whence, by Theorem 8.4, the functional $f(u)$ is weakly lower semicontinuous. Thus the functional $f(Av)$ is weakly lower semicontinuous, since A is a bounded operator. The functional $\|P_2v\|^2$ is weakly lower semicontinuous, for its gradient is a monotone operator, and the functional

$(-1)\|P_1 v\|^2$ is weakly continuous, since its gradient is completely continuous. The functional $\varphi(v)$, then, is weakly lower semicontinuous. Let us write

$$\text{grad}\,\varphi(v) = 2A^* hAv - 2(P_1 - P_2)v.$$

Arguing as in VM, pp. 277 — 278, we obtain

$$\varphi(v) \geqslant (v, v)^{\alpha/2}\left[(v, v)^{(2-\alpha)/2} - 2\beta\right] + 2\int_G c(x)\,dx,$$

where

$$\beta = \left(\int_G |b(x)|^\gamma\right)^{1/\gamma}.$$

It follows that there is a sphere $\|v\| = r > 0$ on which $\varphi(v) > \varphi(0)$. Thus, the weakly lower semicontinuous functional $\varphi(v)$ has the m-property, so by Theorem 9.4 there exists a vector $v_0(\|v_0\| < r)$ such that grad $\varphi(v_0) = 0$, i. e.,

$$A^* hAv_0 - (P_1 - P_2)v_0 = 0.$$

Hence, as in the proof of the preceding theorem, we conclude that $Bhu_0 - u_0 = 0$, where $u_0 = Av_0$. Q. E. D.

24.3. Equations in Orlicz spaces*

We begin with a brief account of the theory of Orlicz spaces.** Let $\varphi(\xi), 0 \leqslant \xi < +\infty$, be a nonnegative nondecreasing function (which may assume infinite values but is neither identically zero nor identically infinity on $[0, +\infty)$), which is right continuous for any ξ. Set $\psi(\eta) = 0$ if $0 \leqslant \eta < \varphi(0)$ and $\psi(\eta) = \sup\{\xi : \varphi(\xi) \leqslant \eta\}$ if $\eta \geqslant \varphi(0)$. The function $\psi(\eta)$, which is the inverse of $\varphi(\xi)$, has the same properties as $\varphi(\xi)$. The functions

$$M(u) = \int_0^u \varphi(\xi)\,d\xi \quad \text{and} \quad N(v) = \int_0^v \psi(\eta)\,d\eta$$

are known as *complementary Young functions*.

Let G be a set in finite-dimensional euclidean space, of positive (finite or infinite) Lebesgue measure. For $\alpha > 0$, set

* See Vainberg and Shragin /3, 4/.

** See Zaanen /1/ and Luxemburg /1/.

$$L_M^{\alpha}=\left\{u(x):\int_G M(\alpha|u(x)|)\,dx<+\infty\right\},$$

where $u(x)$ is measurable on G. The Orlicz space L^M is defined as the union of the classes L_M^{α} for all $\alpha>0$, with norm*

$$\|u\|_M=\inf\left\{p>0:\int_G M(p^{-1}|u(x)|)\,dx\leqslant 1\right\}.$$

It can be shown out that this is a Banach space. The space L^N is similarly defined. The intersection of all the classes L_M^{α} is a closed separable subspace of L^M, denoted by L_M^f. $L_M^f=L^M$ if and only if $M(u)$ is finite for all u and satisfies the Δ_2-condition: $M(2u)\leqslant CM(u)$ for $u\geqslant u_0\geqslant 0$, where $C=\text{const}$ and $u_0=0$ if $\operatorname{mes}G=\infty$. Note that if $M(u)$ is finite for all u (only in this case does L_M^f contain nonzero vectors), then the general form of a linear functional on L_M^f is

$$lu=\int_G u(x)\,v(x)\,dx \qquad (u=u(x)\in L_M^f),$$

where $v=v(x)$ is a fixed vector in L^N. Moreover,

$$\|v\|_N\leqslant\|l\|\leqslant 2\|v\|_N.$$

Thus the dual $(L_M^f)^*$ of L_M^f may be identified with L^N (up to equivalent norms). In this sense, we write $(L_M^f)^*=L^N$. Note that L^p is an Orlicz space: if $M(u)=u^p$, $1\leqslant p<\infty$, then $L^M=L^p$; if $M(u)=0$ for $0\leqslant u\leqslant 1$ and $M(u)=+\infty$ for $u>1$, then $L^M=L^{\infty}$.

We shall assume henceforth that the Young function $M(u)$ is finite for all u and that $N(v)$ satisfies the Δ_2-condition, so that $L^N=L_N^f$. Then $(L_M^f)^*=L^N$ and $(L^N)^*=(L_N^f)^*=L^M$. Suppose, furthermore, that $L^M\subset L^2$. Then the embedding operator of L^M into L^2 is bounded.** Thus $L_M^f\subset L^M\subset L^2\subset L^N=L_N^f$. Examples of such Young functions are $M(u)=e^u-1$ and $M(u)=e^u-u-1$. In these cases L_M^f is dense in L^2, since the set of all functions bounded on G is a subset of L_M^f (if $\operatorname{mes}G=\infty$, a bounded function must vanish off a bounded subset of G). L^2 is dense in $L^N=L_N^f$, since L_N^f is the closure (in the L^N-metric) of the set of bounded functions (see Luxemburg /1/, p. 55).

Now consider the linear integral operator B defined by (24.1), assuming that $\operatorname{mes}G<\infty$ and that B maps all of L^N into L_M^f, i. e.

* For the definition of another equivalent norm, see Luxemburg /1/ or Krasnosel'skii and Rutitskii /1/.

** Luxemburg /1/, p.51 or Krasnosel'skii and Rutitskii /1/, p.132.

(see Banach /1/, p. 75) it is bounded from L^N to L_M^f and its restriction B_0 to L^2 is a bounded operator. Then, if B is a positive operator, it can be represented as

$$B = A_0 A, \tag{24.2}$$

where $A_0 = B_0^{1/2}: L^2 \to L^M$ and A is a continuous extension of A_0 from L^N to L^2. Then $\langle w, A_0 v\rangle = (Aw, v)$ for any $v \in L^2$, $w \in L^N$, where

$$\langle w, A_0 v\rangle = \int_G w(x) A_0 v(x)\, dx.$$

If B_0 is a quasinegative operator, we have a representation of type (24.6).

Finally, we shall assume that the Nemytskii operator h generated by $g(u, x)$ is defined in L_M^f with range in $L^N = L_N^f$. Then (see Vainberg and Shragin /1, 2 /) h is continuous and

$$hu = \operatorname{grad} f(u); \quad f(u) = \int_G dy \int_0^{u(y)} g(v, y)\, dv, \quad u \in L_M^f.$$

Theorem 24.6. Let B_0 be a positive selfadjoint operator and $h: L_M^f \to L^N$, and suppose that

$$\langle hu - hv, u - v\rangle \leqslant a\| u - v\|^2 (u, v \in L_M^f, a > 0, a\| B\| < 1).$$

Then the Hammerstein equation has a unique solution.

Proof. Consider the mapping $\psi(x) = x - AhA_0 x$ in L^2. It is continuous and, as in the proof of Theorem 24.2, one can show that

$$(\psi(x) - \psi(y), x - y) \geqslant c\| x - y\|_2^2 \quad (c = 1 - a\| B\| > 0).$$

Consequently, by Theorem 18.5, there exists a unique solution z_0 to the equation $\psi(x) = 0$, i. e., $z_0 - AhA_0 z_0 = 0$. Applying the operator A_0, we obtain $u_0 - Bhu_0 = 0$, where $u_0 = A_0 z_0 \in L_M^f$. By Lemma 6.1, the solution u_0 is unique. Q. E. D.

Theorem 24.7. Let B_0 be a positive selfadjoint operator, and suppose that the Nemytskii operator $h: L_M^f \to L^N$ and its generating function satisfy (24.4) and (24.5).

Then the equation $u = Bhu$ has a solution in L_M^f.

The proof is similar to that of Theorem 24.3.

Theorem 24.8. Let B_0 be a quasinegative selfadjoint operator in L^2 and let the Nemytskii operator h from L_M^f to L^N satisfy the inequality

* See Vainberg and Shragin /4/.

$$\langle hu - hv,\ u - v\rangle \geqslant \frac{1+\alpha}{m}\| u - v \|_2^2 \qquad (\forall u,\ v \in L_M^f),$$

where $\alpha > 0$ *and* m *is the smallest eigenvalue of* B_0. *Then the equation* $u - Bhu = 0$ *has a unique solution in* L_M^f.

Proof. Let P_1 be the orthogonal projection from L^2 onto the invariant subspace H_1 of the operator B_0 corresponding to its positive spectrum; let P_2 be the projection from L^2 onto $H_2 = L^2 \ominus H_1$.

Consider the mapping $\psi(y) = AhA_0y - (P_1 - P_2)y$ in L^2. It is continuous and, as in the proof of Theorem 24.4, one shows that

$$(\psi(x) - \psi(y),\ x - y) \geqslant \alpha \| x - y \|_2^2.$$

Hence, by Theorem 18.5, the equation

$$AhA_0y - (P_1 - P_2)\,y = 0$$

has a unique solution. Using the equivalence results of § 6, we may now prove our theorem as in the case of Theorem 24.4.

The assertion of the theorem remains valid if the operator B is regular rather than B_0 being quasinegative.

Theorem 24.9. Let B_0 *be a quasinegative selfadjoint operator; suppose that the function* $g(u,\ x)$ *generating the Nemytskii operator* $h: L^f \to L^N$ *is nondecreasing in* $u \in (-\infty, +\infty)$ *for almost every* $x \in G$, *and that*

$$\int_0^u g(v,\ x)\,dv \geqslant \frac{1}{m} u^2 + b(x)\,|\,u\,|^{\alpha} + c(x)$$

for almost every $x \in G$, *where* m *is the smallest positive eigenvalue of the operator* B_0, $0 < \alpha < 2$, $0 \geqslant b(x) \in L^{\gamma}$, $\gamma = 2(2-\alpha)^{-1}$, $0 > c(x) \in L$. *Then the equation* $u = Bhu$ *has a solution in* L_M^f.

The proof is similar to that of Theorem 24.5.

24.4. General theorem for systems

Let G be a measurable set of positive (finite or infinite) Lebesgue measure in euclidean s-space, and let $g_i(u_1,\ u_2,\ \ldots,\ u_n,\ x)$ be real functions generating a continuous Nemytskii operator $h = (h_1,\ h_2,\ \ldots,\ h_n)$ from $L_{p,n}$ to $L_{q,n}$ $(p > 1,\ p^{-1} + q^{-1} = 1)$, i. e. (see VM, § 19),

$$|\,g_i(u_1,\ u_2,\ \ldots,\ u_n,\ x)\,| \leqslant a_i(x) + b\sum_{k=1}^{n} |\,u_k\,|^{p-1},$$

$$a_i(x) \in L^q, \quad b > 0.$$

Consider the system of nonlinear integral equations ($i=1, 2, \ldots, n$)

$$u_i(x)+\sum_{j=1}^{n}\int_G K_{ij}(x, y)\, g_j(u_1(y), u_2(y), \ldots, u_n(y), y)\, dy, \qquad (24.8)$$

where each kernel $K_{ij}(x, y)$ generates a bounded integral operator

$$B_{ij}(v)=\int_G K_{ij}(x, y)\, v(y)\, dy$$

from $L^q(G)$ to $L^p(G)$, $p>1$, $p^{-1}+q^{-1}=1$.

Theorem 24.10. Suppose that the matrix $K(K_{ij}(x, y))$ generates a positive operator from $L_{q,n}$ to $L_{p,n}$, i.e.,*

$$\sum_{i,j=1}^{n}\int_G\int_G K_{ij}(x, y)\, v_i(x)\, v_j(y)\, dx\, dy \geqslant 0, \qquad v_i \in L^q,$$

and the functions g_i, which generate a continuous Nemytskii operator h from $L_{p,n}$ to $L_{q,n}$, satisfy the inequalities

$$\sum_{i=1}^{n}[g_i(u_1, u_2, \ldots, u_n, x)- g_i(v_1, v_2, \ldots, v_n, x)](u_i-v_i)\geqslant 0, \qquad (24.9)$$

$$u_i g_i(u_1, u_2, \ldots, u_n, x)\geqslant \alpha |u_i|^p - \sum_{k=1}^{n}|\beta_{ik}(x)||u_k|^{\gamma}-|c_i(x)| \qquad (24.10)$$

for almost every $x\in G$ and all $u_i, v_i \in(-\infty, +\infty)$, where

$$a>0,\ 1\leqslant\gamma<p,\ \beta_{ik}(x)\in L^r,\ r=p/(p-\gamma),\ c_i(x)\in L(G).$$

Then system (24.8) has a solution in $L_{p,n}(G)$. If K generates a strictly positive operator or if (24.9) is a strict inequality, this solution is unique.

Proof. We first observe that the spaces $L_{p,n}$ satisfy property (π), since the operator $A=(A_1, A_2, \ldots, A_n)$, where

$$A_i u = |u_i(x)|^{q-1}\operatorname{sign} u_i(x),$$

$$u=(u_1(x), u_2(x), \ldots, u_n(x))\in L_{q,n}$$

meets the requirements of Definition 19.1. Indeed, by a theorem on Nemytskii operators (see VM, § 19), A is a continuous operator

* See A. M. Vainberg /1/.

from $L_{q,n}$ to $L_{p,n}$ and, since the function $\varphi(t) = |t|^{q-1} \operatorname{sign} t$ is strictly increasing, we have that for any $u, v \in L_{q,n}$

$$\langle Au - Av, u - v\rangle = \sum_{i=1}^{n} \int_G [|u_i(x)|^{q-1} \operatorname{sign} u_i(x) - \\ - |v_i(x)|^{q-1} \operatorname{sign} v_i(x)] [u_i(x) - v_i(x)]\, dx \geqslant 0,$$

i. e., A is a monotone operator. Finally

$$\langle Au, u\rangle = \sum_{i=1}^{n} \int_G u_i(x) |u_i(x)|^{q-1} \operatorname{sign} u_i(x)\, dx = \| u \|_q^q.$$

The existence of the operator A in $L_{p,n}$ clearly follows from the general reasoning of § 21. Inequality (24.9) is equivalent (see Example 5.5) to the assumption that the operator $h = (h_1, h_2, \ldots, h_n)$ is monotone, and inequality (24.10) gives

$$\langle hu, u\rangle = \sum_{i=1}^{n} \int_G u_i(x) g_i(u_1(x), u_2(x), \ldots, u_n(x), x)\, dx \geqslant \\ \geqslant \alpha \sum_{i=1}^{n} \| u_i \|^p - \sum_{i=1}^{n} \int_G | \beta_i(x) u_i | |(x)|^\gamma\, dx - \sum_{i=1}^{n} \int_G |c_i(x)|\, dx.$$

But by the Hölder inequality $(p_1 = p/\gamma,\ q_1 = p/(p-\gamma))$, we have

$$\int_G |\beta_i(x)| |u_i(x)|^\gamma\, dx \leqslant \\ \leqslant \left(\int_G |u_i(x)|^p\, dx \right)^{\gamma/p} \left(\int_G |\beta_i(x)|^{p/(p-\gamma)}\, dx \right)^{(p-\gamma)/p} = \| u_i \|^\gamma b_i^{(p-\gamma)/p}.$$

Using the Hölder inequality for sums,

$$\sum_{i=1}^{n} \| u_i \|^\gamma b_i^{(p-\gamma)/p} \leqslant \left(\sum_{i=1}^{n} \| u_i \|^p \right)^{\gamma/p} \left(\sum_{i=1}^{n} b_i \right)^{(p-\gamma)/p} = \| u \|^\gamma b,$$

we get $\langle hu, u\rangle \geqslant \| u \|^\gamma (\alpha \| u \|^{p-\gamma} - b) - c$, where

$$c = \sum_{i=1}^{n} \int_G |c_i(x)|\, dx, \quad b = \left(\sum_{i=1}^{n} \int_G |\beta_i(x)|^{p/(p-\gamma)}\, dx \right)^{(p-\gamma)/p}.$$

Hence there exists $M > 0$ such that $\langle hu, u\rangle > 0$ for $\|u\| \geqslant M$. The space $E = L_{p,n}$ and the operators B and $F = h$ therefore satisfy the hypotheses of Theorem 19.1, whence follow the assertions of our theorem.

Note that Theorem 24.10 is a generalization of Theorem 24.1 to spaces of vector-functions. Similarly, the other theorems proved in subsections 24.1 — 24.3 can be carried over to spaces of vector-functions. Moreover, all the theorems of §§ 17, 19, 20 possess analogs valid for ordinary Hammerstein equations, with the operator F replaced by a Nemytskii operator h and the operator B by a similar integral operator.

§ 25. BOUNDARY-VALUE PROBLEMS FOR QUASILINEAR PARTIAL DIFFERENTIAL EQUATIONS

Here we shall demonstrate the application of our general theorems on monotone operators to some boundary-value problems for partial differential equations, whose theory was elaborated by Vishik [1 — 6]. This approach was developed by Browder /1 — 7, 11, 16, 18/, being later used by Leray and Lions /1/, as well as other authors.* As a matter of fact, special cases of the method appear in earlier work of Engel'son /2, 4/ and Zeragiya /3/.

25.1. Quasilinear elliptic differential operators

Let ω be an open bounded set in euclidean n-space with sufficiently smooth boundary; we denote points of ω by $x=(x_1, x_2, \ldots, x_n)$. We consider real vector-functions $u(x) = (u_1(x), u_2(x), \ldots, u_r(x))$ defined on ω and belonging to the Sobolev space $W_p^m(\omega)$ (see Sobolev /1/), where m is a natural number and $p>1$ a real number. We recall the definition of W_p^m (see Smirnov /1/). Let α be a multi-index, $\alpha=(\alpha_1, \alpha_2, \ldots, \alpha_n)$, where α_i are nonnegative integers, $|\alpha|=\alpha_1+\alpha_2+\ldots+\alpha_n$. Set $D_k=\frac{\partial}{\partial x_k}$, $D^\alpha=D_1^{\alpha_1}\ldots D_n^{\alpha_n}$ and $D^\alpha u=(D^\alpha u_1, D^\alpha u_2, \ldots, D^\alpha u_r)$. The derivatives are understood in the generalized sense.** We write $u\in L^p(\omega)$ if every component u_k of the vector u lies in $L^p(\omega)$. The Sobolev space is defined by

$$W_p^m=\{u\colon\ u\in L^p,\ D^\alpha u\in L^p \text{ for all } |\alpha|\leqslant m\},$$

with norm given by

$$\|u\|_{m,p}=\Bigl(\sum_{|\alpha|\leqslant m}\int_\omega |D^\alpha u|^p\,dx\Bigr)^{1/p}.$$

* See the review by Dubinskii /2/, where the work of Browder, Vishik and others is discussed.

** See Sobolev /1/, Smirnov /1/ or Lyusternik and Sobolev /1/.

Let $C_C^\infty(\omega)$ be the family of infinitely differentiable functions with compact support in ω. The closure of $C_C^\infty(\omega)$ in the metric of $W_p^m(\omega)$ is denoted by $\overset{\circ}{W}{}_p^m(\omega)$. Clearly, $\overset{\circ}{W}{}_p^m \subset W_p^m$, and, if Γ is the boundary of ω, then for any function $u \in \overset{\circ}{W}{}_p^m$ we have* $D^\gamma u|_\Gamma = 0$ for $|\gamma| \leqslant m-1$. $W_p^m(\omega)$ is a separable reflexive Banach space.

Let the real vector-functions $A_\alpha(x, v_1, v_2, \ldots, v_{s\alpha})$, $A_\alpha = (A_\alpha^{(1)}, A_\alpha^{(2)}, \ldots, A_\alpha^{(r)})$ be measurable in $x \in \omega$ for fixed $v_i \in (-\infty, +\infty)$ and continuous in v_i for almost every $x \in \omega$. Consider the following quasilinear elliptic differential operator of order $2m$ in generalized divergence form:

$$Au = \sum_{|\alpha| \leqslant m} (-1)^{|\alpha|} D^\alpha A_\alpha(x, u, \ldots, D^m u). \tag{25.1}$$

The functions A_α may be written in another form. Set

$$\tilde{D}^k u = \{D^\nu u, \ |\nu| = k\},$$
$$\delta u = \{u, \ Du, \ \ldots, \ D^{m-1} u\}.$$

Then

$$A_\alpha(x, u, \ldots, D^m u) = A_\alpha(x; \delta u; \tilde{D}^m u).$$

We assume the validity of

Condition 25.1. For any $u, v \in W_p^m(\omega)$,

$$A_\alpha(x; \delta u; \tilde{D}^m v) : L^p(\omega) \to L^q(\omega), \ p > 1, \ p^{-1} + q^{-1} = 1.$$

Using a theorem (VM, Theorem 19.2; see also Example 5.2 in this book) giving necessary and sufficient conditions for a Nemytskii operator to be defined and continuous in an L-space, we see that Condition 25.1 holds if

$$A_\alpha^i(x, v_1, v_2, \ldots, v_{s\alpha})| \leqslant$$
$$\leqslant a_{i\alpha}(x) + b \sum_{k=1}^{s\alpha} |v_k|^{p-1}, \ a_{i\alpha}(x) \in L^q(\omega).$$

Actually, using certain inequalities due to Sobolev,** this estimate can be sharpened (raising the degree of those terms v_k corresponding to δu; see Browder /5, 11/). Thus, Condition 25.1 imposes a restriction on the rate of increase of $A_\alpha(x, v_1, v_2, \ldots, v_{s\alpha})$ with respect to v_k, namely, it must have polynomial increase. Vishik /6/ has shown that one can define spaces similar to $W_p^m(\omega)$ but based on Orlicz spaces, in such a way that this restriction is

* Smirnov /1/, pp.356–357.

** Sobolev /1/.

relaxed. The reason for this is that the necessary and sufficient condition for Nemytskii operators to be defined in Orlicz spaces allow the functions A_α to increase with v_k more rapidly than a polynomial (see Vainberg, Shragin /1, 2/).

25.2. The boundary-value problem

In order to formulate various boundary value problems for the equation $Au = f$, where A is defined by (25.1), we define the space of admissible functions.

Let V be a closed subspace of $W_p^m(\omega)$ containing $\mathring{W}_p^m(\omega)$. This space, too, is separable and reflexive, since $V \subset W_p^m(\omega)$. We construct a nonlinear operator $G: V \to V^*$, using the nonlinear Dirichlet form

$$a(u, v) = \sum_{|\alpha| \leqslant m} \int_\omega A_\alpha\left(x;\ \delta u;\ \tilde{D}^m u\right) D^\alpha v\, dx. \tag{25.2}$$

This form is defined for any $u, v \in V$, since $D^\alpha v \in L^p(\omega)$, $A_\alpha(x;\ \delta u,\ \tilde{D}^m u) \in L^q(\omega)$. This form is linear and continuous in $v \in V$, so that it uniquely determines an operator $G(u) \in V^*$ via the formula

$$a(u, v) = \langle G(u), v\rangle. \tag{25.3}$$

Definition 25.1. A function $u_0(x) \in V$ is said to be a *weak solution of the boundary-value problem* for the equation

$$Au = f \qquad (f \in V^*),$$

corresponding to the subspace V, if for any $v \in V$

$$a(u_0, v) = \langle f, v\rangle, \tag{25.4}$$

or, equivalently,

$$G(u_0) = f. \tag{25.4'}$$

This boundary-value problem is a Dirichlet problem if $V = \mathring{W}_p^m(\omega)$, a Neumann problem if $V = W_p^m(\omega)$, or a mixed boundary-value problem if $\mathring{W}_p^m \subset V \subset W_p^m(\omega)$ (proper inclusion!).

25.3. Auxiliary propositions

We cite some propositions, due essentially to Browder /5, 11/.

Lemma 25.1. If Condition 25.1 holds, the operator $G: V \to V^*$ is demicontinuous.

Proof. Let $\|u_n - u_0\|_{m,p} \to 0$ as $n \to \infty$. Then for any $|\gamma| \leqslant m$, we have $D^\gamma u_n \to D^\gamma u_0$ in the L^p-metric. Since the Nemytskii operator generated by the function $A_\alpha(x, v_1, v_2, \ldots, v_{s\alpha})$ is continuous from L^p to L^q $(p^{-1} + q^{-1} = 1)$, it follows that as $n \to \infty$

$$A_\alpha(x; \delta u_n, \tilde{D}^m u_n) \to A_\alpha(x; \delta u_0, D^m u_0)$$

in the L^q-metric. Hence $a(u_n, v) \to a(u, v)$ for any $v \in V$, i.e., $G(u_n) \rightharpoonup G(u_0)$ as $n \to \infty$. Q. E. D.

Note that if the order of increase of A_α with respect to the variables v_k corresponding to δu exceeds $p-1$ (this is admissible, by virtue of Sobolev's inequalities; see Browder /5, 11/), the argument may be modified as follows. Since the Nemytskii operator is continuous in measure (VM, Theorem 18.6), it follows that as $u_n \to u_0$ we have

$$A_\alpha(x; \delta u_n; \tilde{D}^m u_n) \to A_\alpha(x; \delta u_0; \tilde{D}^m u_0)$$

in measure, so there exists a subsequence $\{n_k\}$ such that

$$A_\alpha(x; \delta u_{n_k}; \tilde{D}^m u_{n_k}) \to A_\alpha(x; \delta u_0, \tilde{D}^m u_0)$$

almost everywhere. Thus, $a(u_{n_k}, v) \to a(u_0, v)$. Hence, by Lemma 16.1, $G(u)$ is demicontinuous.

We now consider subordinate operators. Let

$$Bu = \sum_{|\beta| \leqslant m-1} (-1)^{|\beta|} B_\beta(x, u, \ldots, D^{m-1}u), \tag{25.5}$$

where the functions $B_\beta(x, v_1, v_2, \ldots, v_{s\beta})$ generate continuous Nemytskii operators from L^p to L^q $(p^{-1} + q^{-1} = 1)$. The operator B is said to be subordinate to the operator A defined by (25.1), or of lower order than A. As before, the operator B generates a Dirichlet form

$$b(u, v) = \sum_{|\beta| \leqslant m-1} \int_\omega B_\beta(x, u, \ldots, D^{m-1}u) D^\beta v \, dx \tag{25.6}$$

$$(\forall u, v \in V),$$

which determines a nonlinear operator $C: V \to V^*$ via the formula

$$b(u, v) = \langle C(u), v\rangle.$$

Since B is a subordinate operator, it follows from the Sobolev embedding theorems,* via Hölder's inequality and the properties of Nemytskii operators, that

$$|\langle C(u), v\rangle| \leqslant g(\|u\|_{m,p})\|v\|_{m-1,r},$$

where $r^{-1} > p^{-1} - n^{-1}$ and $g(t)$ is a continuous function; moreover, the embedding operator J of $W_p^m(\omega)$ into $W_r^{m-1}(\omega)$ is completely continuous.

Let V_1 be the space consisting of the same vectors as V, but with the relative topology as a subspace of $W_r^{m-1}(\omega)$. Then, as before,

$$b(u_1, v_1) = \langle C_1(u_1), v_1\rangle \quad (\forall u_1, v_1 \in V_1),$$

where the operator C_1 is demicontinuous by Lemma 25.1. Setting $u_1 = Ju$, where J is the embedding operator, we have

$$C = J^* C_1 J.$$

Since J and J^* are completely continuous linear operators and C_1 is demicontinuous, we obtain the following proposition:

Lemma 25.2. The operator $C = J^*C_1J$ is strongly continuous from V to V^*.

25.4. Existence theorems

Using Theorem 18.4 and Lemmas 25.1, 25.2, we have the following proposition**:

Theorem 25.1. Suppose that:

1. A is an operator of type (25.1) and B is an operator subordinate to A, of type (25.2), both satisfying Condition 25.1.

2. V is a closed subspace of $W_p^m(\omega)$ containing $C_c^\infty(\omega)$.

3. The nonlinear Dirichlet forms $a(u,v)$ and $b(u,v)$ defined by (25.2) and (25.6) satisfy the inequalities

$$a(u, u-v) - a(v, u-v) + b(u, u-v) - b(v, u-v) \geqslant 0,$$
$$a(u, u) \geqslant c(\|u\|_{m,p})\|u\|_{m,p},$$

* See Sobolev /1/, Smirnov /1/, Kondrashov /1/, /2/.

** See Browder /1/.

for all $u, v \in V$, *where* $c(t)$ *is a real function such that* $c(t) \to +\infty$ *as* $t \to +\infty$.

Then the boundary-value problem for the equation $Au = f$ *has at least one solution.*

This theorem implies the following assertion. If an operator A of type (25.1) is representable as $Au = A_1 u + A_2 u$ where A_1 is of order m and A_2 is of order less than m, and if $a_1(u, v)$ and $a_2(u, v)$ are the corresponding Dirichlet forms, then only the operator A_1 need be monotone, i. e., only the operator G_1 defined by $a_1(u, v) = \langle G_1(u), v\rangle$ (see (25.3)) need be monotone. By Theorem 18.5 and Lemma 25.1, we then have the following proposition for strongly elliptic systems.

Theorem 25.2. Suppose that:

1. A *is an operator of type (25.1) satisfying Condition 25.1.*
2. V *is a closed subspace of* $W_p^m(\omega)$ *containing* $C_c^\infty(\omega)$.
3. *The nonlinear Dirichlet form defined by (25.2) satisfies the inequality*

$$a(u, u - v) - a(v, u - v) \geqslant \gamma(\| u - v \|_{m,p}) \| u - v \|_{m,p}$$

where $\gamma(t)$ *is an increasing function such that* $\gamma(0) = 0$ *and* $\gamma(t) \to +\infty$ *as* $t \to +\infty$.

Then the boundary-value problem for the equation $Au = f$ *has a unique solution for any* $f \in V^*$.

25.5. The boundary-value problem for parabolic equations

Let ω be an open bounded set in euclidean n-space, $t \in [0, b] = S$ and Q the cylinder $\omega \times S$. As before, we consider the operator

$$A(t)u = \sum_{|\alpha| \leqslant m} (-1)^{|\alpha|} D^\alpha A_\alpha(x, t; u, \ldots, D^m u)$$

on the assumption that the r-dimensional vector-functions A_α are measurable in $(x, t) \in Q$, continuous in the other variables, and generate a continuous Nemytskii operator from $L^p(Q)$ to $L^q(Q)$ $(p > 1, p^{-1} + q^{-1} = 1)$, i. e. (VM, Theorem 19.2),

$$| A_\alpha^{(i)}(x, t; v_1, v_2, \ldots, v_{s\alpha}) | \leqslant a(x, t) + c \sum_{k=1}^{s\alpha} | v_k |^{p-1},$$

where $c > 0$, $a(x, t) \in L^q(Q)$. As before, let V be a closed subspace of $W_p^m(\omega)$ containing $C_c^\infty(\omega)$. Consider the space $L^p(\dot{S}, V)$ of abstract functions $u(t)$ from S to V, with norm given by

$$\| u \| = \left\{ \int_0^b \| u(t) \|_{m,p}^p \, dt \right\}^{1/p}.$$

$L^p(S, V)$ is a separable reflexive Banach space. By virtue of our assumptions, for any $u, v \in L^p(S, V)$ there exists a nonlinear Dirichlet form

$$l(u, v) = \sum_{|\alpha| \leqslant m} \int_0^b \langle A_\alpha(x, t; u(t), \ldots, D^m u(t)), D^\alpha v(t) \rangle \, dt.$$

Consider the set of all continuous and piecewise-differentiable functions $F = \{u(t)\}$ in the intersection

$$L^p(S, V) \cap L^q(S, V)$$

and set $F_0 = \{u \in F : u(0) = 0\}$. Given $u \in F_0$ and $v \in L^p(S, V)$, we take the bilinear form

$$l_0(u, v) = \int_0^b \left\langle \frac{du(t)}{dt}, v(t) \right\rangle dt,$$

which defines a linear operator δ_0 from F_0 to

$$(L^p(S, V))^* : \langle \delta_0 u, v \rangle = l_0(u, v);$$

δ_0 has closure δ.

Definition 25.2. A function $u \in D(\delta)$ is said to be a solution of the boundary-value problem for the parabolic equation

$$\frac{\partial u}{\partial t} + A(t) u = f(u, t), \ f \in (L^p(S, V))^* \tag{25.7}$$

if, for any $v \in L^p(S, V)$,

$$\langle \delta u, v \rangle + l(u, v) = \langle f, v \rangle.$$

Note that this problem may also be formulated in terms of the classical parabolic boundary-value problem.

Using an analog of Theorem 20.6, one can prove the following proposition.*

Theorem 25.3. Suppose that the Dirichlet form is monotone, in the sense

$$l(u, u-v) - l(v, u-v) \geqslant 0,$$

and coercive, in the sense

$$l(u, u) \geqslant c(\|u\|)\|u\|,$$

where $c(t) \geqslant 0$ *and* $c(t) \to +\infty$ *as* $t \to +\infty$. *Then the problem posed for equation (25.7) has a unique solution* $u_0(t)$. *This solution is continuous from* S *to* $L^2(\omega)$ *and* $u_0(0) = 0$.

§ 26. NONLINEAR DIFFERENTIAL EQUATIONS IN BANACH SPACES**

The theory of differential equations in linear spaces is comparatively young.† We only wish to point out the role of monotone operators in this theory. The parabolic equation (25.6) considered in subsection 25.5 may be viewed as an equation $u' + A(t)u = f$, where the operator $A(t)u$ is monotone in u. A similar remark is valid for the equations of type $Bx + F(x) = y$, as considered in subsection 20.3, where F is a monotone operator and B a densely defined linear operator, since if other assumptions are adopted B may be regarded, in particular, as a differentation operator. However, differential equations with monotone operators in Hilbert and Banach spaces were first studied explicitly by Browder /8, 14, 20/, Kato /3/, Lions /1/, Lions and Strauss /1/, and later by other authors.

26.1. Equations with bounded monotone operators

Let H be a separable Hilbert space with inner product (u, v), $S = [0, b]$, and let $f(t, u)$ be a bounded demicontinuous mapping from $S \times H$ to H. Consider the initial-value problem

$$\frac{du}{dt} = f(t, u), \qquad u(0) = u_0 \, (t \in S) \tag{26.1}$$

* Browder /7/.

** This section was written in collaboration with T. Kh. Enikeeva.

† See review by Nemytskii, Vainberg and Gusareva /1/.

and the corresponding integral equation

$$u(t)=u_0+\int_0^t f(s, u(s))\,ds, \qquad t\leqslant b. \tag{26.2}$$

If $f(t, u)$ is a continuous function from $S\times H$ to H, then problem (26.1) and equation (26.2) are equivalent in the class of continuous functions from S to H. This is proved as in the classical case. If $f(t, u)$ is not continuous, problem (26.1) and equation (26.2) need not be equivalent. A solution of the integral equation is then known as a generalized solution of problem (26.1).

Along with H, we shall consider the space $C(S, H)$ of continuous functions from S to H, the space of continuously differentiable functions $C^1(S, H)$, and the Hilbert space $L^2=L^2(S, H)$ with the usual norm and inner product:

$$\|u\|_2=\left(\int_0^b \|u(t)\|^2\,dt\right)^{1/2}; \qquad (u, v)_2=\int_0^b (u, v)\,dt.$$

The norm and inner product in H are denoted by $\|u\|$, (u, v), respectively, the norm in $C(S, H)$ by $\|u\|_1$.

Theorem 26.1. Let* $f(t, u)$ *be a continuous bounded mapping from* $S\times H$ *to* H, *such that*

$$\operatorname{Re}(f(t, u)-f(t, v), u-v)\leqslant M\|u-v\|^2 \tag{26.3}$$

for $t\in S$ *and any* $u, v\in H$. *Then problem (26.1) has a unique solution* $u=u(t)\in C^1(S,H)$ *for any* $u_0\in H$, *and the mapping* $u_0\to u$ *is continuous from* H *to* $C(S,H)$.

Proof. We first observe that without loss of generality we may set $M=0$ in (26.3), i. e., inequality (26.3) may be replaced by the assumption that the mapping $(-f)$ is monotone. Indeed, the substitution $u(t)=v(t)e^{Mt}$ reduces equation (26.1) to

$$\frac{dv}{dt}+Mv=e^{-Mt}f(t, e^{Mt}v), \quad v(0)=u_0,$$

where the function $F(t, v)=e^{-Mt}f(t, e^{Mt}v)-Mv$ satisfies (26.3) with $M=0$. We thus assume for the proof that $(-f)$ is a monotone mapping from $S\times H$ to H. We may then weaken the continuity requirements, assuming only that $f(t, u)$ is a demicontinuous mapping from $S\times H$ to H.

* Browder /8/ and Kato /3/.

Let L' be the differentiation operator in L^2, $(L'u)(t) = u'(t)$, with domain

$$D(L') = \{u(t) \in C^1(S, H) : u(0) = 0\}.$$

This operator has a closure in L^2, which we denote by L.

Lemma 26.1. The closure L of L' is a maximal monotone operator, with domain $D(L) \subset C(S, H)$ dense in L^2, and has a bounded inverse L^{-1} in L^2.

Proof. $D(L)$ is dense in L^2, since $D(L) \supset D(L')$ and $D(L')$ is dense in L^2. Let $u \in D(L)$. Then there exists a sequence $\{u_n\} \subset D(L')$ such that $u_n \to u$, $u_n'(t) \to Lu$, in L^2. But then

$$\| u_n - u_m \|_1 = \max_t \left\| \int_0^t (u_n'(s) - u_m'(s))\, ds \right\| \leqslant$$
$$\leqslant \int_0^b \| u_n'(s) - u_m'(s) \| \, ds \leqslant \sqrt{b} \left(\int_0^b \| u_n'(s) - u_m'(s) \|^2 \, ds \right)^{1/2} =$$
$$= \sqrt{b} \| u_n' - u_m' \|_2 \to 0,$$

as $m, n \to \infty$. Hence $\{u_n(t)\}$ is a fundamental sequence in $C(S, H)$, and so $u_n \to v \in C(S, H)$. Therefore, $u_n \to v$ in L^2, so that $u = v$ almost everywhere. By modifying $u(t)$ on a set of measure zero, we may assume that $u(t) \in C(S, H)$. Thus $D(L) \subset C(S, H)$.

Now let $u \in D(L)$, $u_n \in D(L')$ and suppose that $u_n \to u$, $u_n'(t) \to Lu$ in L^2. Then

$$\operatorname{Re}(Lu, u)_2 = \lim_{n \to \infty} \operatorname{Re} \left(\frac{du_n}{dt}, u_n \right)_2 =$$
$$= \lim_{n \to \infty} \int_0^b \operatorname{Re} \left(\frac{du_n}{dt}, u_n \right) dt = \lim_{n \to \infty} \int_0^b \frac{1}{2} \frac{d}{dt} (u_n, u_n)\, dt =$$
$$= \frac{1}{2} \lim_{n \to \infty} (\| u_n(b) \|^2 - \| u_n(0) \|^2) = \frac{1}{2} \| u(b) \|^2 \geqslant 0,$$

so that L is a monotone operator.

In order to prove that L is maximal and its inverse L^{-1} bounded, it suffices to verify* that for any function $v \in L^2$

$$(L^{-1} v)(t) = \int_0^t v(s)\, ds. \qquad (26.4)$$

Since this equality holds for any function in $C(S, H)$, the fact that $C(S, H)$ is dense in L^2 implies that the equality is true throughout L^2.

* See Lumer and Phillips /1/.

Consider the mapping

$$(Gu)t = -f(t, u(t)) \tag{26.5}$$

from $C(S, H)$ to L^2; it is, by assumption,* bounded in $C(S, H)$, so that for any function $u(t)$ continuous on S we have $Gu(t) \in L^2$.

Lemma 26.2. The mapping G is bounded, monotone and demicontinuous from $C(S, H)$ to L^2.

Proof. By assumption, the mapping G has all these properties as a mapping from H to H. Hence it is certainly bounded and monotone from $C(S,H)$ to L^2. To prove that G is demicontinuous, we consider an arbitrary sequence such that $u_n \to u$ in the metric of $C(S, H)$. For any fixed $t \in S$, we have $f(t, u_n(t)) \rightharpoonup f(t, u(t))$ in the metric of H. Hence, for any vector $v \in L^2$,

$$|(Gu_n - Gu, v)_2| = \left| \int_0^b (f(t, u_n(t)) - f(t, u(t)), v(t))\, dt \right| \leqslant$$

$$\leqslant \int_0^b |(f(t, u_n(t)) - f(t, u(t)), v(t))|\, dt = I_n.$$

But since $u_n(t) \to u(t)$ in $C(S, H)$ and f is a bounded mapping in H, it follows that

$$|(f(t, u_n(t)) - f(t, u(t)), v(t))| \leqslant M_1 \| v(t) \|, \quad v(t) \in L^2.$$

Thus, by Lebesgue's theorem, $I_n \to 0$ as $n \to \infty$, i. e., $(Gu_n - Gu, v)_2 \to 0$ or $Gu_n \rightharpoonup Gu$ in L^2.

Let $H_1 \subset H_2 \subset \ldots$ be an increasing sequence of subspaces of H such that $\bigcup_n H_n$ is dense in H, and let P_n be the orthogonal projections of H onto H_n, so that $P_n \to I$ strongly. Each P_n defines a projection in L^2, which we again denote by P_n: $(P_n u)t = P_n u(t)$ for any function $u(t) \in L^2$. Consider the initial-value problems

$$v_n' = P_n f(t, v_n + u_0), \quad v_n(0) = 0 \quad (n = 1, 2, 3, \ldots), \tag{26.6}$$

where $v_n(t) \in H_n$. For every fixed n, problem (26.6) is a classical differential equation in a finite-dimensional space H_n, with continuous right-hand side,** and so the initial-value problem (26.6) has at least one local solution $v_n = v_n(t)$. This solution may be extended to the entire interval S, since the norms $\|v_n(t)\|$ are bounded by a constant independent of both n and the length of the interval on

* The mapping G is a Nemytskii operator in abstract spaces.

** In finite-dimensional spaces demicontinuity is equivalent to continuity; see Remark 1.4.

which $v_n(t)$ was originally defined. Indeed, setting $f(t, v+u_0) = f_1(t, v)$ for brevity and noting that $P_n v_n = v_n$, we have from the monotonicity of f that

$$\frac{d}{dt}\| v_n(t)\|^2 = 2\,\mathrm{Re}\,(v_n'(t), v_n(t)) =$$
$$= 2\,\mathrm{Re}\,(P_n f_1(t, v_n(t)), v_n(t)) = 2\,\mathrm{Re}\,(f_1(t, v_n(t)), v_n(t)) =$$
$$= 2\,\mathrm{Re}\,(f_1(t, v_n(t)) - f_1(t, 0), v_n(t)) + 2\,\mathrm{Re}\,(f_1(t, 0), v_n(t)) \leqslant$$
$$\leqslant 2\,\mathrm{Re}\,(f_1(t, 0), v_n(t)) \leqslant 2\| f_1(t, 0)\| \| v_n(t)\|,$$

i. e. $d\| v_n(t)\|/dt \leqslant \| f_1(t, 0)\|$, provided $\| v_n(t)\| \neq 0$. Consequently,

$$\| v_n(t)\| \leqslant \int_0^t \| f_1(s, 0)\| ds \leqslant \int_0^b \| f_1(s, 0)\| ds = C_1. \tag{26.7}$$

This inequality is trivially true for $t \in [0, b]$ at which $v(t) = 0$. Thus, $v_n \in C(S, H)$ and $\{v_n\}$ is a bounded sequence in $C(S, H)$. Hence, by Lemma 26.2, the sequence $\{G(v_n + u_0)\}$ is bounded in L^2. Setting $G(v+u_0) = G_1 v$, we may then state our result as a

Lemma 26.3. For each n, there exists a solution $v_n \in C(S, H)$ to problem (26.6), i.e.,

$$(L + P_n G_1) v_n = 0, \quad v_n(0) = 0 \tag{26.8}$$

and $\{G_1 v_n\}$ is a bounded sequence in L^2.

Next, since the sequence $\{G_1 v_n\}$ is bounded in L^2, we may select a subsequence such that $(-G_1 v_{n_k}) \rightharpoonup z$ in L^2 as $k \to \infty$. Hence, by (26.8) and the fact that $P_n^* = P_n \to I$ as $n \to \infty$, we get $L v_{n_k} = -P_{n_k} G_1 v_{n_k} \rightharpoonup z$, and since L^{-1} is bounded it follows that

$$v_{n_k} \rightharpoonup L^{-1} z = v \in D(L) \subset D(G).$$

We claim that this vector v satisfies the equation $(L + G_1) u = 0$. Indeed, let $w \in D(L)$. Since the operators L and G_1 are monotone (Lemmas 26.1 and 26.2), it follows that

$$\mathrm{Re}\,(L(w - v_n) + G_1 w - G_1 v_n, w - v_n)_2 \geqslant 0.$$

But, by Lemma 26.3,

$$((L + G_1) v_n, v_n)_2 = ((L + G_1) v_n, P_n v_n)_2 =$$
$$= (P_n (L + G_1) v_n, v_n)_2 = ((L + P_n G_1) v_n, v_n)_2 = 0,$$

since $(P_nLv_n, v_n)_2 = (Lv_n, P_nv_n)_2 = (Lv_n, v_n)_2$. The previous results now give

$$\operatorname{Re}((L+G_1)w, w-v_{n_k})_2 - \operatorname{Re}((L+G_1)v_{n_k}, w)_2 \geqslant 0;$$

now letting $k\to\infty$, we get

$$\operatorname{Re}((L+G_1)w, w-v)_2 \geqslant 0, \tag{26.9}$$

since $(Lv_{n_k}, w)_2 \to (z, w)_2$ and $(G_1v_{n_k}, w)_2 \to -(z, w)_2$. Consider any vector $x \in D(L)$ and set $w = w_n = v + x/n$ in (26.9). Then

$$\operatorname{Re}\left(Lv + \frac{1}{n}Lx + G_1w_n, x\right)_2 \geqslant 0.$$

Letting $n\to\infty$, we see by Lemma 26.2 that

$$\operatorname{Re}(Lv + G_1v, x)_2 \geqslant 0.$$

Hence, since x is arbitrary, it follows that $Lv+G_1v=0$ or $Lv + G(v+u_0) = 0$, where $v(0) = 0$.* Setting $u = v + u_0$ and then applying L^{-1} to the last equality, but one, we get

$$u(t) = u_0 + \int_0^t f(s, u(s))\,ds,$$

i. e., $u = v + u_0$ is a solution of the integral equation (26.2).

Up to this point we have assumed that f is demicontinuous, proving the existence of a generalized solution to problem (26.1). However, if $f(t, u)$ is continuous from $S\times H$ to H, then

$$L(u-u_0) = -Gu \in C(S, H), \tag{26.10}$$

and so u is a solution of (26.1). Its uniqueness and continuity with respect to the initial values follow from the following proposition.

Lemma 26.4. Let u_1, u_2 be solutions of (26.10) with initial values u_{01}, u_{02}, respectively. Then $\|u_1(t) - u_2(t)\| \leqslant \|u_{01} - u_{02}\|$ for any $t \in S$.

Proof. By assumption $L(u_i - u_{0i}) + G(u_i) = 0$ $(i = 1, 2)$. Set $u = u_1 - u_2$, $u_0 = u_{01} - u_{02}$. Since G is monotone, it follows that

$$\operatorname{Re}(L(u-u_0), u)_2 = -\operatorname{Re}(Gu_1 - Gu_2, u_1 - u_2)_2 \leqslant 0.$$

* This follows from the form of the operator L^{-1} and the equality $v = L^{-1}z$.

But

$$\operatorname{Re}(L(u-u_0), u)_2 =$$
$$= \operatorname{Re}(L(u-u_0), u-u_0)_2 + \operatorname{Re}(L(u-u_0), u_0)_2 =$$
$$= \frac{1}{2}\int_0^b \frac{d}{dt}(u-u_0, u-u_0)\,dt + \operatorname{Re}\int_0^b \left(\frac{d}{dt}(u-u_0), u_0\right)dt =$$
$$= \frac{1}{2}\|u(b)-u_0\|^2 + \operatorname{Re}(u(b)-u_0, u_0) = \frac{1}{2}(\|u(b)\|^2 - \|u_0\|^2).$$

Consequently, $\|u(b)\| \leqslant \|u_0\|$, i. e.,

$$\|u_1(b)-u_2(b)\| \leqslant \|u_{01}-u_{02}\|.$$

Repeating these arguments for an interval $[0, t] \subset [0, b]$, we arrive at the assertion of lemma. Thus the proof of Theorem 26.1 is complete.

Remark 26.1. Inequality (26.7) remains valid for the solution $v(t)$ of the equation $(L+G_1)v=0$. Indeed, let

$$F = \{u(t) \in L^2 : \|u(t)\| \leqslant C_1 \text{ almost everywhere}\}.$$

This set is convex. Now, if a sequence $\{u_n\}$ in F converges in L^2 to $u \in L^2$, then, selecting a subsequence $\{u_{n_k}\}$ which converges almost everywhere to u, we get $\|u(t)\| \leqslant C_1$ almost everywhere. Since F is convex and closed, it is weakly closed.* But, as we have seen, $v(t)$ is the weak limit of a sequence in F, and thus $v(t) \in F$.

26.2. Statement of the problem and auxiliary propositions

We now proceed to our examination of the initial-value problem with unbounded operators

$$\frac{du}{dt} = A(t)u + f(t, u), \quad u(0) = u_0, \quad t \in [0, b] = S \tag{26.11}$$

in a separable Hilbert space H. We shall assume that the following conditions hold.

I. The mapping f is bounded and demicontinuous from $S \times H$ to H and satisfies the inequality

$$\operatorname{Re}(f(t, u) - f(t, v), u - v) \leqslant M\|u-v\|^2 \quad (t \in S;\ u, v \in H).$$

* Dunford and Schwartz /1/, Theorem 5.3.13.

II. For each $t \in S, (-1)A(t)$ generates a strongly continuous contraction semigroup,* i. e., $A(t)$ is a closed maximal monotone operator in H with dense domain.** Hence,

$$\|A(t)+\lambda I)^{-1}\| \leqslant \lambda^{-1} \quad (\lambda > 0).$$

III. The operator-valued function $(A(t)+\lambda I)^{-1}$ is strongly continuously differentiable with respect to t for $\lambda \geqslant 0$, and

$$\|A(t)+\lambda I)^{-1}\| \leqslant (1+\lambda)^{-1}, \quad \lambda \geqslant 0. \tag{26.12}$$

This inequality is obtained from the preceding one by replacing $A(t)$ by $A(t)+I$.

IV. There exists a family $U(t, s)$ of bounded operators from H to H, defined and strongly continuous in (s, t) for $0 \leqslant s \leqslant t \leqslant b$ and having the following properties:

1. If $u(t)$ is an exact solution of the problem

$$\frac{du}{dt} + A(t)u = f(t), \quad u(0) = u_0 \tag{26.13}$$

on the interval $[0, b]$, i. e., $u(t) \in D(A(t))$ for every $t \in S$, $A(t)u(t)$ and $u'(t)$ are continuous and $u(t)$ satisfies (26.13), then this solution is given by

$$u(t) = U(t, 0)u_0 + \int_0^t U(t, s) f(s)\, ds. \tag{26.14}$$

2. Conversely, formula (26.14) defines an exact solution of problem (26.13) on the interval S, provided $f \in C^1(S, H)$ and $u_0 \in D(A(0))$.

Note that condition IV holds if, for example, $A(t)$ is a family of operators in H with domain independent of t, satisfying conditions II and III (see Kato /1/).

Note, moreover, that (as in subsection 26.1) we may set $M=0$ in condition I, i. e., we may assume that the mapping $(-f)$ is monotone.

Definition 26.1. A continuous solution of the equation

$$u(t) = U(t, 0)u_0 + \int_0^t U(t, s) f(s, u(s))\, ds \tag{26.15}$$

is called a *generalized solution* of problem (26.11).

* See Hille and Phillips /1/ or Nemytskii, Vainberg and Gusarova /1/.

** See Yosida /1/, p.250.

As in subsection 26.1, we introduce the Hilbert space $L^2 = L^2(S, H)$ and the linear operator L'. Let $A(t)$ be an operator in L^2 such that

$$D(A(t)) = \{u \in L^2 : A(t)u(t) \in C(S, H)\}.$$

Lemma 26.5. The operator $T' = L' + A$ is densely defined in L^2, is monotone, and has a closure.

Proof. Let $v \in C^1(S, H)$, $v(0) = 0$ and $u(t) = A^{-1}(t)v(t)$. By condition III, $u \in D(L') \cap D(A) = D(T')$. Since $C^1(S,H)$ is dense in L^2, it follows that $A^{-1}C^1(S, H)$ is dense in L^2. In fact, if there exists a nonzero vector $y \in L^2$ such that $(A^{-1}C^1(S,H), y)_2 = 0$, then $(A^{-1})^* y = (A^*)^{-1} y = 0$, i. e., the operator $(A^*)^{-1}$ is not invertible. This contradicts condition III. Next, since the operators L' and A are monotone, so is T'. Finally, in order to prove that T' has a closure, it suffices to show that $(T')^*$ is defined on a dense set. Let $x \in C^1(S, H)$ and let $x(b) = 0$. Define functions $y(t)$ and $z(t)$ by

$$y(t) = (A^{-1}(t))^* x(t);$$
$$z(t) = -\left(\frac{d}{dt} A^{-1}(t)\right)^* x(t) - (A^{-1}(t))^* x'(t) + x(t).$$

A direct check shows that $(u, z)_2 = (T'u, y)_2$ for all $u \in D(T')$. Hence $y \in D(T'^*)$ and $(T')^* y = z$. Just as we proved that $D(T')$ is dense, we can now prove that the set of all functions $y(t)$ is dense in L^2.

Define an operator V in L^2 by

$$(Vx)(t) = \int_0^t U(t, s)x(s)\,ds.$$

Lemma 26.6. V is a bounded linear operator from L^2 to itself and from L^2 to $C(S, H)$.

Proof. Condition IV implies that $\|U(t, s)\| \leqslant c$ for $0 \leqslant s \leqslant t \leqslant b$, so that

$$\|Vx\| \leqslant c\sqrt{b}\,\|x\|_2 \qquad (26.16)$$

and $Vx \in C(S, H)$ for $x \in C^1(S, H)$. Now, the set $C^1(S, H)$ is dense in L^2; thus, for any $y \in L^2$, we see by considering (26.16) for any sequence $x_n \in C^1(S,H)$ converging to y that $Vy \in C(S,H)$. Observe, moreover, that (26.16) implies

$$\|V\|_{L^2 \to C(S,H)} \leqslant c\sqrt{b}.$$

Q. E. D.

In light of this lemma, we may rewrite condition IV thus:

$$\left.\begin{aligned} VT'u = u \quad &\text{for} \quad u \in D(T'), \\ T'Vx = x \quad &\text{for} \quad x \in C^1(S, H). \end{aligned}\right\} \tag{26.17}$$

In fact, since $D(T') \subset D(L')$, it follows that $u(0) = 0$. We thus see by setting $u_0 = 0$ in (26.14) that

$$u(t) = \int_0^t U(t, s) f(s)\, ds,$$

and therefore

$$VT'u = \int_0^t U(t, s) \left[\frac{du(s)}{ds} + A(s) u\right] ds = u(t).$$

The second relation in (26.17) follows from the definition.

Lemma 26.7. The closure T of T' is maximal, monotone and invertible, $D(T) \subset C(S, H)$ and $T^{-1} = V$.

Proof. Let $u \in D(T)$. Then there exists a sequence $\{u_n\} \subset D(T')$ such that $u_n \to u$, $T'u_n \to Tu$ in L^2. From Lemma 26.6 and (26.17), we have

$$VTu = \lim_{n\to\infty} VT'u_n = \lim_{n\to\infty} u_n = u.$$

Consequently, if $Tu = 0$, then $u = 0$, i. e., the inverse T^{-1} exists and $T^{-1} \subset V$. (Hence, in particular, $D(T) \subset R(V) \subset C(S, H)$.)

Since V is bounded, T^{-1} is also bounded. Since the set $R(T)$ is closed, we have $R(T) = L^2$, as (26.17) implies that $R(T')$ is dense in L^2 and $T^{-1} = V$. The operator T is monotone, being the closure of the monotone operator T', and T is maximal because $R(T) = L^2$. Q. E. D.

Lemma 26.8. The operator $\lambda(T + \lambda I)^{-1}$ from $L^\infty(S, H)$ to $C(S, H)$ is uniformly bounded for $\lambda \geqslant 0$, the operator $(T + \lambda I)^{-1} = U_\lambda$ is bounded from L^2 to L^2 for all $\lambda \geqslant 0$, and

$$(U_\lambda x)(t) = \int_0^t e^{-\lambda(t-s)} U(t, s) x(s)\, ds. \tag{26.18}$$

Proof. Let $(B_\lambda x)(t) = e^{\lambda t} x(t)$. Then

$$T'B_\lambda x = B_\lambda (T' + \lambda I) x \quad \text{for any} \quad x \in D(T').$$

Hence, by continuity,

$$TB_\lambda x = B_\lambda (T + \lambda I) x$$

for any $x \in D(T)$. Going over to inverses, we get

$$(T + \lambda I)^{-1} = B_\lambda^{-1} V B_\lambda,$$

since $T^{-1} = V$, so that (26.18) holds. This equality shows that

$$\lambda \| U_\lambda x(t) \| \leqslant c \|x\|_\infty$$

for all t, where the constant c is independent of t and x, and $\|x\|_\infty$ is the norm of x in L^∞. Q. E. D.

Let u be a generalized solution of problem (26.11), i. e., a solution of the integral equation (26.15), $g(t) = U(t, 0)u_0$ and G the Nemytskii operator defined by (26.5). Set $G_2 v = G(v + g)$. Using Lemma 26.7 and the expression for the operator V, we see that for the solution u in question formula (26.15) takes the form

$$u - g + T^{-1} G u = 0.$$

Setting $v = u - g$, we have $v + T^{-1} G_2, \ v = 0$ or

$$(T + G_2) v = 0. \tag{26.19}$$

Since $g \in C(S, H)$, it follows that G_2 has the same properties as the mapping G in Lemma 26.2.

Definition 26.2. Let A and B be operators from a Banach space E to E. The operator A is said to be B-demicontinuous, if, for any sequence $\{u_n\}$ such that $Bu_n \to Bu$, we have $Au_n \rightharpoonup Au$.

Lemma 26.9. The Nemytskii operator G is T-demicontinuous.

Proof. If $Tu_n \to Tu$, Lemma 26.7 (see (26.17)) entails that the following holds in $C(S, H)$:

$$u_n = VTu_n \to VT.$$

By Lemma 26.2, then, $Gu_n \rightharpoonup Gu$.

26.3. Quasilinear equations in Hilbert space

Theorem 26.2. Suppose that conditions I - IV of subsection 26.2 hold. Then problem (26.11) has a unique generalized solution u and $u_0 \to u$ is a continuous mapping of H into $C(S, H)$.*

* Browder /8/ and Kato /3/.

Proof. Let $A_n(t) = A(t)(I + n^{-1}A(t))^{-1}$, $n = 1, 2, 3, \ldots$. Using $A_n(t)$, we define an operator A_n in L^2, in the same way as the operator A was defined in subsection 26.2 with the help of $A(t)$. It can be shown (see Kato /2/) that condition III implies the estimate $\|A_n(t)\| \leqslant 2n$. The operator A_n is bounded and monotone, since the operator $A_n(t)$ has these properties; hence the mapping

$$f^{(n)}(t, v) = A_n(t)v + f_2(t, v) \quad (f_2 = -G_2)$$

is monotone and demicontinuous. Thus, as in subsection 26.1, there exists a solution v_n of the equation

$$Lv_n + (A_n + G_2)v_n = 0. \tag{26.20}$$

Using (26.7), Remark 26.1 and the fact that $A_n 0 = 0$, we now see that

$$\|v_n(t)\| \leqslant \int_0^b \|f^{(n)}(t, 0)\| dt = \int_0^b \|f_2(t, 0)\| dt = C_2,$$

so we can select a subsequence such that $v_{n_k} \rightharpoonup v$ in L^2 as $k \to \infty$, and moreover $\|v(t)\| \leqslant C_2$.

In order to prove that v is a solution of equation (26.19), i. e., that $u = v + g$ is a generalized solution of problem (26.11), we set $T_n = L + A_n$ and write (26.11) in the form

$$(T_n + G_2)v_n = 0. \tag{26.21}$$

Lemma 26.10. For any $T_n w \to Tw$, $w \in D(T')$ in L^2 as $n \to \infty$.
The proof follows (via Lebesgue's theorem) from the relation

$$(T_n w - Tw)(t) = (A_n(t) - A(t))w(t) = \\ = ((I + n^{-1}A(t))^{-1} - I)A(t)w(t) \to 0, \quad n \to \infty,$$

since $A(t)w(t)$ is continuous on S and therefore bounded in t.

Lemma 26.11. If v is the weak limit of $\{v_{n_k}\}$ in L^2, then

$$\mathrm{Re}((T + G_2)w, w - v)_2 \geqslant 0 \ (\forall w \in D(T)) \tag{26.22}$$

and $v \in D(T)$.

The proof of inequality (26.22) is similar to that of Lemma 26.3.

To prove that $v \in D(T)$, we set $w = v_\lambda = \lambda(T + \lambda I)^{-1}v$ in (26.22). We then have from (26.22) that

$$\lambda \|v_\lambda - v\|_2^2 \leqslant \mathrm{Re}(G_2 v_\lambda, v_\lambda - v)_2 \leqslant \|G_2 v_\lambda\|_2 \|v_\lambda - v\|_2,$$

or $\lambda\|v_\lambda - v\|_2 \leqslant \|G_2 v_\lambda\|_2$. Since $\|v(t)\| \leqslant C_2$ (see the proof of Lemma 26.9), it follows via Lemma 26.8 that v_λ is uniformly bounded with respect to $\lambda \geqslant 0$. Consequently, by Lemma 26.2, $G_2 v_\lambda$ is uniformly bounded in L^2 with respect to λ, and thus $\lambda\|v_\lambda - v\|_2 \leqslant c = \text{const}$. The preceding argument implies $\lambda(v_\lambda - v) = -Tv_\lambda$, so that $\|Tv_\lambda\| \leqslant c$. But $v_\lambda \to v$ in L^2 as $\lambda \to \infty$ and T is a closed operator in L^2; thus we can select a subsequence v_{λ_k} such that $Tv_{\lambda_k} \to Tv$. Indeed, $Tv_{\lambda_k} \to u \in L^2$. But $v_{\lambda_k} - v = -\lambda_k^{-1} Tv_{\lambda_k} \to 0$, i. e., $v_{\lambda_k} \to v$. Now, $T = (T^*)^*$, so that for any $g \in D(T^*)$ we have $(Tv_{\lambda_k}, g) = (v_{\lambda_k}, T^*g)$. Thus, letting $k \to \infty$, we get $(u, g) = (v, T^*g)$, i. e., $v \in D(T)$. Q. E. D.

Proceeding as in the proof of the equality $(L + G_1)v = 0$ in subsection 26.1, we see from inequality (26.22) and the fact that $v \in D(T)$ that v is a solution of (26.19).

We now prove the uniqueness of the solution and its continuity with respect to the initial values. Let u_1, u_2 be generalized solutions of equation (26.11), with initial values u_{01}, u_{02}, respectively. Setting $g_i = U(t, 0)u_{0i}$ $(i = 1, 2)$ $u = u_1 - u_2$, $u_0 = u_{01} - u_{02}$, $g = g_1 - g_2$ and using (26.19) and the equality $u_i = v_i + g_i$, we have

$$T(u_i - g_i) + Gu_i = 0 \qquad (i = 1, 2).$$

Since the operator G is monotone, we conclude that

$$\operatorname{Re}(T(u - u_0), u)_2 = \operatorname{Re}(Gu_1 - Gu_2, u_1 - u_2)_2 \leqslant 0.$$

We now show that for all u, $u_0 \in D(T)$

$$2\operatorname{Re}(T(u - u_0), u)_2 \geqslant \|u(b)\|^2 - \|u_0\|^2, \tag{26.23}$$

which, by previous results, will imply that $\|u(b)\| \leqslant \|u_0\|$ and, similarly, $\|u(t)\| \leqslant \|u_0\|$ for all $t \in [0, b]$. To prove (26.23), we take any vector $v \in D(T')$ and a vector $u_0 \in D(A(0))$. Then $U(t, 0)u_0 = g(t)$ and, by condition IV, $g' + Ag = 0$. Consequently,

$$Tv = T'v = v' + Av = (v + g)' + A(v + g) = u' + Au,$$
$$2\operatorname{Re}(Tv, u)_2 = 2\operatorname{Re}(u' + Au, u)_2 \geqslant 2\operatorname{Re}(u', u)_2 =$$
$$= \int_0^b \frac{d}{dt}\|u(t)\|^2\, dt = \|u(b)\|^2 - \|u(0)\|^2.$$

This proves (26.23) for $v \in D(T')$. Let $v \in D(T)$. Let $\{v_n\} \subset D(T')$ and $\{u_0^{(n)}\} \subset D(A(0))$ be sequences such that $v_n \to v$, $Tv_n \to Tv$ in L^2 and $u_0^{(n)} \to u_0$ as $n \to \infty$. Inequality (26.23) now follows by a suitable limit passage. Q. E. D.

26.4. Periodic solutions

Let H be a separable Hilbert space. Consider the equation

$$\frac{du}{dt} + A(t)u = f(t, u), \quad u(0) = u_0, \tag{26.24}$$

where $A(t)$ is a family of closed linear operators in H and f maps $E^1 \times H$ into H $(E^1 = (-\infty, +\infty))$.

Theorem 26.3. Let $A(t)$ and $f(t, u)$ be periodic in t with common period T, such that the conditions of Theorem 26.2 hold with $M = 0$. Suppose, moreover, that there exists $R > 0$ such that*

$$\mathrm{Re}\,(f(t, u), u) < 0$$

for $\|u\| = R$ and all $t \in [0, T]$. Then there exists $u_0 \in H$ $\|u_0\| \leqslant R$, such that the generalized solution of equation (26.24) with initial value $u(0) = u_0$ is periodic in t with period T.

We prove Theorem 26.3 using the following fixed-point theorem:

Theorem 26.4. Let S be a nonexpansive contractive mapping of the Hilbert space H into H (i.e., $\|Su - Sv\| \leqslant \|u - v\|$ for all $u, v \in H$). Suppose, moreover, that there exists $R > 0$ such that $\|Su\| \leqslant \|u\|$ for any u, $\|u\| = R$. Then S has a fixed point in the ball $\|u\| \leqslant R$.

Proof of Theorem 26.3 (assuming Theorem 26.4). By Theorem 26.2, for any $u_0 \in H$ there exists a generalized solution $u(t)$ of equation (26.24) with initial value $u(0) = u_0$, defined on $[0, T]$. Let S be the mapping of H into H taking every $u_0 \in H$ $(\|u_0\| \leqslant R)$ to the value of the corresponding solution $u(t)$ at the point T. To a fixed point of S corresponds a T-periodic solution of equation (26.24). Thus the problem of periodic solutions to equation (26.24) reduces to the fixed-point problem for the operator S.

Now, for any generalized solution,

$$\frac{1}{2}\frac{d}{dt}\{\|u(t)\|^2\} = -\mathrm{Re}\,(A(t)u(t), u(t)) + \\ + \mathrm{Re}\,(f(t, u(t)), u(t)) \leqslant \mathrm{Re}\,(f(t, u(t)), u(t)),$$

and so it follows from the hypotheses of Theorem 26.3 that $\frac{d}{dt}\{\|u(t)\|^2\} < 0$ for any $t \in [0, T]$ such that $\|u(t)\| = R$. Consequently, $\|u(T)\| \leqslant R$, i. e., $\|Su\| \leqslant \|u\|$ for $\|u\| = R$.

If $u_{01}, u_{02} \in H$ and u_1, u_2 are the corresponding solutions of equation (26.24), then

$$\frac{1}{2}\frac{d}{dt}(\|u_1(t) - u_2(t)\|^2) = -\mathrm{Re}\,(A(t)(u_1 - u_2), u_1 - u_2) + \\ + \mathrm{Re}\,(f(t, u_1) - f(t, u_2), u_1 - u_2) \leqslant 0,$$

* Browder /20/.

i. e., $\|u_1(T)-u_2(t)\|\leqslant\|u_1(0)-u_2(0)\|$ and $\|Su_{01}-Su_{02}\|\leqslant\|u_{01}-u_{02}\|$, i. e., the assumptions of Theorem 26.4 are satisfied. Since for $\|u\|=R$ we have $\|Su\|\leqslant\|u\|$, it follows that the fixed point of S lies in the ball $\|u\|\leqslant R$. Q. E. D.

Instead of Theorem 26.4, we shall prove a more general assertion.

Theorem 26.5. Let S be a demicontinuous mapping of H into H; suppose that for all $u, v\in H$

$$\mathrm{Re}(Su-Sv,\ u-v)\leqslant\|u-v\|^2,$$

and, moreover, there exists $R>0$ such that $\mathrm{Re}(Su,\ u)\leqslant\|u\|^2$ for all u with $\|u\|=R$. Then S has a fixed point in H.

Proof. Let $Q=I-S$. Q is a demicontinuous mapping of H into H, and for all $u, v\in H$

$$\mathrm{Re}(Qu-Qv,\ u-v)=\|u-v\|^2-\mathrm{Re}(Su-Sv,\ u-v)\geqslant 0.$$

For all u with $\|u\|=R$ we have

$$\mathrm{Re}(Qu,\ u)=\|u\|^2-\mathrm{Re}(Su,\ u)\geqslant 0.$$

By Theorem 22.1, there exists $u\in H\,(\|u\|\leqslant R)$ such that $Qu=0$, i. e., $Su=u$. Q. E. D.

26.5. Nonlinear equations in Banach spaces

Let E be a real reflexive separable Banach space. We assume that E is embeddable and that the sequence $\{\varphi_k\}$ is a basis in E. We consider the initial-value problem

$$\frac{du}{dt}+f(t,\ u)=0,\quad u(0)=0\ \ (t\in[0,\ b]=S), \tag{26.25}$$

where $f(t,\ u)$ is a mapping from $S\times E$ into E, and the corresponding integral equation

$$u(t)+\int_0^t f(s,\ u(s))\,ds=0,\qquad t\in S. \tag{26.26}$$

A continuous solution of equation (26.26) will be called a *generalized solution* of problem (26.6).

We shall assume henceforth that the function $f(t,\ u(t)$ is measurable for any measurable $u(t)$ and that the following condition holds:

Condition 26.1. There exists a constant C_1 such that, for any interval $[0, a]$, $0 < a \leqslant b$, and any continuous function $v(t) \in E$ on $[0, a]$, if the inequality $\langle f(t, v(t)), v(t) \rangle \leqslant 0$ holds for almost all t, then $\|v(t)\| \leqslant C_1$, where $\langle y, x \rangle$ is the value of $y \in E^x$ at the vector x. This condition holds, in particular, if $f(t, u)$ is coercive in u uniformly for almost all fixed $t \in S$.

Theorem 26.6. Let $f(t, u)$ be a bounded demicontinuous mapping from $S \times E$ into E, which is monotone from E to E^* for almost every fixed $t \in S$. Suppose, moreover, that Condition 26.1 holds. Then problem (26.25) has a generalized solution.

Proof. Some of the steps in this proof are similar to the corresponding steps in the proof of Theorem 26.1, and we shall therefore omit a few details. Consider the space $L^p = L^p(S, E)$ and its dual $L^q = L^q(S, E^*)$, where $p > q > 1$, $p^{-1} + q^{-1} = 1$. We denote the norms in these spaces by $\|\cdot\|_p$ and $\|\cdot\|_q$, respectively. The space of continuous functions $C(S, E)$ is clearly dense in L^p and L^q. The norm in $C(S, E)$ is denoted by $\|\cdot\|_1$. We let $\|\cdot\|$ and $\|\cdot\|_*$ denote the norms in E and E^*, respectively, and $\langle u, v \rangle$ denotes the value of a linear functional $v \in E^*$ at a vector $u \in E$. Note (see Definition 15.2) that if $v \in E$, then $\langle u, v \rangle = (u, v)$, where $(\cdot, \cdot)$ is the inner product in the corresponding Hilbert space. Set

$$\{u, v\} = \int_0^b \langle u(t), v(t) \rangle \, dt \qquad (\forall u \in L^p, \ v \in L^q).$$

We consider the differentiation operators $(L_0 u)(t) = u'(t)$, $(L_b v)(t) = -v'(t)$ from L^p to $L^q(S, E)$ with domains

$$D(L_0) = \{u(t) \in C^1(S, E) : u(0) = 0\},$$
$$D(L_b) = \{v(t) \in C^1(S, E) : v(b) = 0\},$$

and write

$$\{L_0 u, v\} = \int_0^b \langle u'(t), v(t) \rangle \, dt = - \int_0^b \langle u(t), v'(t) \rangle \, dt = \{u, L_b v\},$$

i. e., $D(L_b) \subset D(L_0^*)$. But $D(L_b)$ is dense in L^q and so $D(L_0^*)$ is also dense in L^q. Since $D(L_0^*)$ is dense in L^q, the operator $(L_0 \, !)(t)$ has a closure, which we denote by L.

We claim that $D(L)\subset C(S, E)$. Indeed, let $u\in D(L)$. Then there exists a sequence $\{u_n\}\subset D(L_0)$ such that $u_n\to u$ in L^p and $u_n'\to Lu$ in L^q as $n\to\infty$. Hence,

$$\| u_n - u_m \|_1 = \max_t \left\| \int_0^t (u_n'(s) - u_m'(s))\, ds \right\| \leqslant$$
$$\leqslant \int_0^b \| u_n'(s) - u_m'(s) \|\, ds \leqslant b^{1/p} \left(\int_0^b \| u_n'(s) - u_m'(s) \|^q\, ds \right)^{1/q} =$$
$$= b^{1/p} \| u_n' - u_m' \|_q \to 0$$

as $m, n\to\infty$. Thus $\{u_n(t)\}$ is a fundamental sequence in $C(S, E)$ and so $u_n\to v\in C(S, E)$. Now, the convergence of $\{u_n(t)\}$ in $C(S, E)$ implies that $u_n\to v$ in L^q, i. e., $u=v$ almost everywhere. Thus, by modifying $u(t)$ on a set of measure zero, we may assume that $u(t)\in C(S, E)$, i. e., $D(L)\subset C(S, E)$. Now, for the same sequence $\{u_n\}\subset D(L_0)$, we have

$$\{u, Lu\} = \lim_{n\to\infty} \{u_n, u_n'\} = \lim_{n\to\infty} \int_0^b \langle u_n(t), u_n'(t)\rangle\, dt =$$
$$= \frac{1}{2} \lim_{n\to\infty} \int_0^b \frac{d}{dt} \langle u_n(t), u_n(t)\rangle\, dt =$$
$$= \frac{1}{2} \lim_{n\to\infty} \int_0^b \frac{d}{dt} (u_n(t), u_n(t))\, dt =$$
$$= \frac{1}{2} \lim_{n\to\infty} [(u_n(b), u_n(b)) - (u_n(0), u_0(0))] = \frac{1}{2} (u(b), u(b)) \geqslant 0,$$

i. e., L is a monotone operator. Finally, as in subsection 26.1, we find that L has an inverse given by (26.4), i. e.,

$$(L^{-1}v)(t) = \int_0^t v(s)\, ds \qquad (\forall v\in R(L)).$$

Since

$$\int_0^t v(s)\, ds \leqslant t^{1/p} \left(\int_0^t | v(s) |^q\, ds \right)^{1/p} \leqslant b^{1/p} \| v \|_q,$$

it follows that $\| L^{-1} v \|_p \leqslant b^{2/p} \| v \|_q$, i. e., L^{-1} is bounded.

We now define a mapping from $C(S, E)$ to L^q by

$$(Gu)(t) = f(t, u(t)).$$

By hypothesis, it is bounded from $C(S, E)$ to E, so that for any function $u(t) \in C(S, E)$ we have $G(u) \in L^q$. Hence, it at once follows that the operator G is monotone and bounded from $C(S, E)$ to L^q. Consider an arbitrary sequence $\{u_n\} \subset C(S, E)$, converging to $u(t)$ in the metric of $C(S, E)$. By hypothesis, $f(t, u_n(t)) \rightharpoonup f(t, u(t))$ in the metric of E, for almost every fixed $t \in S$. Hence, for any $v \in L^p$ and almost every fixed $t \in S$,

$$I_n = \langle f(t, u_n(t)) - f(t, u(t)), v(t)\rangle \to 0 \quad \text{as} \quad n \to \infty.$$

Since $u_n(t) \to u(t)$ in $C(S, E)$ and f is a bounded mapping from $S \times E$ to E^*, it follows that

$$|\langle f(t, u_n(t)) - f(t, u(t)), v(t)\rangle| \leqslant M \| v(t) \|, \quad M = \text{const},$$

i. e., $I_n(t)$ is dominated by a summable function and so, by Lebesgue's theorem, $\{Gu_n - Gu, v\} \to 0$ or $Gu_n \rightharpoonup Gu$ in L^q. Thus, G is a bounded demicontinuous monotone operator from $C(S, E)$ to L^q.

Let E_n be the subspace of E spanned by the basis vectors $\varphi_1, \varphi_2, \ldots, \varphi_n$, and P_n the projection from E onto E_n. Consider the sequences $\{E_k\}$, $\{P_k\}$ and $\{P_k^*\}$. By standard theorems (see Dunford and Schwartz /1/, VI. 3. 3. and VI. 9. 19), the adjoint P_k^* projects E^* onto a k-dimensional subspace E_k', so that the subspaces E_k and E_k' $(k = 1, 2, 3, \ldots)$ may be identified. As in subsection 26.1, we consider the initial-value problems

$$v_n' + P_n^* f(t, v_n) = 0, \quad v_n(0) = 0 \qquad (n = 1, 2, 3, \ldots),$$

where $v_n \in E_n$. Since f is demicontinuous, it follows from Remark 1.4 that $P_n^* f(t, v_n)$ is a continuous function in a finite-dimensional space, so for any fixed n the above initial-value problem has at least one local solution $v_n = v_n(t)$, which can be extended to the entire interval S, as the norms $\|v_n(t)\|$ are bounded by a constant depending neither on n nor on the length of the interval on which $v_n(t)$ was originally defined.

Indeed,

$$\begin{aligned}\langle v_n'(t) + f(t, v_n(t)), v_n(t)\rangle &= \langle v_n'(t) + f(t, v_n(t)), P_n v_n(t)\rangle = \\ &= \langle v_n'(t) + P_n^* f(t, v_n(t)), v_n(t)\rangle = 0\end{aligned}$$

Hence, since L is monotone,

$$\langle f(t, v_n(t)), v_n(t)\rangle = -\langle Lv_n, v_n\rangle \leqslant 0,$$

so that, by Condition 26.1, we have*

$$\| v_n(t) \| \leqslant C_1 \qquad (C_1 = \text{const}), \tag{26.27}$$

implying the boundedness of the sequence $\{Gv_n\}$ in L^q. This being so, we can select a subsequence such that $Gv_n \rightharpoonup z$ in L^q as $k \to \infty$. Since the sequence $\{P_n^*\}$ converges strongly to the identity operator, we have $P_{n_k}^* G v_{n_k} \rightharpoonup z \in L^q(SE)$. Thus, $Lv_{n_k} = -P_{n_k}^* G v_{n_k} \rightharpoonup -z$. Hence

$$v_{n_k} \rightharpoonup -L^{-1} z = v \in D(L) \subset C(S, E),$$

and it follows from the representation of L^{-1} that $v(0) = 0$.

We now prove that v satisfies the equation $(L + G)u = 0$. In fact, let $w \in D(L)$. Since L and G are monotone, it follows that

$$\{L(w - v_{n_k}) + Gw - Gv_{n_k},\ w - v_{n_k}\} \geqslant 0.$$

But

$$\begin{aligned}\{(L + G)\, v_{n_k},\ v_{n_k}\} = \{(L + G)\, v_{n_k},\ P_{n_k} v_{n_k}\} = \\ = \left\{\left(L + P_{n_k}^* G\right) v_{n_k},\ v_{n_k}\right\} = 0,\end{aligned}$$

whence

$$\{(L + G)\, w,\ w - v_{n_k}\} - \{(L + G)\, v_{n_k},\ w\} \geqslant 0.$$

Letting $k \to \infty$ and taking into account that

$$\{Lv_{n_k},\ w\} \to \{-z,\ w\}, \quad \{Gv_{n_k},\ w\} \to \{z,\ w\},$$

we have

$$\{(L + G)\, w,\ w - v\} \geqslant 0.$$

Setting $w = v + x/n$, where x is any vector in $D(L)$, and dividing by n, we get

$$\left\{ Lv + \frac{1}{n} Lx + G\left(v + \frac{1}{n} x\right),\ x \right\} \geqslant 0.$$

Letting $n \to \infty$ and using the demicontinuity of G, we obtain

$$\{Lv + Gv,\ x\} \geqslant 0.$$

* This is the only point at which Condition 26.1 is used. If the condition is dropped, one can prove inequality (26.27) as in subsection 26.1.

Hence, since x is an arbitrary vector in a dense set, we have $Lv+Gv=0$. Applying the operator L^{-1}, we finally obtain the assertion of our theorem.

A concluding remark: Theorems of this kind and more general theorems, whose proofs use the monotonicity of $f(t, u)$, may be found in Browder /14/, Lions /14/ and Dubinskii /2/.

BIBLIOGRAPHY

Agaev, G.N.

1. On the theory of nonlinear operator equations in Hilbert space. — Izv. Akad. Nauk Azerbaidzhan. SSR Ser. Fiz.-Mat. Tekhn. Nauk, No. 5 (1966). (Russian)
2. On the solvability of nonlinear operator equations in a Banach space. — Dokl. Akad. Nauk SSSR **174**, No. 6 (1968). (Russian)
3. On the solvability of nonlinear operator equations in locally convex topological vector spaces. — Izv. Akad. Nauk Azerbaidzhan. SSR Ser. Fiz.-Mat. Tekhn. Nauk, No. 4 (1968). (Russian)

Aizengendler, P.G.

1. On a problem in the theory of potential operators. — Uchen. Zap. Moskov. Obl. Ped. Inst. **110**, No. 7 (1962). (Russian)

Akhiezer, N.I.

1. Integral operators with Carleman kernels. — Uspekhi Mat. Nauk **2**, No. 5 (1947). (Russian)

Akhiezer, P.G. and I.M. Glazman

1. Theory of Linear Operators in Hilbert Space. — Moscow, Nauka, 1966.(Russian)

Aleksandrov, P.S.

1. Combinatorial Topology. — Moscow, Gostekhizdat, 1947. (Russian)

Altman, M.

1. Generalized gradient methods of minimizing a functional. — Bull. Acad. Polon. Sci. **14**, No. 6 (1966).

Amann, H.

1. Über die Existenz und Eindeutigkeit einer Lösung der Hammersteinschen Gleichung in Banachräumen. — J. Math. Mech. **19**, No. 2 (1969).
2. Zum Galerkin-Verfahren für die Hammersteinsche Gleichung. — Arch. Rational Mech. Anal. **35**, No. 2 (1969).
3. Hammersteinsche Gleichungen mit kompakten Kernen. — Math. Ann. **186**, No. 4 (1970).

Asplund, E.

1. Positivity of duality mappings. — Bull. Amer. Math. Soc. **73**, No. 2 (1967).

Asplund, E. and R.T. Rockafellar

1. Gradients of convex functions. — Trans. Amer. Math. Soc. **139** (1969).

Averbukh, V.I. and O.G. Smolanov

1. Theory of differentiation in linear topological spaces. — Uspekhi Mat. Nauk **22**, No. 6 (1967). (Russian)
2. Different definitions of the derivative in linear topological spaces. — Uspekhi Mat. Nauk **23**, No. 4 (1968). (Russian)

Banach, S.

1. A Course of Functional Analysis. — Kiev, 1948. [Ukrainian translation]

Bennett, A.A.

1. Newton's method in general analysis. — Proc. Nat. Acad. Sci. USA **2**, No. 10 (1916).

Bonic, R.A.

1. Four brief examples concerning polynomials on certain Banach spaces. — J. Differential Geometry **2**, No. 4 (1968).

Bourbaki, N.
1. Eléments de mathématique, Fasc. XV: Espaces vectoriels topologiques. — Paris, Hermann, 1949.
2. Eléments de mathématique, Fasc. IX: Fonctions d'une variable réelle. — Paris, Hermann, 1953.
3. Eléments de mathématique, Fasc. II: Topologie générale. — Paris, Hermann, 1940.

Brezis, H.
1. Une généralisation des opérateurs monotones. — C.R. Acad. Sci. Paris **264**, No. 15 (1967).
2. Les opérateurs monotones. — Seminar Choquet, Fac. Sci. Paris 5, No. 2 (1968).

Brezis, H. and M. Sibony
1. Méthodes d'approximation et d'itération pour les opérateurs monotones. — Arch. Rational Mech. Anal. 28, No. 1 (1968).

Brouwer, L. E. J.
1. Über Abbildung von Mannigfaltigkeiten. — Math. Ann. **71**, No. 2. (1912).

Browder, F. E.
1. Variational boundary value problems for quasilinear elliptic equations of arbitrary order. — Proc. Nat. Acad. Sci. USA 50, No. 1 (1963).
2. Variational boundary value problems for quasilinear elliptic equations, II. — Proc. Nat. Acad. Sci. USA **50**, No. 4 (1963).
3. Variational boundary value problems for quasilinear elliptic equations. III. — Proc. Nat. Acad. Sci. USA **50**, No. 5 (1963).
4. Nonlinear parabolic boundary value problems of arbitrary order. — Bull. Amer. Math. Soc. 69, No. 6 (1963).
5. Nonlinear elliptic boundary value problems. — Bull. Amer. Math. Soc. **69**, No. 6 (1963).
6. Nonlinear elliptic problems, II. — Bull. Amer. Math. Soc. **70**, No. 2 (1964).
7. Strongly nonlinear parabolic boundary value problems. — Amer. J. Math. **86**, No. 2 (1964).
8. Nonlinear equations of evolutions. — Ann. of Math. 80, No. 3 (1964).
9. Continuity properties of monotone nonlinear operators in Banach spaces. — Bull. Amer. Math. Soc. 70, No. 4 (1964).
10. Remarks on nonlinear functional equations. — Proc. Nat. Acad. Sci. USA **51**, No. 6 (1964).
11. Nonlinear elliptic boundary value problems, II. — Trans. Amer. Math. Soc. **117**, No. 2 (1965).
12. Remarks on nonlinear functional equations, II. — Illinois J. Math. **9**, No. 4 (1965).
13. Remarks on nonlinear functional equations, III. — Illinois J. Math. **9**, No. 4 (1965).
14. Nonlinear initial value problems. — Ann of Math. **82**, No. 1 (1965).
15. On a theorem of Beurling and Livingston. — Canad. J. Math. **17** (1965).
16. Variational methods for nonlinear elliptic eigenvalue problems. — Bull. Amer. Math. Soc. **71**, No. 1 (1965).
17. Multivalued monotone nonlinear mappings and duality mappings in Banach spaces. — Trans. Amer. Math. Soc. **118**, No. 6 (1965).
18. Existence and uniqueness theorems for solutions of nonlinear boundary value problems. — Proc. Symp. Appl. Math. 17 (1965).
19. Nonlinear monotone operators and convex sets in Banach spaces. — Bull. Amer. Math. Soc. **71**, No. 5 (1965).
20. Existence of periodic solutions for nonlinear equations of evolution. — Proc. Nat. Acad. Sci. USA **53**, No. 5 (1965).
21. Fixed point theorems for noncompact mappings in Hilbert space. — Proc. Nat. Acad. Sci. USA 53, No. 6 (1965).

22. Mapping theorems for noncompact nonlinear operators in Banach space. — Proc. Nat. Acad. Sci. USA 54, No. 2 (1965).
23. Nonexpansive nonlinear operators in a Banach space. — Proc. Nat. Acad. Sci. USA 54, No. 4 (1965).
24. Nonlinear elliptic functional equations in nonreflexive Banach spaces. — Bull. Amer. Math. Soc. 72, No. 1 (1966).
25. Problèmes nonlinéaires. — University of Montreal Press, 1966.
26. On the unification of the calculus of variations and the theory of monotone nonlinear operators in Banach spaces. — Proc. Nat. Acad. Sci. USA 56, No. 2 (1966).
27. Existence and approximation of solutions of nonlinear variational inequalities. — Proc. Nat. Acad. Sci. USA 56, No. 4 (1966).
28. Nonlinear operators in Banach spaces. — Math. Ann. 162, No. 2 (1966).
29. Fixed point theorems for nonlinear semicontractive mappings in Banach spaces. — Arch. Rational Mech. Anal. 21, No. 4 (1966).
30. Further remarks on nonlinear functional equations. — Illinois J. Math. 10, No. 2 (1966).
31. Convergence of approximants to fixed points of nonexpansive nonlinear mappings in Banach spaces. — Arch. Rational Mech. Anal. 24, No. 1 (1967).
32. Nonlinear accretive operators in Banach spaces. — Bull. Amer. Math. Soc. 73, No. 3 (1967).
33. Existence and perturbation theorems for nonlinear maximal monotone operators in Banach spaces. — Bull. Amer. Math. Soc. 73, No. 3 (1967).
34. Approximation-solvability of nonlinear functional equations in normed linear spaces. — Arch. Rational Mech. Anal. 26, No. 1. (1967).
35. Nonlinear equations of evolution and nonlinear accretive operators in Banach spaces. — Bull. Amer. Math. Soc. 73, No. 6. (1967).
36. Nonlinear mappings of nonexpansive and accretive type in Banach spaces. — Bull. Amer. Math. Soc. 73, No. 6 (1967).
37. Nonlinear functional analysis and nonlinear partial differential equations. — In: Proc. of Conference on Differential Equations and their Applications Equadiff II (1966), Bratislava, 1967.
38. Semicontractive and semiaccretive nonlinear mappings in Banach spaces. — Bull. Amer. Math. Soc. 74, No. 4 (1968).
39. Nonlinear monotone and accretive operators in Banach spaces. — Proc. Nat. Acad. Sci. USA 61, No. 2 (1968).
40. Nonlinear maximal monotone operators in Banach spaces. — Math. Ann. 175, No. 2 (1968).

Browder, F.E. and D.G. de Figueiredo
1. J-monotone nonlinear operators in Banach spaces. — Nederl. Akad. Wetensch. Indag. Math. 69, No. 4 (1966).

Browder, F.E. and C.P. Gupta
1. Monotone operators and nonlinear integral equations of Hammerstein type. — Bull. Amer. Math. Soc. 75, No. 6 (1969).

Clarkson, J.A.
1. Uniformly convex spaces. — Trans. Amer. Math. Soc. 40, No. 1 (1936).

Cudia, D.F.
1. The geometry of Banach spaces, smoothness. — Trans. Amer. Math. Soc. 110, No. 2 (1964).

Day, M.M.
1. Normed Linear Spaces, Second edition. — New York, Academic Press, 1962.

Dem'yanov, V.F. and A.M. Rubinov
1. Approximate Methods for Solution of Extremum Problems. — Leningrad, Izdatel'stvo Leningradskogo Universiteta, 1969. (Russian)

Dolph, C.L. and G.J. Minty

1. On nonlinear integral equations of Hammerstein type. — In: Nonlinear Integral Equations (ed. by P. Anselone), Madison, Wis., University of Wisconsin Press, 1964.

Dubinskii, Yu.A.

1. Nonlinear Elliptic and Parabolic Equations of Arbitrary Order. — Author's Summary of Dissertation, Moscow State University, 1965. (Russian)
2. Quasilinear elliptic and parabolic equations of arbitrary order. — Uspekhi Mat. Nauk 23, No. 1 (1968). (Russian)

Dubinskii, Yu.A. and S.I. Pokhozhaev

1. On a certain class of operators and the solvability of quasilinear elliptic equations. — Mat. Sb. 72 (114), No. 2 (1967). (Russian)

Dubovitskii, A.Ya. and A.A. Milyutin

1. Extremum problem with constraints. — Zh. Vychisl. Mat. i Mat. Fiz. 5, No. 3 (1965). (Russian)

Dunford, N. and J.T. Schwartz

1. Linear Operators, Part 1: General Theory. — New York, Interscience Publishers, 1958.
2. Linear Operators, Part 2: Spectral Theory. — New York, Interscience Publishers, 1963.

Engel'son, Ya.L.

1. On the square root of linear operators in linear topological spaces. — Uchen. Zap. Latv. Univ. 8, No. 2 (1956). (Russian)
2. On the variational theory of nonlinear equations in locally convex spaces. — Nauchn. Dokl. Vysshei Shkoly, No. 4 (1958). (Russian)
3. On potential operators in linear topological spaces. — Uchen. Zap. Latv. Univ. 20, No. 3 (1958). (Russian)
4. Some questions in the variational theory of nonlinear equations in locally convex spaces. — Uchen. Zap. Latv. Univ., No. 4 (1959). (Russian)
5. Application of variational theory to a class of nonlinear integrodifferential equations. — In: Latv. Mat. Ezhegodnik, Riga, 1966. (Russian)

Gavurin, M.K.

1. Analytical methods for the study of nonlinear functional transformations. — Uchen. Zap. Leningrad. Univ. Ser. Mat. 19 (1950). (Russian)

Gel'fand, I.M. and S.V. Fomin

1. Calculus of Variations. — Moscow, Fizmatgiz, 1961. (Russian)

Gel'fand, I.M. and A.G. Kostyuchenko

1. Eigenfunction expansion of differential and other operators. — Uchen. Zap. Leningrad. Univ. Ser. Mat. 19 (1960). (Russian)

Gel'fand, I.M. and G.E. Shilov

1. Generalized Functions, Vol. 2. — Moscow, Fizmatgiz, 1958. (Russian)

Gel'fand, I.M. and N.Ya. Vilenkin

1. Generalized Functions, Vol. 4. — Moscow, Fizmatgiz, 1961. (Russian)

Giles, J.R.

1. Classes of semi-inner-product spaces. — Trans. Amer. Math. Soc. 129, No. 3 (1967).

Giusti, E.

1. Regolarita parziale delle soluzioni di sistemi ellittici quasi-lineari di ordine arbitrario. — Ann. Scuola Norm. Sup. Pisa 23, No. 1 (1969).

Gossez, J.P.

1. Remarques sur les opérateurs monotones. — Bull. Cl. Sci. Acad. Roy. Belg. 52, No. 9. (1966).

Graves, L.M.

1. Riemann integration and Taylor's theorem in general analysis. — Trans. Amer. Math. Soc. 29 (1927).

Hartman, P. and G. Stampacchia

1. On some nonlinear elliptic differential-functional equations. Acta Math. Ser.3 115, Nos. 3–4 (1966).

Hille, E. and R.S. Phillips

1. Functional Analysis and Semi-Groups (revised edition). – Providence, R.I., American Mathematical Society, 1957.

James, R.C.

1. Orthogonality and linear functionals in normed linear spaces. – Trans. Amer. Math. Soc. 61, No.2 (1947).

Kachurovskii, R.I.

1. On monotone operators and convex functionals. – Uspekhi Mat. Nauk 15, No.4 (1960). (Russian)
2. On the variational theory of operator equations. – Uchen. Zap. Moskov. Obl. Ped. Inst. 96, No.6 (1960). (Russian)
3. On the existence of a solution to a certain variational problem. – Uchen. Zap. Moskov. Obl. Ped. Inst. 110, No.7 (1962). (Russian)
4. Nonlinear monotone operators in Banach spaces. – Uspekhi Mat. Nauk 23, No.2 (1968). (Russian)
5. Approximate methods for solution of nonlinear operator equations. – Izv. Vyssh. Uchebn. Zaved. Matematika 12 (1968). (Russian)

Kadets, M.I.

1. Conditions for differentiability of the norm of a Banach spaces. – Uspekhi Mat. Nauk 20, No.3 (1965). (Russian)

Kantorovich, L.V.

1. On a new method for approximate solution of partial differential equations. – Dokl. Akad. Nauk SSSR 4, Nos. 8–9 (1934). (Russian)
2. Functional analysis and applied mathematics. – Uspekhi Mat. Nauk 3, No. 6 (1948). (Russian)
3. Newton's method for functional equations. – Dokl. Akad. Nauk SSSR 60, No.7 (1948). (Russian)
4. On Newton's method. – Trudy Mat. Inst. Steklov 28 (1949). (Russian)
5. Some further applications of Newton's method. – Vestnik Leningrad. Univ. 2, No.7 (1957). (Russian)

Kantorovich, L.V. and G.P. Akilov

1. Functional Analysis in Normed Spaces. – Moscow, Fizmatgiz, 1959. (Russian)

Kato, T.

1. Integration of the equation of evolution in a Banach space. – J. Math. Soc. Japan 5 (1953).
2. Abstract evolution equations of parabolic type in Banach spaces. – Nagoya Math. J. 19 (1961).
3. Nonlinear evolution equations in Banach spaces. – Proc. Symp. Appl. Math. 17 (1965).
4. Demicontinuity, hemicontinuity and monotonicity, II. – Bull. Amer. Math. Soc. 73, No.6 (1967).
5. Nonlinear semigroups and evolution equations. – J. Math. Soc. Japan 19, No.4 (1967).

Keldysh, M.V.

1. On Galerkin's method for the solution of boundary-value problems. – Izv. Akad. Nauk SSSR Ser. Mat. 6 (1942). (Russian)

Kirpotina, N.V.

1. On the theory of systems of nonlinear integral equations. – In: Trudy V Vsesoyuznoi Konferentsii po Funktsional'nomu Analizu i ego Primeneniyu (1959), Baku, 1961. (Russian)

2. Variational methods for solution of certain systems of nonlinear integral equations. — Uchen. Zap. Moskov. Obl. Ped. Inst. 110, No. 7 (1962). (Russian)

Kolmogorov, A. N. and S. V. Fomin

1. Elements of the Theory of Functions and Functional Analysis. — Moscow, Nauka, 1968. (Russian)

Kolodner, I. I.

1. Equations of Hammerstein type in Hilbert spaces. — J. Math. Mech. 13, No. 5 (1964).

Kolomy, J.

1. Applications of some existence theorems for the solution of Hammerstein integral equations. — Comment. Math. Univ. Carolinae 7, No. 4 (1966).
2. The solvability of nonlinear integral equations. — Comment. Math. Univ. Carolinae 8, No. 2 (1967).

Kondrashov, V. I.

1. On some properties of functions in the space L_p. — Dokl. Akad. Nauk SSSR 48, No. 8 (1945). (Russian)
2. On the Theory of Boundary-Value Problems and Eigenvalue Problems for Variational and Differential Equations in Regions with Degenerate Contours. — Author's Summary of Dissertation, Moscow, Mat. Inst. Steklov, 1950. (Russian)

Kositskii, M. E.

1. Nonlinear equations of Hammerstein type with a monotone operator. — Dokl. Akad. Nauk SSSR 190, No. 1 (1970). (Russian)
2. On Nonlinear Equations with Monotone Operators. — Author's summary of dissertation, Moscow, MOPI, 1970. (Russian)
3. Galerkin approximations for nonlinear equations with unbounded and monotone operators. — Uchen. Zap. Moskov. Obl. Ped. Inst. 269, 1969 (1970). (Russian)

Krachkovskii, S. N. and A. A. Vinogradov

1. On a criterion for uniform convexity of a space of type B. — Uspekhi Mat. Nauk 7, No. 3 (1952). (Russian)

Krasnosel'skii, M. A.

1. Topological Methods in the Theory of Nonlinear Integral Equations. — Moscow, Gostekhizdat, 1956. (Russian)

Krasnosel'skii, M. A. and S. G. Krein

1. Criteria for continuity and complete continuity of a linear operator, in terms of the properties of its square. — Trudy Sem. Funktsional. Anal. (Voronezh) 5, 1957. (Russian)

Krasnosel'skii, M. A. and Ya. B. Rutitskii

1. Convex Functions and Orlicz Spaces. — Moscow, Fizmatgiz, 1958. (Russian)

Krasnosel'skii, M. A. and V. I. Sobolev

1. On splitting of linear operators. — Uspekhi Mat. Nauk 12, No. 4 (1957). (Russian)

Krasnosel'skii, M. A., G. M. Vainikko et al.

1. Approximate Solution of Operator Equations. — Moscow, Nauka, 1969. (Russian)

Krasnosel'skii, M. A., P. P. Zabreiko et al.

1. Integral Operators in Spaces of Summable Functions. — Moscow, Nauka, 1966. (Russian)

Langenbach, A.

1. On the application of a variational principle to certain nonlinear integral equations. — Dokl. Akad. Nauk SSSR 121, No. 2 (1958). (Russian)
2. On the application of the method of least squares to nonlinear equations. — Dokl. Akad. Nauk SSSR 148, No. 1 (1962). (Russian)
3. Über Lösungsverzweigungen bei Potentialoperatoren. — Math. Nachr. 42, Nos. 1—3. (1969).

Lavrent'ev, I.M.
1. On the variational theory of nonlinear equations. — Dokl. Akad. Nauk SSSR **166**, No. 2 (1966). (Russian)
2. On the solvability of nonlinear equations. — Dokl. Akad. Nauk SSSR **175**, No. 6 (1967). (Russian)
3. On the Theory of Nonlinear Equations. — Author's Summary of Dissertation, Moscow State University, 1967. (Russian)

Lavrent'ev, M.A. and L.A. Lyusternik
1. Elements of the Calculus of Variations, Vol. I, Part II. — Moscow-Leningrad, ONTI, 1935. (Russian)

Lavrent'ev, M.M.
1. On Some Ill-Posed Problems of Mathematical Physics. — Novosibirsk, 1962. (Russian)

Leray, J. and J.L. Lions
1. Quelques résultats de Visik sur les problèmes elliptiques non linéaires par les méthodes de Minty-Browder. — Bull. Soc. Math. France 93, No. 1 (1965).

Leray, J. and J. Schauder
1. Topologie et équations fonctionelles. — Ann. Sci. Ecole Norm. Sup. (3) **51** (1934).

Levchenko, V.I. and I.V. Shragin
1. The Nemytskii operator from a space of continuous functions into an Orlicz-Nakano space. — Mat. Issledovaniya (Akad. Nauk Mold. SSR) 3, No. 3 (1968). (Russian)

Levin, V.L.
1. On some properties of support functionals. — Mat. Zametki **4**, No. 6 (1968). (Russian)

Levitin, E.S. and B.T. Polyak
1. Methods of minimization with constraints. — Zh. Vychisl. Mat. i Mat. Fiz. **6**, No. 5 (1966). (Russian)

Lezanski, T.
1. Über das Minimumproblem für Funktionale in Banachschen Räumen. — Math. Ann. **152**, No. 4 (1963).

Lin'kov, E.I.
1. On some properties of the norm in Banach spaces with a convex sphere. — Uchen. Zap. Moskov. Obl. Ped. Inst. **110**, No. 7 (1962). (Russian)

Lions, J.L.
1. Sur certaines équations paraboliques non linéaires. — Bull. Soc. Math. France **93** (1969).
2. Quelques méthodes de résolution des problèmes aux limites non linéaires. — Paris, 1969.

Lions, J.L. and G. Stampacchia
1. Variational inequalities. — Comm. Pure Appl. Math. **20**, No. 3 (1967).

Lions, J.L. and W.A. Strauss
1. Some nonlinear evolution equations. — Bull. Soc. Math. France **93**, No. 1 (1965).

Lumer, G.
1. Semi-inner-product spaces. — Trans. Amer. Math. Soc. **100**, No. 1 (1961).

Lumer, G. and R.S. Phillips
1. Dissipative operators in a Banach space. — Pacific J. Math. **11** (1961).

Luxemburg, W.A.J.
1. Banach Function Spaces. — Assen, van Gorcum, 1955.

Lyubich, Yu.I. and G.D. Maistrovskii
1. On stability of relaxation processes. — Dokl. Akad. Nauk SSSR **191**, No. 1 (1970). (Russian)
2. General theory of relaxation processes for convex functionals. — Uspekhi Mat. Nauk 25, No. 1 (1970). (Russian)

Lyusternik, L.A.

1. On conditional extrema of functionals. Mat. Sb. 41, No. 3 (1934). (Russian)

Lyusternik, L.A. and V.I. Sobolev

1. Elements of Functional Analysis. — Moscow, Nauka, 1965. (Russian)

Marinescu, G.

1. Différentielles de Gâteaux et Fréchet dans les espaces localement convexes. — Bull. Math. Soc. Sci. Math. Phys. RPR 1, No. 1 (1957).

Mazur, S.

1. Über schwache Konvergenz in den Räumen (L^p). — Studia Math. 4 (1933).
2. Über konvexe Mengen in linearen normierten Räumen. — Studia Math. 4 (1933).

Mikhlin, S.G.

1. The Problem of the Minimum of a Quadratic Functional. — Moscow-Leningrad, Gostekhizdat, 1952. (Russian)
2. Variational Methods in Mathematical Physics. — Moscow, Gostekhizdat, 1957. (Russian)
3. Numerical Realization of Variational Methods. — Moscow, Nauka. 1966. (Russian)

Mil'man, D.P.

1. On some criteria for regularity of spaces of type B. — Dokl. Akad. Nauk SSSR 20 (1938). (Russian)

Minty, G.J.

1. Monotone (nonlinear) operators in Hilbert space. — Duke Math. J. 29, No. 3 (1962).
2. On a "monotonicity" method for the solution of nonlinear equations in Banach spaces. — Proc. Nat. Acad. Sci. USA 50, No. 6 (1963).
3. Two theorems on nonlinear functional equations in Hilbert spaces. — Bull. Amer. Math. Soc. 69, No. 5 (1963).
4. On the solvability of nonlinear functional equations of "monotonic" type. — Pacific J. Math. 14, No. 1 (1964).
5. On the "monotonicity" of the gradient of a convex function. — Pacific J. Math. 14, No. 1 (1964).
6. A theorem on maximal monotonic sets in Hilbert spaces. — J. Math. Anal. Appl. 11, Nos. 1—3 (1965).
7. On the generalization of a direct method of the calculus of variations. — Bull. Amer. Math. Soc. 75, No. 3 (1967).

Mosolov, P.P.

1. Variational methods in nonstationary problems (parabolic case). — Izv. Akad. Nauk SSSR Ser. Mat. 34, No. 2 (1970). (Russian)

Mysovskikh, I.P.

1. On convergence of the method of L.V. Kantorovich for solution of functional equations and its applications. — Dokl. Akad. Nauk SSSR 70, No. 4 (1950). (Russian)

Naimark, M.A.

1. Normed Rings. — Moscow, Nauka, 1968. (Russian)

Natanson, I.P.

1. Theory of Functions of a Real Variable. — Moscow, Gostekhizdat, 1957. (Russian)

Nečas, J.

1. Sur l'alternative de Fredholm pour les opérateurs nonlinéaires avec applications aux problèmes aux limites. — Ann. Scuola Norm. Sup. Pisa Sci. Fiz. Mat. 23, No. 2 (1969).
2. Les équations elliptiques non linéaires. Czechslovak Math. J. 19, No. 2 (1969).

Nemytskii, V.V.

1. The fixed point method in analysis. — Uspekhi Mat. Nauk, No. 1 (1936). (Russian)

Nemytskii, V.V., M.M. Vainberg and R.S. Gusarova

1. Operator differential equations. — In: Matematicheskii Analiz, 1964 (Itogi Nauki), Moscow, 1966. (Russian)

Nguyen-Phong-Chau
1. Remarques sur deux théorèmes de G.J. Minty et de F.E. Browder. — C.R. Acad. Sci. Paris 265, No.12 (1967).

Nikol'skii, S.M.
1. Approximations of Functions of Several Variables and Embedding Theorems. — Moscow, Nauka, 1969. (Russian)

Orlicz, W.
1. Über eine gewisse Klasse von Räumen von Typus B. — Bull. Intern. Acad. Polon. Ser.A, Nos. 8—9 (1932).
2. Über Räume (L^M). — Bull. Intern. Acad. Polon. Ser. A, Nos. 3—4 (1936).

Palmer, K.J.
1. On the complete continuity of differentiable mappings. — J. Austral. Math. Soc. 9, Nos. 3—4 (1969).

Petryshyn, W.V.
1. On the extension and the solution of nonlinear operator equations. — Illinois J. Math. 10, No.2 (1966).

Piontkovskii, O.V.
1. On a corollary of a theorem of M.M. Vainberg. — Uchen. Zap. Moskov. Obl. Ped. Inst. 225, No.12 (1969). (Russian)

Pokhozhaev, S.I.
1. On the solvability of nonlinear equations with odd operators. — Funktsional. Anal. i Pril. 1, No.3 (1967). (Russian)
2. Normal solvability of nonlinear equations in uniformly convex spaces. — Funktsional. Anal. i Pril.3, No.2 (1969). (Russian)

Polyak, B.T.
1. Theorems on existence and convergence of minimizing sequences for extremum problems with constraints. — Dokl. Akad. Nauk SSSR 166, No.2 (1966). (Russian)
2. Semicontinuity of integral functionals and existence theorems in extremum problems. — Mat. Sb. 78, No.1 (1969). (Russian)
3. Minimization of nonsmooth functionals. — Zh. Vychisl. Mat. i Mat. Fiz. 9, No.3 (1969). (Russian)
4. The method of conjugate gradients in extremum problems. — Zh. Vychisl. Mat. i Mat. Fiz. 9, No.4 (1959). (Russian)

Raikov, D.A.
1. On the complete continuity of the adjoint operator. — Dokl. Akad. Nauk SSSR 119, No.3 (1958). (Russian)
2. Completely continuous spectra of locally convex spaces. — Trudy Moskov. Mat. Obshch.7 (1958). (Russian)

Ritz, W.
1. Über eine neue Methode zur Lösung gewisser Variationsprobleme der mathematischen Physik. — J. Reine Angew. Math. 135 (1908).
2. Theorie der Transversalschwingungen einer quadratischen Platte mit freien Rändern. — In: Gesammelte Werke, Paris, 1911.

Rockafellar, R.T.
1. Characterization of the subdifferentials of convex functions. — Pacific J. Math. 17, No.3 (1966).
2. Convex Analysis. — Princeton, N.J., Princeton University Press, 1970.
3. Convex functions, monotone operators and variational inequalities. — In: Theory and Applications of Monotone Operators (Proc. NATO Advanced Study Institute, Venice, 1968), Gubbio, "Oderisi," 1969.

Rosenbloom, P.
1. The method of steepest descent. — Proc. Symp. Appl. Math. 6 (1956).
Schwartz, L.
1. Théorie des distributions, Vol. 1. — Paris, 1950.
Sebastião e Silva, J.
1. Le calcul différentiel et intégral dans les espaces localement convexes, réels ou complexes, I, II. — Atti Accad. Naz. Lincei Rend. Cl. Sci. Fis. Mat. Natur. 20, No. 6 (1956); 21, Nos. 1—3 (1956).
Shimogaki, T.
1. A generalization of Vainberg's theorem, I. — Proc. Japan Acad. 34, No. 8 (1958).
2. A generalization of Vainberg's theorem, II. — Proc. Japan Acad. 34, No. 10 (1958).
Shmul'yan, V.L.
1. On some geometric properties of the unit sphere in a space of type B. — Mat. Sb. 6 (1939). (Russian)
2. On different topologies in a Banach space. — Dokl. Akad. Nauk SSSR 23 (1939). (Russian)
3. On differentiability of the norm in a Banach space. — Dokl. Akad. Nauk SSSR 27 (1940). (Russian)
4. Sur la structure de la sphère unitaire dans l'espace de Banach. — Mat. Sb. 9 (51) (1941).
Shragin, I.V.
1. On the weak continuity of the Nemytskii operator. — Uchen. Zap. Moskov. Obl. Ped. Inst. 57 (1957). (Russian)
2. On some operators in generalized Orlicz spaces. — Dokl. Akad. Nauk SSSR 117, No. 1 (1957). (Russian)
3. On a certain nonlinear operator. — Nauchn. Dokl. Vyssh. Shkoly, Fiz.-Mat. Nauki, No. 2 (1958). (Russian)
4. On the Hammerstein operator and Hammerstein equation. — Nauchn. Dokl. Vyssh. Shkoly, Fiz.-Mat. Nauki, No. 3 (1958). (Russian)
5. The Nemytskii operator from C to L^M. — Uchen. Zap. Moskov. Obl. Ped. Inst. 77 (1959). (Russian)
6. On the weak continuity of the Nemytskii operator in generalized Orlicz spaces. — Uchen. Zap. Moskov. Obl. Ped. Inst. 77 (1959). (Russian)
7. On a certain nonlinear operator in Orlicz spaces. — Uspekhi Mat. Nauk 14, No. 4 (1959). (Russian)
8. On continuity of the Nemytskii operator. — In: Funktsional'nyi Analiz i ego Primenenie (Trudy V Vsesoyuznoi Konferentsii po Funktsional'nomu Analizu i ego Primeneniyu), Baku, 1961. (Russian)
9. On continuity of the Nemytskii operator in Orlicz spaces. — Dokl. Akad. Nauk SSSR 140, No. 3 (1961). (Russian)
10. On the boundedness of the Nemytskii operator in Orlicz spaces. — Uchen. Zap. Kishinev. Univ. 50 (1962). (Russian)
11. On the question of the continuity of the Nemytskii operator. — Uchen. Zap. Kishinev, Univ. 70 (1964). (Russian)
12. Some properties of the Nemytskii operator. — Mat. Sb. 65, No. 3 (1964). (Russian)
13. The Hammerstein equation in the space of continuous functions. — Uchen. Zap. Kishinev. Univ. 82 (1965). (Russian)
14. Generalization of a theorem of M. M. Vainberg. — Trudy Tambov. Inst. Khim. Mashinostr., No. 3 (1969). (Russian)
15. The Nemytskii operator in spaces generated by gen-functions. — Dokl. Akad. Nauk SSSR 189, No. 1 (1969). (Russian)
16. Measurability of superposition of discontinuous functions. — Trudy Tambov. Inst. Khim. Mashinostr., No. 3 (1969). (Russian)

Smirnov, V.I.
1. Course of Higher Mathematics, Vol.V. — Moscow, Fizmatgiz, 1959. (Russian)
Sobolev, S.L.
1. Some Applications of Functional Analysis in Mathematical Physics. — Izdatel'stvo Leningrad. Gos. Univ., 1950. (Russian)
Sobolev, V.I.
1. On the splitting of linear operators. — Dokl. Akad. Nauk SSSR **111**, No.5 (1956). (Russian)
Stampacchia, G.
1. Formes bilinéaires coercitives sur les convexes. — C.R. Acad. Sci. Paris 258, No.18 (1964).
2. Le problème de Dirichlet pour les équations elliptiques du second ordre à coéfficients discontinus. — Ann. Inst. Fourier (Grenoble) **15**, No.1 (1965).
3. Variational inequalities. — In: Theory and Applications of Monotone Operators (Proc. NATO Advanced Study Institute, Venice, 1968), Gubbio, "Oderisi," 1969.
Strauss, W.A.
1. Evolution equations nonlinear in the time derivative. — J. Math. Mech. **15**, No.1 (1966).
Sunderasan, K.
1. Smooth Banach spaces. — Math. Ann. **173**, No.3 (1967).
Tikhonov, A.N.
1. On methods for regularization of optimal control problems. — Dokl. Akad. Nauk SSSR **162**, No.4 (1965). (Russian)
Tsitlanadze, E.S.
1. On differentiability of functionals. — Mat. Sb. **29**, No.1 (1951). (Russian)
Vainberg, A.M.
1. Nonlinear equations with monotone operators. — Dokl. Akad. Nauk SSSR **188**, No.3 (1969). (Russian)
Vainberg, M.M.
1. On continuity of some operators of a special type. — Dokl. Akad. Nauk SSSR **73**, No.2 (1950). (Russian)
2. On the differential and gradient of functionals. — Uspekhi Mat. Nauk **7**, No.3 (1952). (Russian)
3. Some questions of differential calculus in linear spaces. — Uspekhi Mat. Nauk **7**, No.4 (1952). (Russian)
4. On the structure of a certain operator. — Dokl. Akad. Nauk SSSR **92**, No.2 (1953). (Russian)
5. On the solvability of certain operator equations. — Dokl. Akad. Nauk SSSR **92**, No.3. (1953). (Russian)
6. On some properties of quadratic forms in L^q spaces ($q \le 2$). — Dokl. Akad. Nauk SSSR **100**, No.5 (1955). (Russian)
7. The Nemytskii operator. — Ukrain. Mat. Zh. **7** (1955). (Russian)
8. On the variational theory of nonlinear operator equations. — Uspekhi Mat. Nauk **10**, No.4 (1955). (Russian)
9. Variational Methods for the Study of Nonlinear Operators. — Moscow, Gostekhizdat, 1956. [English translation: San Fransisco, Holden-Day, 1964.]
10. New theorems for nonlinear operators and equations. — Uchen. Zap. Moskov. Obl. Ped. Inst. **77**, No.5 (1959). (Russian)
11. On the convergence of a steepest-descent method for nonlinear equations. — Dokl. Akad. Nauk SSSR **130**, No.1 (1960). (Russian)
12. On some new principles in the theory of nonlinear equations. — Uspekhi Mat. Nauk **15**, No.1 (1960). (Russian)

13. On the convergence of a steepest-descent method for nonlinear equations. — Sibirsk. Mat. Zh. 2, No. 2 (1961). (Russian)
14. On a class of potential operators. — Uchen. Zap. Moskov. Obl. Ped. Inst. 110, No. 7 (1962). (Russian)
15. On the minimum of convex functionals. — Uspekhi Mat. Nauk 20, No. 1 (1965). (Russian)
16. Nonlinear equations with potential and monotone operators. — Dokl. Akad. Nauk SSSR 183, No. 4 (1968). (Russian)
17. On the unconditional extremum of functionals and the convergence of minimizing sequences. — Dokl. Akad. Nauk SSSR 183, No. 6 (1968). (Russian)
18. Metodo varizionale e metodo di Caccioppoli nella teoria delle equazioni funzionali non lineari. — In: Symposia Mathematica II (Istituto nazionale di alta matematica, 1968), New York, Academic Press. 1969.
19. On the minimum of certain nonlinear functionals. — Uchen. Zap. Moskov. Obl. Ped. Inst. 225, No. 12 (1969). (Russian)
20. On the question of uniqueness of the solution of nonlinear equations of Hammerstein type. — Uchen. Zap. Moskov. Obl. Ped. Inst. 269 (1969—1970). (Russian)
21. On application of the variational method in the theory of differential equations. — Časopis Pěst. Mat. 95, No. 2 (1970). (Russian)
22. Equations with nonlinear accretive and monotone operators and Galerkin-Petrov approximations. — Dokl. Akad. Nauk SSSR 197, No. 4 (1971). (Russian)
23. Le problème de la minimisation des fonctionelles non linéaires. — In: Problems in Non-Linear Analysis, C.I.M.E. (IV Ciclo, Varenna, 1970), Rome, Ed. Cremonese, 1971.

Vainberg, M.M. and Ya.L. Engel'son
1. On the square root of a linear operator in locally convex spaces. — Dokl. Akad. Nauk SSSR 122, No. 5 (1958). (Russian)
2. On the conditional extremum of functionals in linear topological spaces. — Mat. Sb. 45, No. 4 (1958). (Russian)

Vainberg, M.M. and R.I. Kachurovskii
1. On the variational theory of nonlinear operators and equations. — Dokl. Akad. Nauk SSSR 129, No. 6 (1959). (Russian)

Vainberg, M.M. and I.M. Lavrent'ev
1. Equations with monotone and potential operators in Banach spaces. — Dokl. Akad. Nauk SSSR 187, No. 4 (1969). (Russian)
2. On the square root of a linear operator in certain Banach spaces. — Vestnik Moskov. Univ. Mat. Mekh., No. 6 (1970). (Russian)
3. Nonlinear equations of Hammerstein type with potential and monotone operators in Banach spaces. — Mat. Sb. 87, No. 3 (1972). (Russian)

Vainberg, M.M. and E.I. Lin'kov
1. On some approximate methods for solution of nonlinear equations. In: Funtsional'nyi Analiz i ego Primeneniya. Trudy Vsesoyuznoi Konferentsii po Funktsional'nomu Analizu i ego Primeneniyu (1959), Baku, 1961. (Russian)

Vainberg, M.M. and I.V. Shragin
1. The Nemytskii operator and its potential in Orlicz spaces. — Dokl. Akad. Nauk SSSR 120, No. 5 (1958). (Russian)
2. The Nemytskii operator in generalized Orlicz spaces. — Uchen. Zap. Moskov. Obl. Ped. Inst. 77 (1959). (Russian)
3. Nonlinear operators and Hammerstein equations in Orlicz spaces. — Dokl. Akad. Nauk SSSR 128, No. 1 (1959). (Russian)
4. The Hammerstein operator in Orlicz spaces, I. — Izv. Vyssh. Uchebn. Zaved. Matematika, No. 1 (1965). (Russian)

5. The Hammerstein operator in Orlicz spaces, II. — Izv. Vyssh. Uchebn. Zaved. Matematika, No. 3 (1965). (Russian)

Vishik, M.I.

1. Solution of a system of quasilinear equations in divergence form with periodic boundary conditions. — Dokl. Akad. Nauk SSSR **137**, No. 3 (1961). (Russian)
2. Boundary-value problems for quasilinear strongly elliptic systems of equations in divergence form. — Dokl. Akad. Nauk SSSR **138**, No. 3 (1961). (Russian)
3. Quasilinear elliptic systems of equations containing subordinate terms. — Dokl. Akad. Nauk SSSR **144**, No. 1 (1962). (Russian)
4. On the solvability of boundary-value problems for quasilinear parabolic equations of higher orders. — Mat. Sb. 59 (add.) (1962). (Russian)
5. Quasilinear strongly elliptic systems of differential equations in divergence form. — Trudy Moskov. Mat. Obshch. 12 (1963). (Russian)
6. On the solvability of the first boundary-value problem for quasilinear equations with rapidly increasing coefficients in Orlicz classes. — Dokl. Akad. Nauk SSSR **151**, No. 4 (1963). (Russian)

Vishik, M.I. and O.A. Ladyzhenskaya

1. Boundary-value problems for partial differential equations and some classes of operator equations. — Uspekhi Mat. Nauk **11**, No. 6. (1956). (Russian)

von Neumann, J.

1. Zur Algebra der Funktionaloperationen und Theorie der normalen Operatoren. — Math. Ann. **102**, No. 3 (1929).

Wang Shêng-wang

1. Some remarks on solutions of certain nonlinear differential equations. — Dokl. Akad. Nauk SSSR **150**, No. 5 (1963). (Russian)

Webb, J.R.L.

1. Mapping and fixed-point theorems for nonlinear operators in Banach spaces. — Proc. London Math. Soc. **20**, No. 3 (1970).

Yakovlev, M.N.

1. On some methods for solution of nonlinear equations. — Trudy Mat. Inst. Akad. Nauk SSSR 84 (1965). (Russian)

Yosida, K.

1. Functional Analysis. — Berlin, Springer, 1966.

Zaanen, A.C.

1. Linear Analysis. — New York, Humanities Press, 1964.

Zabreiko, P.P. and A.I. Povolotskii

1. Theorems on existence and uniqueness of solutions for the Hammerstein equation. — Dokl. Akad. Nauk SSSR **176**, No. 4 (1967). (Russian)

Zarantonello, E.H.

1. Solving functional equations by contractive averaging. — Univ. Wisconsin Math. Res. Center, Tech. Rep. No. 160, 1960.
2. The closure of the numerical range contains the spectrum. — Bull. Amer. Math. Soc. **70**, No. 6 (1964).
3. The closure of the numerical range contains the spectrum. — Pacific J. Math. **22**, No. 3 (1967).

Zeragiya, D.P.

1. On the variational theory of nonlinear equations. — Soobshch. Akad. Nauk Gruz. SSR **29**, No. 2 (1962). (Russian)
2. On the variational theory of nonlinear equations. — Tbilisi, Trudy Univ. 84 (1962). (Russian)
3. On solution of the Dirichlet problem for certain nonlinear equations of elliptic type. — Soobshch. Akad. Nauk Gruz. SSR **31**, No. 1 (1963). (Russian)
4. On the Variational Theory of Nonlinear Equations. — Author's Summary of Dissertation, Tbilisi, 1963. (Russian)

SUBJECT INDEX